Food Science Text Series

The Food Science Text Series provides faculty with the leading teaching tools. The Editorial Board has outlined the most appropriate and complete content for each food science course in a typical food science program and has identified textbooks of the highest quality, written by the leading food science educators.

Series Editor
Dennis R. Heldman

Editorial Board
David A. Golden, Ph.D., Professor of Food Microbiology, Department of Food Science and Technology, University of Tennessee

Richard W. Hartel, Professor of Food Engineering, Department of Food Science, University of Wisconsin

Hildegarde Heymann, Professor of Food Sensory Science, Department of Food Science and Technology, University of California-Davis

Joseph H. Hotchkiss, Professor, Institute of Food Science and Institute for Comparative and Environmental Toxicology, and Chair, Food Science Department, Cornell University

Michael G. Johnson, Ph.D., Professor of Food Safety and Microbiology, Department of Food Science, University of Arkansas

Joseph Montecalvo, Jr., Professor, Department of Food Science and Nutrition, California Polytechnic and State University-San Luis Obispo

S. Suzanne Nielsen, Professor and Chair, Department of Food Science, Purdue University

Juan L. Silva, Professor, Department of Food Science, Nutrition and Health Promotion, Mississippi State University

For further volumes:
http://www.springer.com/series/5999

Omar A. Oyarzabal · Steffen Backert
Editors

Microbial Food Safety

An Introduction

Springer

Editors
Omar A. Oyarzabal
Department of Biological Sciences
Alabama State University
Montgomery, AL 36101, USA
oaoyarzabal@gmail.com

Steffen Backert
University College Dublin
Belfield Campus
School of Biomolecular and Biomedical Science
Dublin-4, Ireland
steffen.backert@ucd.ie

ISSN 1572-0330
ISBN 978-1-4614-1176-5   e-ISBN 978-1-4614-1177-2
DOI 10.1007/978-1-4614-1177-2
Springer New York Dordrecht Heidelberg London

Library of Congress Control Number: 2011941615

© Springer Science+Business Media, LLC 2012
All rights reserved. This work may not be translated or copied in whole or in part without the written permission of the publisher (Springer Science+Business Media, LLC, 233 Spring Street, New York, NY 10013, USA), except for brief excerpts in connection with reviews or scholarly analysis. Use in connection with any form of information storage and retrieval, electronic adaptation, computer software, or by similar or dissimilar methodology now known or hereafter developed is forbidden.
The use in this publication of trade names, trademarks, service marks, and similar terms, even if they are not identified as such, is not to be taken as an expression of opinion as to whether or not they are subject to proprietary rights.

Printed on acid-free paper

Springer is part of Springer Science+Business Media (www.springer.com)

Preface

For many centuries humans have used empirical knowledge to cook and prepare foods, and although we have known for a long time about many different hazards inherent to food products, our understanding of infectious agents transmitted by foods did not materialize until the theory of germs was well established, approximately 150 years ago. Food hazards are classified as physical, chemical, and biological. By far, the biological hazards – primarily bacteria and viruses – pose the greatest risk in modern food safety. Like other infectious diseases, foodborne diseases repeat themselves, in part because we still do not fully understand their epidemiology to prevent their appearance, and in part because we do not always apply the acquired knowledge consistently. Therefore, there is always a need to revisit basic concepts to better understand food safety hazards. This book is intended to provide a review of the most prevalent biological hazards in the most common food categories.

In general, books related to food safety deal with a detailed description of known physical, chemical, and biological agents, emphasize the normative related to the presence of pathogens in foods, or review how these pathogens can be detected. More recently, some books have attempted to review our current knowledge of control strategies to reduce foodborne diseases. However, it appears that a general training tool for undergraduate and graduate students pursuing careers in food science, animal science, microbiology, and similar fields is still missing. Therefore, this book attempts to provide a study tool to advanced undergraduate and graduate students who need or wish to take a class on food safety. Nevertheless, any student with some basic knowledge in microbiology will find additional information related to different food safety topics in this book.

From the three major components that make up food safety – perception, regulations, and science – this book attempts to summarize the current scientific understanding of the most common biological hazards by food commodity. The book then provides an overview of the current regulations related to food safety in the United States. The first part includes a chapter that briefly describes our current understanding of the evolution of foodborne pathogens. The other chapters in this first part describe the basic microbiology concepts applied to food safety, the methodology used to identify microbial hazards transmitted by foods, the clinical presentations and pathogenicity of foodborne diseases, foodborne viruses, and the methodology used to type microbial pathogens for epidemiological studies. We have included a separate chapter for foodborne viruses because fewer scientists are working with viruses than are studying with bacterial agents. The methodologies that we have developed so far for viruses do not allow for an easy reproduction of viruses under laboratory conditions; thus, our studies of viruses depend heavily on molecular techniques. We have also added a chapter on molecular techniques for typing bacterial pathogens because these techniques provide unique tools to better understand the epidemiology of foodborne agents. We now know that strains from the same bacterial species have different pathogenicity potentials to humans. Therefore, as the methodologies for molecular studies become more simplified and available, we will be able to better understand the risk posed by certain bacterial strains in food commodities.

The second part of the book summarizes the major food commodities and the major biological hazards associated with these products. Several concepts may overlap in these chapters, such as the definition of certain bacterial pathogens. We believe that each of these chapters should be able to "stand alone"; if readers do skip some food commodity chapters, they will still get the basic concepts for the food commodities of interest.

The third part includes the chapters related to risk analysis, interventions, and regulations. Several books have already been written about interventions for those interested in this topic. Similarly, several books have recently emerged on the application of the risk analysis model to food safety. However, these two topics either are relatively new to food safety (risk assessment) or have undergone many different changes in the last few decades (interventions) to warrant some attention among food safety professionals. These areas of food safety are expanding rapidly, and as the world population will reach 10 billion in a few decades according to the United Nations's predictions, food safety and the control of food safety hazards will become increasingly important in the near future. The current regulations for food safety described in this area are all related to the United States and its federal agencies. Without food laws and guidelines addressing the presence of specific biological agents in food, little would be done to control these agents. As the international trade of food commodities becomes more complex, we will see more consolidation of food safety standards for an ever-expanding international market.

The last part of this book includes a list of other books and Internet resources related to food safety. Throughout the book, there is an assumption that the reader has a basic knowledge in microbiology, such as the way bacteria grow and multiply, the effect of temperature on the survival or destruction of bacteria, and the composition of viruses. For those interested in a more in-depth review of microbiology concepts, a list of microbiology books and Internet resources is also provided. It is important to highlight that many regulations and most of the documents generated by regulatory agencies in the United States are published mainly online. Therefore, the Internet can be a useful resource for food safety information. Throughout the book, there are italicized terms and words whose definitions are found in the Glossary.

We hope this book brings a new resource to undergraduate and graduate students, food professionals, biologists, and microbiologists interested in food safety. We also hope this book will expand the resources for those food safety professionals already working for the food industry, in academia, or in regulatory agencies. We welcome any feedback to improve future editions.

Montgomery, AL, USA Omar A. Oyarzabal
Dublin, Ireland Steffen Backert

Contents

Part I Microorganisms and Food Contamination

Emerging and Reemerging Foodborne Pathogens 3
Omar A. Oyarzabal

**Clinical Presentations and Pathogenicity Mechanisms of Bacterial
Foodborne Infections** 13
Nicole Tegtmeyer, Manfred Rohde, and Steffen Backert

Microbiology Terms Applied to Food Safety 33
Anup Kollanoor-Johny, Sangeetha Ananda Baskaran, and Kumar Venkitanarayanan

Methods for Identification of Bacterial Foodborne Pathogens 45
Ramakrishna Nannapaneni

Methods for Epidemiological Studies of Foodborne Pathogens 57
Omar A. Oyarzabal

Foodborne Viruses 73
Daniel C. Payne, Umid Sharapov, Aron J. Hall, and Dale J. Hu

Part II Safety of Major Food Products

Safety of Produce 95
Maha N. Hajmeer and Beth Ann Crozier-Dodson

Safety of Fruit, Nut, and Berry Products 109
Mickey Parish, Michelle Danyluk, and Jan Narciso

Safety of Dairy Products 127
Elliot T. Ryser

Safety of Meat Products 147
Paul Whyte and Séamus Fanning

Safety of Fish and Seafood Products 159
Kenneth Lum

Part III Risk Analysis, Interventions and Regulations

Food Risk Analysis 175
Thomas P. Oscar

Interventions to Inhibit or Inactivate Bacterial Pathogens in Foods 189
P. Michael Davidson and Faith M. Critzer

Food Regulation in the United States 203
Patricia Curtis

Role of Different Regulatory Agencies in the United States 217
Craig Henry

Part IV List of Other Food Safety Resources

Food Safety Resources 235
Omar A. Oyarzabal and Steffen Backert

Glossary 241

Index 253

Contributors

Steffen Backert Belfield Campus, School of Biomolecular and Biomedical Science, University College Dublin, Dublin, Ireland

Sangeetha Ananda Baskaran Department of Animal Science, University of Connecticut, Storrs, CT, USA

Faith M. Critzer Department of Food Science and Technology, University of Tennessee, Knoxville, TN, USA

Beth Ann Crozier-Dodson Food Safety Consulting, LLC, Manhattan, KS, USA

Patricia Curtis Department of Poultry Science, Auburn University, Auburn, AL, USA

Michelle Danyluk Citrus Research and Education Center, University of Florida, Lake Alfred, FL, USA

P. Michael Davidson Department of Food Science and Technology, University of Tennessee, Knoxville, TN, USA

Séamus Fanning Centre for Food Safety & Institute of Food and Health, School of Public Health, Physiotherapy and Population Science, University College Dublin, Ireland

Maha N. Hajmeer Food and Drug Branch, California Department of Public Health, Sacramento, CA, USA

Aron J. Hall National Center for Immunization and Respiratory Diseases, Division of Viral Diseases, Epidemiology Branch, U.S. Centers for Disease Control and Prevention, Atlanta, GA, USA

Craig Henry Grocery Manufacturers Association (GMA), Washington, DC, USA

Dale J. Hu National Center for Immunization and Respiratory Diseases, Division of Viral Hepatitis, Epidemiology and Surveillance Branch, U.S. Centers for Disease Control and Prevention, Atlanta, GA, USA

Anup Kollanoor-Johny Department of Animal Science, University of Connecticut, Storrs, CT, USA

Kenneth Lum Seafood Products Association, Seattle, WA, USA

Jan A. Narciso USDA/ARS/CSPRU, US Horticultural Research Laboratory, Fort Pierce, FL, USA

Thomas P. Oscar U. S. Department of Agriculture, Microbial Food Safety Research Unit, University of Maryland Eastern Shore, Princess Anne, MD, USA

Ramakrishna Nannapaneni Department of Food Science, Nutrition and Health Promotion, Mississippi State University, Mississippi State, MS, USA

Omar A. Oyarzabal Department of Biological Sciences, Alabama State University, Montgomery, AL, USA

Mickey E. Parish U. S. Food and Drug Administration, College Park, MD, USA

Daniel C. Payne Division of Viral Diseases, National Center for Immunization and Respiratory Diseases, Epidemiology Branch, U. S. Centers for Disease Control and Prevention, Atlanta, GA, USA

Manfred Rohde Helmholtz Centre for Infection Research, Braunschweig, Germany

Elliot T. Ryser Department of Food Science and Human Nutrition, Michigan State University, East Lansing, MI, USA

Umid Sharapov National Center for Immunization and Respiratory Diseases, Division of Viral Hepatitis, Epidemiology and Surveillance Branch, U.S. Centers for Disease Control and Prevention, Atlanta, GA, USA

Nicole Tegtmeyer Belfield Campus, School of Biomolecular and Biomedical Science, University College Dublin, Dublin, Ireland

Kumar Venkitanarayanan Department of Animal Science, University of Connecticut, Storrs, CT, USA

Paul Whyte Centre for Food Safety & Institute of Food and Health, School of Veterinary Medicine, University College Dublin, Ireland

Part I
Microorganisms and Food Contamination

Emerging and Reemerging Foodborne Pathogens

Omar A. Oyarzabal

1 Introduction

Emerging and "reemerging" pathogens are mainly zoonoses, and emerging foodborne diseases are not the exception. The interface between humans and food animals, the potential for new infectious diseases to emerge, and the adaptation of bacteria to infect humans by the species jump concept will be examined in this chapter. However, to understand how pathogens evolve and spread, it is important to remember that the microbiology events that happened in the last 200 years have consolidated our view of food as a source of microbial contamination and have helped us to recognize some of the events that result in the emergence of new pathogens, or the reemergence of known pathogens in food products. This chapter will focus mainly on bacterial foodborne pathogens and will review our current understanding of emerging foodborne pathogens.

2 Emerging and Reemerging Infectious Diseases

The term "emerging infectious diseases" is used to define those infections that newly appear in a population or have existed but are rapidly increasing in incidence or spreading in geographic range (Morse 1995). Emerging or reemerging *pathogens* appear because of a series of circumstances that favor their spread. In the case of foodborne pathogens, the factors that play an important role include those related to the pathogen itself, the environment, food production and distribution, and the consumers (Altekruse et al. 1997; Smith and Fratamico 2005). The World Health Organization (WHO) associates the appearance of foodborne diseases with factors that include changes in microorganisms, change in the human population and lifestyle, the globalization of the food supply, the inadvertent introduction of pathogens into new geographic areas, and exposure to unfamiliar foodborne hazards while abroad (Anonymous 2002).

There are approximately 1,415 species of microorganisms known to produce disease to humans. From this total, 60% of the species are *zoonotic* and the majority (72%) originates in wildlife. Approximately 175 pathogenic species are associated with diseases considered to be emerging, and approximately 54% of emerging infectious diseases are caused by bacteria or rickettsia (Tables 1 and 2).

O.A. Oyarzabal (✉)
Department of Biological Sciences, Alabama State University, Montgomery, AL, USA
e-mail: oaoyarzabal@gmail.com

Table 1 Species of microorganisms known to be pathogenic to humans[a]

Category	Number of infectious organisms
Bacteria and rickettsia	538
Helminths	287
Viruses and prions	217
Protozoa	66
Fungi	30

[a]Adapted from Taylor et al. (2001)

Table 2 Examples of emerging infection diseases caused by bacteria and the probable factors explaining their appearance[a]

Infection or agent	Disease	Possible factors contributing to emergence
Haemophilus influenza (biotype aegyptius)	Brazilian purpuric fever	Probably new strain
Vibrio cholera	Cholera	Probably introduced from Asia to South America. Spread facilitated by reduced water chlorination
Helicobacter pylori	Gastric ulcers	Probably long widespread but just recently recognized
Escherichia coli O157:H7	Hemolytic-uremic syndrome	Mass food processing allowing point contamination of large amounts of meat
Legionella pneumophila	Legionnaires' disease	Cooling and plumbing systems
Borrelia burgdorferi	Lyme disease	Reforestation around homes and conditions favoring the expansion of deer (secondary reservoir host)
Streptococcus, group A	Necrotizing skin disease	Unclear

[a]Adapted from Morse (1995)

In general, zoonotic pathogens are more likely to be associated with emerging diseases than nonzoonotic pathogens, although there are variations among taxa, with protozoa and viruses more likely to emerge than helminthes. Presently, no association between the transmission route and the type of emerging infectious diseases has been found (Jones et al. 2008; Taylor et al. 2001). The U. S. National Institute of Allergies and Infectious Diseases has published a list of emerging and reemerging infectious agents; the different foodborne and waterborne pathogens are included in Category B. Within bacteria, this list includes *Escherichia coli* O157:H7, *Campylobacter jejuni*, *Listeria monocytogenes*, *Shigella* spp., *Salmonella* spp., and *Yersinia enterocolitica*. Several *protozoa* species (e.g., *Cryptosporidium parvum*, *Cyclospora cayatanensis*, *Giardia lamblia*, and *Entamoeba histolytica*) as well as viruses (Caliciviruses and Hepatitis A) also appear on the list. For instance, the *hemolytic-uremic syndrome* caused by certain strains of *E. coli* O157:H7 in the United States is an example of an emerging foodborne pathogen that was not reported prior to 1980. On the other hand, the increase in the number of human listeriosis cases in the 1980s was due to the concentration of food production that allowed for a known pathogen, *L. monocytogenes*, to disseminate in a novel way.

3 The Origin of Human Pathogens

It is important to remember that many species closely related to us, such as chimpanzees, have donated many zoonotic diseases. There are different reasons why an animal species that serves as host for a pathogen may become a source of contamination for humans. In the case of chimpanzees, although they have few and infrequent encounters with humans, they may have donated several *zoonoses*. For example, molecular studies of hepatitis B viruses from chimpanzees and humans show that these viruses have a high phylogenetic relationship and therefore may have been donated from chimpanzees to humans. In addition, the emergence of agriculture and the domestication of livestock animals in the last 10,000 years have also favored the appearance of the major human infectious diseases (Wolfe et al. 2007). It has been theorized that in temperate regions of the world, these infectious diseases originated from animals and arrived at humans through what is defined as *species jumps*, which means that a pathogen that was originally confined to animal species evolved to infect humans. Figure 1 shows the proposed five stages in the evolutionary adaptation of a pathogen from being only an animal pathogen to becoming a pathogen that infects only humans (Wolfe et al. 2007). The second category depicted in this figure appears to be the right category in which most of the bacterial and viral foodborne pathogens would fall. Yet we have to recognize that our understanding of some of these diseases increases with time and that these disease agents and their host (humans) are evolving and, therefore, the degree of host–pathogen interaction is continuously in flux.

4 Modern Views of Disease Agents, Evolution, and Epidemiology

Until the 1670s, when Anton van Leeuwenhoek used high-quality lenses to observe living microorganisms (Black 1996), the prevalent theory was spontaneous generation, the idea that living organisms arise from nonliving molecules. The work of Ignaz Semmelweis, who demonstrated that the washing of hands could prevent the spread of childbirth fever; Louis Pasteur, who dismissed the theory of spontaneous generation and developed the *pasteurization* method to make milk safe, among other things; Joseph Lister, who combined the work by Semmelweis and Pasteur to develop and promote antiseptic surgery by the use of chemical compounds; and Robert Koch, who developed a series of postulates (*Koch's postulates*) to directly correlate a microorganism with a specific disease, consolidated the germ theory of disease (Rothman et al. 1995a, b, c).

These events happened in the last 150 years, and the germ theory of disease may be the most important contribution by the science of microbiology to medicine. This theory opened up the possibility for the treatment of diseases by antimicrobials. This theory is also the most important concept to explain biological hazards present in foods because the contamination of foods by pathogenic microorganisms is by far the most important hazard among the three hazard categories (physical, chemical, and biological).

At the same time, the theory of evolution by Charles Darwin provided the platform by which natural processes, including the reproduction, survival, and spread of bacteria, could be studied in an objective fashion. However, it has been within the last 50 years that our tools to study pathogenic microbes flourished to the point where we could interrogate different bacteria and the environment for clues on how these organisms spread and produce disease. Foodborne pathogens are not an exception when compared to other infectious disease agents. However, the systematic study of foodborne disease agents did not appear in a formalized curriculum until just 30–40 years ago.

Another important event that took place in England about 150 years ago allowed for scientists to think about disease agents as "transmissible" agents. When John Snow's request to close a water pump resulted in the control of a cholera outbreak in Soho, England, in 1854 (Porter 1997), the discipline that we now know as epidemiology started. This simple event appears almost an anecdote when compared to the complex epidemiological studies needed to understand modern foodborne outbreaks, in which just the simple association of a food product to a bacterial pathogen during an outbreak investigation becomes a real challenge. The variety of infectious agents

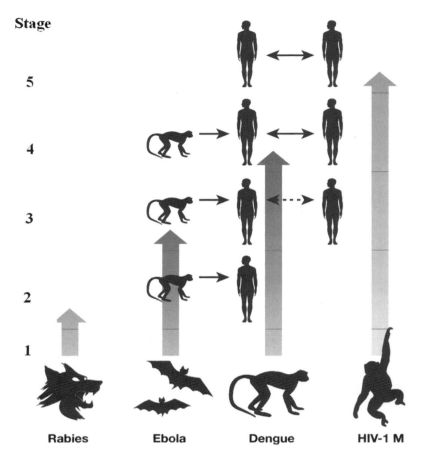

Fig. 1 Different stages of pathogen evolution and adaptation to human infection. In stage 1, the pathogen is constrained to infecting animals only. In stage 2, a "species jump" has occurred and the pathogen can now infect humans. However, humans act as terminal hosts. This second stage is the most common for bacterial foodborne pathogens. It is not clear why animal pathogens that have survived the initial species jump to infect humans do not evolve past stages 2 and 3. Pathogens that make the transition to stages 4 and 5 have a global impact in human populations (Adapted by permission from Macmillan Publishers Ltd., Wolfe et al. 2007)

and the variety of the immunology status of the hosts create a problem that is very difficult to study with current models. For instance, the incubation period of some of these foodborne diseases is measured in days and even weeks, and by the time the first symptoms appear, most of the contaminated foods have been distributed through the retail channels and have spread across vast geographical areas.

5 How Bacteria Evolve

Bacteria, like other prokaryotes, are unicellular organisms that divide using an asexual reproduction scheme called binary fission. In this process, a living bacterium (plural = bacteria) replicates its internal components and organelles and divides itself into two new daughter cells. Although there is no exchange of genetic material from different parents, as is the case with sexual reproduction, bacteria have adopted a series of successful mechanisms to guarantee a degree of DNA variability for their progeny. These key mechanisms include: *mutations* in the DNA mismatch repair system, which increase mutation and recombination rate, and genome rearrangements and horizontal DNA transfer, which ensure the acquisition of survival and/or pathogenicity traits.

Homology-dependent recombination and horizontal (lateral) gene transfer are important mechanisms for the acquisition of DNA diversity (Gogarten et al. 2002). In general terms, genetic recombination in bacteria refers to the occurrence of *mutations* and *horizontal DNA transfer* to change the genetic makeup of a cell. The uptake and acquisition of "foreign" DNA comprise mechanisms such as genetic transformation, bacteriophage transduction, or conjugation. However, it is important to highlight that our understanding of the plasticity of the bacterial genome is limited. It is believed that only half of the bacterial genes from those bacterial species for which we have the complete genomes have known biological functions, and only half of those genes appear to be species-specific (Wren 2006). In addition, the simple uptake of DNA may not explain the pathogenicity potential in bacterial species. The *cadA* gene from *Escherichia coli*, which is missing in *Shigella flexneri*, can reduce virulence when heterologously expressed in *S. flexneri* (Maurelli et al. 1998). Independent losses of the *cadA* gene and other genes in different *Shigella* spp. have provided additional evidence for what is called negative selection, or "purifying" selection, in which deleterious alleles are hindered from being spread (Day et al. 2001; Prunier et al. 2007).

But all these scientific findings are still controversial in their explanation of the relationship of *gene loss* or gene inactivation with pathogenicity. For instance, mice infected with four *Mycobacterium tuberculosis* mutants died more rapidly than those infected with wild-type bacteria (Parish et al. 2003). Yet some data suggest that disruption of some genes leads to attenuation in a mouse aerosol model using the more resistant BALB/c and C57BL/6 mouse strains (Converse et al. 2009). Therefore, we are still missing some key knowledge to understand how bacteria may increase or decrease the activity of certain genes to become more pathogenic for their hosts.

Genetic transformation refers to the acquisition, or uptake, of foreign DNA by bacterial cells. This definition encompasses the acquisition of DNA, usually as single-stranded, that will produce heritable changes. In the majority of the cases, the exchange of DNA occurs among homologous genes, although heterologous genes can also be associated.

The capacity of some bacteria to acquire DNA from the environment is called genetic competence. Some bacteria are competent in natural environments and are naturally prone to the uptake of single-stranded DNA from the surroundings. These bacteria are usually more successful in acquiring linear DNA.

The term "transduction" refers to the passage of DNA from bacteriophages, or viral particles, into bacteria when these viruses infect bacterial cells. Although the main goal of this event is for viruses to perpetuate by using the reproduction machinery of the host cells, some cells can acquire DNA from other bacteria by the viral vector. *Bacteriophages* can also leave other molecules in the infected cell, such as RNA and proteins that make up the coat of the virions.

In the case of bacterial *conjugation*, cell-to-cell contact is necessary for DNA exchange. For DNA to be transferred through conjugation, the presence of an appendage called a pilus (plural=pili) in the membrane of the donor cell is necessary. This appendage probably acts like a tubelike device that connects the donor cell with the recipient cells for DNA exchange to happen. Sometimes *pilus* is used as a synonym of fimbria (plural=fimbriae). However, the latter term refers to small, hairlike appendages that are involved in the attachment of bacteria to surfaces and in the production of biofilms. The mechanism of conjugation is complex and involves different proteins that form what is called a *type IV secretion system*. The most common DNA molecules exchanged during conjugation are *plasmids*, which are extrachromosomal DNA molecules that replicate independently from the chromosome. The secretion systems that allow for conjugation are important for the transfer of plasmids from one bacterium to another. Plasmids can eventually be integrated into the chromosome of the recipient bacterium by genetic recombination and can bring some extrachromosomal DNA that may confer specific traits to the bacteria that now carry those plasmids. For instance, some plasmids carry genetic material that provides antimicrobial resistance to the new host cell.

Another group of mobile genetic elements, called *pathogenicity islands* (PAIs), can move from cell to cell probably using conjugative transfer systems and can contribute essential elements for virulence in the pathogens of both animals and plants. PAIs are frequently part of complex regulatory networks that include regulators encoded by genetic material in the chromosome or by plasmids. PAIs themselves can act as regulators of genes located outside the PAI (Schmidt and Hensel 2004).

Some pathogens also have the ability to reversibly alternate or change between two genetic phenotypes, a phenomenon called *phase variation*, which results in two different phenotypic appearances according to the level of expression of one or several proteins among the different cells of a bacterial population. The occurrence of phase variation can be in one cell per 10 cells per generations, but it is more frequently on the order of one change per 10^{-5} cells per generation (Villemur and Deziel 2005). If phase variation results in changes to the surface pathogenic factors of infectious bacteria, such as pili or glycoproteins, which are recognized by the immune system of host cells, the mechanism is known as antigenic variation. The major benefit of antigenic variation is that pathogenic bacteria can alter their surface proteins to create clonal populations that are antigenically distinct and therefore can evade the hosts' immune responses. This mechanism is the main reason why it is so difficult to create stable vaccines against some pathogenic bacteria (Villemur and Deziel 2005).

6 New Opportunities for Pathogens to Infect Humans

The changes in bacterial pathogens are important evolutionary strategies to create genetic diversity and take advantage of conquering new colonization niches. However, the expansion of humans into new land and changes in human behavior have also created new opportunities for bacterial foodborne pathogens to be exposed to and infect humans. To further complicate this scenario, the exposure of humans to new carriers of foodborne pathogens creates additional new opportunities for these pathogenic bacteria to infect us. An example of the latter scenario is the increase trade of exotic animals as pets, which have increased the risk of introducing some pathogens that otherwise will not be present in certain human populations. Particularly, several foodborne cases of *Salmonella serotypes* in the United States have been linked to reptile pets imported from South America (Anonymous 1995; Mermin et al. 1997).

Several crucial changes have occurred in agricultural practices in the last 50 years. One of these changes, the concentration of massive food production, has created unique food safety concerns. As food distribution has increased to cover large areas, and even different countries, it has become more difficult to keep track of where the food was produced and processed. In some cases, food is transported across different countries; therefore, a bacterial pathogen unique to some specific areas in the world may end up in a completely different area of the world. A good example of the latter is the 2008 outbreak of a virulent *Salmonella* serotype Saintpaul responsible for illnesses associated with the consumption of tomatoes. Suppliers of tomatoes normally rely on more than one grower to fill their orders, and tomatoes are not classified by origin but by ripeness, size, and grades during processing. Thus, tomatoes collected in Florida may be shipped to Mexico for packaging before they are sent back to the United States for final sale. In addition, the incorporation of sliced tomatoes in salad bars, deli counters, or supermarket salsas makes it extremely difficult to track where the tomatoes originated. The investigation into this particular outbreak of *Salmonella* Saintpaul resulted in suspicion that farms from Mexico and Florida were the ones involved in the production of the contaminated tomatoes. However, more than 1,700 samples collected from irrigation sources and packing, washing, and storage facilities were negative, and there was never a clear resolution of the actual source of the outbreak.

The international trade of food commodities and the ease with which people can move from different geographical areas have a long-term effect on food safety. The movement of foods increases the possibility of pathogens traveling in hiding from seemingly remote geographical. But humans also serve as carriers when they get infected in a country but develop the symptoms and suffer the disease in another country. An example is the cases of salmonellosis in Sweden that still remain despite all the efforts to control the domestic cases of salmonellosis (Motarjemi and Adams 2006). Most of these cases are associated with the contamination of travelers who return home with the infectious agents. As international food trade becomes more prevalent, countries that strive to control specific foodborne agents may see their efforts curtailed and therefore will pressure international organizations to adopt more stringent international food safety regulations.

Viruses are also opportunistic agents. The fact that we are still missing reliable techniques to isolate and identify some viruses makes it more difficult to study them than to study bacteria. The most recent examples of noroviruses affecting passengers on recreational cruises highlight the importance of food safety in new settings that were uncommon years ago.

6.1 Changes in Food Production and Processing Practices

The changes in human populations and the way the increased need for more foods has been addressed are historically similar in many industrialized nations. The key for the successful provision of quality foods has depended on the availability of technologies to preserve foods, mainly refrigeration systems to lower the temperature, and the availability to transport the food in an efficient, economic fashion, mainly the development of railroad systems.

Since the 1950s, food manufacturing companies have been consolidating to process more food per unit of land. Until the 1970s, that consolidation related mainly to the processing of meats, but in the last few years the consolidation has expanded into other vegetable food products. As the population expanded, there was a demand for more food, and the basic food needs, such as milk and eggs, were covered by increasing production in suburban areas. However, other food products (corn, meat, etc.) have tended to concentrate in areas where the productivity has been the highest. For instance, in the U.S. Midwest, the fertility of the soils is high enough that the production of corn or soybeans allows for the highest profitability of the land. Therefore, the shipping of foods across different areas has allowed for large human populations to concentrate and get a more steady supply of food products. The meat industry took advantage of these developments; by the end of the 1890s, refrigerated train cars were already in place to transport the cattle stock to central points for processing and to transport processed products to large urban conglomerates across the nation. The consolidation of the meat-packing industry started early, with a large number of animals processed in one location and the opportunity for unsanitary conditions to emerge and contaminate the product, as depicted in 1906 by Upton Sinclair in his novel The Jungle (Sinclair 1906). Therefore, new regulations were put in place to deal with these new challenges. More recently, the increase in the consumption of fresh-cut produce and leafy greens, such as carrots, celery, and spinach, products that are usually consumed raw, have created a similar scenario where the industry and the government have to work on the appropriate minimum set of regulations to be put in place to control the occurrence of foodborne diseases associated with these products. For more details, refer to chapter "Food Regulation in the United States."

As the traditional agricultural systems have evolved from large areas/low-productivity systems to more concentrated, small areas/high-productivity systems, so have some biological agents. Some pathogenic bacteria have adapted to thrive when food animals and their corresponding food products concentrate in small areas. For instance, *Listeria monocytogenes*, a pathogenic bacterium, can colonize a niche within a processing facility and contaminate a large volume of food in a matter of hours.

Listeria monocytogenes is a dangerous pathogen because of the chances of the postprocessing contamination of food products. This example again shows the different opportunities for adaptation and resiliency to survive, replicate, and spread that some bacterial foodborne pathogens have, even when presented with adverse environmental conditions.

7 Recognition of At-Risk Populations

Many important improvements in public health have been achieved in the last century. Most of these improvements, such as the pasteurization of milk or the processing alternatives developed for meat products, are directly related to the control of foodborne pathogens. In the same fashion we have improved our understanding of the immunological limitations that some unique groups of individuals in a given community have. Certain sectors of the population, such as infants, elders, people suffering from debilitating diseases, and pregnant women, may have an immature or compromised immune system that makes them more susceptible to diseases. These populations of individuals, generally referred to as at-risk populations, pose an important challenge in the control of foodborne diseases. More importantly, the demographics of these populations are always in flux. For instance, the proportion of people described as "elders" is increasing, and by the year 2025, more than 20% of the worldwide population is expected to be above 60 years old (Motarjemi and Adams 2006).

For these populations, educational messages are very important for their health, and the four principles promoted to help reduce the risk of contracting a foodborne illness (clean, *s*eparate, cook, and chill) are part of the educational campaigns of several governmental agencies and the food industry. These individuals must develop a strict habit of thoroughly washing their hands before and after eating, and before and after handling or preparing any foods. Keeping raw or uncooked products, such as meat and poultry, away from *ready-to-eat* foods, such as fresh fruits and vegetables, is also an important principle to prevent the cross-contamination of ready-to-eat food with pathogenic bacteria from raw food products.

8 Changes to and Expansion of Our Diets

In industrialized nations and even in urban sectors of developing countries, people have better access to a variety of food products than ever before. And the trend is that more food product choices will be available for the public. Yet, at the same time, urban dwellers have less understanding of how the food is produced and processed than ever before; unfortunately, the trend is that people know less and less about the origin and composition of their foods. About 40–50 years ago, most people knew the basis of how the common foods were produced. Today, more people are not aware of the intricacies of food production and are inclined to believe erroneous concepts about food safety. Some examples are the belief that hormones are commonly used in the raising of commercial broiler chickens, when, in reality, no hormones are used in commercial broiler production in the United States.

The perception of food safety is very important and creates different conflicts among people with different knowledge of food production and processing. For instance, some people that choose to consume raw milk do so because they may believe there are certain advantages of consuming raw milk, such as improved immunological responses. Although there may be some perceived benefits associated with the consumption of raw milk, the risk of acquiring foodborne diseases is much greater by following this practice. The consumption of raw milk has been repeatedly demonstrated in the past to cause outbreaks of *Escherichia coli* O157:H7, which can cause hemolytic-uremic syndrome, a life-threatening complication for children. There are many public health challenges that emerge from the expansion of our food supplies and from choosing to consume high-risk foods. The development of food safety legislation can help protect people, but consumer education and more research on disease epidemiology are also important factors to control foodborne diseases.

9 Summary

Emerging and reemerging pathogens are disproportionately zoonotic, and emerging foodborne diseases are not an exception. New infectious diseases emerge and adapt to infect humans by the "species jump" concept. Although pathogens may have been exposed to similar evolutionary forces, each bacterial pathogen appears to adapt in its own very unique way. Besides the changes associated with the pathogenic organisms themselves, the changes in the human population and lifestyle, the globalization of the food supply, and the inadvertent introduction of pathogens into new geographic areas are some of the most important forces responsible for the creation of new opportunities for pathogens to infect humans. The complexity of foodborne diseases is highlighted by the unpredictable susceptibility of certain individuals to infection by foodborne agents. Similar to other diseases, the complete eradication of the etiological agents responsible for foodborne diseases is not feasible. Finally, continuous research efforts to better understand the conditions necessary to control foodborne pathogens and consistent consumer education efforts will allow for a quick response to react to new, or reemerging, foodborne pathogens when they strike.

References

Altekruse, S.F., M.L. Cohen, and D.L. Swerdlow. 1997. Emerging foodborne diseases. *Emerging Infectious Diseases* 3: 285–293.
Anonymous. 1995. CDC. Reptile-associated salmonellosis – selected states, 1994–1995. *Morbidity and Mortality Weekly Report* 44: 347–350.
Anonymous. 2002. Foodborne diseases, emerging. Fact sheet no.124. [Online] http://www.who.int/mediacentre/factsheets/fs124/en/. Accessed 5 Dec 2010.
Black, J.G. 1996. *Microbiology: Principles and applications*, 3rd ed, 1–25. Upper Saddle River: Prentice Hall.
Converse, P.J., P.C. Karakousis, L.G. Klinkenberg, A.K. Kesavan, L.H. Ly, S.S. Allen, J.H. Grosset, S.K. Jain, G. Lamichhane, Y.C. Manabe, D.N. McMurray, E.L. Nuermberger, and W.R. Bishai. 2009. Role of the *dosR-dosS* two-component regulatory system in *Mycobacterium tuberculosis* virulence in three animal models. *Infection and Immunity* 77: 1230–1237.
Day Jr., W.A., R.E. Fernandez, and A.T. Maurelli. 2001. Pathoadaptive mutations that enhance virulence: genetic organization of the *cadA* regions of *Shigella* spp. *Infection and Immunity* 69: 7471–7480.
Gogarten, J.P., W.F. Doolittle, and J.G. Lawrence. 2002. Prokaryotic evolution in light of gene transfer. *Molecular Biology and Evolution* 19: 2226–2238.
Jones, K.E., N. Patel, M. Levy, A. Storeygard, D. Balk, J.L. Gittleman, et al. 2008. Global trends in emerging infectious diseases. *Nature* 451: 990–993.
Maurelli, A.T., R.E. Fernandez, C.A. Bloch, C.K. Rode, and A. Fasano. 1998. "Black holes" and bacterial pathogenicity: a large genomic deletion that enhances the virulence of *Shigella* spp. and enteroinvasive *Escherichia coli*. *Proceedings of the National Academy of Sciences of the United States of America* 95: 3943–3948.
Mermin, J., B. Hoar, and F.J. Angulo. 1997. Iguanas and *Salmonella* Marina infection in children: a reflection of the increasing incidence of reptile-associated salmonellosis in the United States. *Pediatrics* 99: 399–402.
Morse, S.S. 1995. Factors in the emergence of infectious diseases. *Emerging Infectious Diseases* 1: 7–15.
Motarjemi, Y., and A. Adams. 2006. *Emerging foodborne pathogens*. Boca Raton: CRC Press.
Parish, T., D.A. Smith, S. Kendall, N. Casali, G.J. Bancroft, and N.G. Stoker. 2003. Deletion of two-component regulatory systems increases the virulence of *Mycobacterium tuberculosis*. *Infection and Immunity* 71: 1134–1140.
Porter, R. 1997. Public medicine. In: The *Greatest benefit to mankind: a medical history of humanity*, 412–414. New York: W.W. Norton.

Prunier, A.L., R. Schuch, R.E. Fernandez, and A.T. Maurelli. 2007. Genetic structure of the *nadA* and *nadB* antivirulence loci in *Shigella spp*. *Journal of Bacteriology* 189: 6482–6486.

Rothman, D.J., S. Marcus, and S.A. Kiceluk. 1995a. On the extension of the germ theory to the etiology of certain common diseases. In: *Medicine and western civilization*. 253–257. New Brunswick: Rutgers University Press.

Rothman, D.J., S. Marcus, and S.A. Kiceluk. 1995b. On the antiseptic principle in the practice of surgery. In: *Medicine and western civilization*. 247–252. New Brunswick: Rutgers University Press.

Rothman, D.J., S. Marcus, and S.A. Kiceluk. 1995c. The etiology of tuberculosis. In: *Medicine and western civilization*. 319–329. New Brunswick: Rutgers University Press.

Schmidt, H., and M. Hensel. 2004. Pathogenicity islands in bacterial pathogenesis. *Clinical Microbiology Reviews* 17: 14–56.

Sinclair, U. 1906. The Jungle. Prepared and Published by E-BooksDirectory.com.

Smith, J.L., and P.M. Fratamico. 2005. Emerging foodborne pathogens. In *Foodborne pathogens: Microbiology and molecular biology*, ed. P.M. Fratamico, A.K. Bhunia, and J.L. Smith. Norwich: Caister Academic.

Taylor, L.H., S.M. Latham, and M.E.J. Woolhouse. 2001. Risk factors for human disease emergence. *Philosophical Transactions of the Royal Society of London. Series B* 356: 983–989.

Villemur, R., and E. Deziel. 2005. Phase variation and antigenic variation. In *The dynamic bacterial genome*, ed. P. Mullany, 277–322. New York: Cambridge University Press.

Wolfe, N.D., C. Panosian Dunavan, and J. Diamond. 2007. Origins of major human infectious diseases. *Nature* 447: 279–283.

Wren, B. 2006. How bacterial pathogens evolve. In *Emerging foodborne pathogens*, ed. Y. Motarjemi and A. Adams, 3–22. Boca Raton: CRC Press.

Clinical Presentations and Pathogenicity Mechanisms of Bacterial Foodborne Infections

Nicole Tegtmeyer, Manfred Rohde, and Steffen Backert

1 Introduction

The gastrointestinal (GI) tract is one of the largest and most important organs in humans. In a normal male, the GI tract is approximately 6.5 m (20 ft) long and is covered by the intestinal epithelium, which has a surface of about 400–500 m^2. This epithelium not only exhibits crucial absorptive and digestive properties, it also represents an efficient barrier against commensal microbial flora as well as foodborne pathogens. The gut flora consists of more than 1,000 microbial species, which shape a highly complex and dynamic community (Hooper and Gordon 2001; Eckburg et al. 2005). The exclusion of these microbes is not only a result of the continuous physical barrier formed by the tightly bound epithelial cells; the intestinal epithelium also exhibits crucial host immune functions to recognize commensals and eliminate pathogens (Sansonetti 2004; Tsolis et al. 2008). The immune system controls the resident microflora and defends against infections by foodborne pathogens through two functional arms: the innate immunity and the adaptive immunity. Interestingly, commensal bacteria that colonize the gut also protect the host from intruding pathogens by imposing a colonization barrier, also called the barrier effect (Stecher and Hardt 2008). The recognition of these microbes is commonly based on the identification of pathogen-associated molecular patterns (PAMPs) by defined pattern recognition receptors (PRRs) expressed in a variety of host cells. Typical PAMPs are lipopolysaccharides (LPSs), flagellins, or peptidoglycans that are either present on the bacterial cell surface or spontaneously released from the bacteria upon contact with the target cell. Such factors are commonly recognized at the plasma membrane by PRRs. A classical PRR is the family of Toll-like receptors (TLRs), which consists of 10–15 members in most mammalian species (Beutler et al. 2006; Palm and Medzhitov 2009). The pattern recognition by TLRs is subsequently transduced into proinflammatory signaling pathways that activate numerous transcription factors, including nuclear factor kappa B (NF-κB) and AP-1 (Tato and Hunter 2002; Chen and Greene 2004; Backert and Koenig 2005; Ghosh and Hayden 2008). Most of these signals are transported through dendritic cells (DCs), which deliver pathogen-derived antigens from the tissues to the secondary lymphoid organs and prime T cells by providing costimulation as well as appropriate

N. Tegtmeyer • S. Backert (✉)
University College Dublin, School of Biomolecular and Biomedical Sciences,
Science Center West, Belfield Campus, Dublin 4, Ireland
e-mail: steffen.backert@ucd.ie

M. Rohde
Helmholtz Centre for Infection Research, Braunschweig, Germany

cytokines and other mediators. These mediator molecules also activate macrophages, neutrophils, and mast cells, which are recruited to the site of infection and then eliminate a given pathogen. The functions of the above-mentioned immune and epithelial cells have been reviewed thoroughly (Hornef et al. 2002; Boquet and Lemichez 2003; Liston and McColl 2003; Monack et al. 2004; Backert and Koenig 2005; Pédron and Sansonetti 2008; Tsolis et al. 2008).

Despite the sophisticated immune system, some foodborne pathogens have coevolved with their hosts to overcome protective cell barriers and to establish short- or long-term infections. *Escherichia coli, Campylobacter, Salmonella, Listeria, Shigella,* and other bacterial species as well as some enteric viruses and parasites represent the most common foodborne pathogens (Fang et al. 1991; Salyers and Whitt 1994; Sougioultzis and Pothoulakis 2003; Eckmann and Kagnoff 2005; Lamps 2007). Importantly, infections with these microbes are one of the leading causes of morbidity and death of humans worldwide. Estimations by the World Health Organization (WHO) indicate that the human world population suffered from 4.5 billion incidences of diarrhea, causing 1.8 million deaths, in the year 2002 (WHO 2004). These infections are especially problematic among infants, young children, and immunocompromised persons, whereas the majority of enteric infections in healthy adults seem to be self-limiting. Those patients who undergo endoscopic biopsy often have chronic or debilitating diarrhea, systemic symptoms, or other significant clinical scenarios. Foodborne infections are estimated to affect one in four Americans each year. Most of these infections (67%) are caused by the noroviruses, but *Campylobacter* and nontyphoidal *Salmonellae* together account for about one fourth of the cases of illness in which a pathogen can be detected (Mao et al. 2003). Less common bacterial infections, such as with Shiga toxin-producing *E. coli, Shigella,* or *Listeria* species, cause fewer infections but are also important because of their severe complications or high mortality rate or both (Sougioultzis and Pothoulakis 2003). Upon ingestion, such pathogens commonly pass through the stomach in sufficient numbers to infect the small intestine or colon. To establish and maintain a successful infection in this compartment, microbial pathogens have evolved a variety of strategies to invade the host, avoid or resist the innate immune response, damage the cells, displace the normal flora, and multiply in specific and normally sterile regions. During evolution, several bacterial pathogens developed well-known weapons, such as protein toxins or effector proteins of a specialized type III secretion system (T3SS), which play major roles in these processes (Thanassi and Hultgren 2000; Burns et al. 2003; Alouf and Popoff 2005). Most, but not all, bacterial foodborne pathogens can be classified as so-called "invasive bacteria," which are able to induce their own uptake into gastric epithelial cells that are normally nonphagocytic. According to specific characteristics of the entry process, we distinguish between the "zipper" and "trigger" mechanisms, respectively (Cossart and Sansonetti 2004). The "zipper" mechanism is initiated by a bacterial surface protein (adhesin), which binds to a specific host cell receptor followed by internalization of the bacterium, whereas the "trigger" mechanism involves injected bacterial factors by T3SSs, which often mimic or highjack specific host cell factors to trigger the uptake process (Fig. 1). Typical examples and morphologic features are shown in respective scanning electron micrographs (Fig. 2, top). The invasion process commonly involves rearrangements of the cytoskeleton and/or the microtubule network, which facilitate bacterial uptake at the host cell membrane (Rottner et al. 2005). Other cross-talks alter the trafficking of cellular vesicles and induce changes in the intracellular compartment in which they reside, thus creating niches favorable to bacterial survival and growth. Finally, a variety of strategies also exist to deal with other components of the epithelial barrier, such as macrophages. Prophagocytic, antiphagocytic, and proapoptotic processes seem to be of particular importance in this context. This chapter describes the pathogenicity mechanisms and clinical presentations of selected bacterial foodborne pathogens as well as the associated diseases in humans.

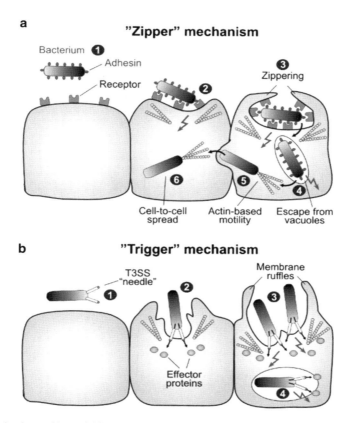

Fig. 1 Primary mechanisms of bacterial invasion into nonphagocytic host cells. Schematic representation of the two different routes of entry by intracellular bacterial pathogens. The pathogens induce their own uptake into target cells by subversion of host cell signaling pathways using the "zipper" and "trigger" mechanism, respectively. (**a**) Bacterial GI pathogens commonly colonize the gastric epithelium (step 1). The zipper mechanism of invasion involves the high-affinity binding of bacterial surface adhesins to their cognate receptors on mammalian cells (step 2), which is required to initiate cytoskeleton-mediated zippering of the host cell plasma membrane around the bacterium (step 3). Subsequently, the bacterium is internalized into a vacuole. Some bacteria developed strategies to survive within or to escape from this compartment (step 4). A well-known example of this invasion mechanism is *Listeria*, which escapes into the cyotsol and triggers actin-based motiliy (step 5) involved in the cell-to-cell spread of the bacteria (step 6). (**b**) The trigger mechanism is used by *Shigella* or *Salmonella* spp., which also colonize the gastric epithelium (step 1). These pathogens use a sophisticated type III secretion system (T3SS) and translocate multiple injected effector proteins into the host cell cytoplasm (step 2). These factors manipulate a variety of signaling events, including the activation of small Rho GTPases and actin-cytoskeletal reorganization, to induce membrane ruffling and subsequently bacterial uptake (step 3). As a consequence of this signaling, the bacteria are internalized into a vacuole (step 4), followed by the induction of different signaling pathways to establish infection including actin-based motility, entry into macrophages, and others. For more details, see text

2 *Salmonella* spp.

Salmonella spp. are Gram-negative bacilli that are members of the enterobacteriaceae family. Due to old nomenclature, the genus was originally split into three species: *Salmonella typhi* (the cause of typhoid fever), *Salmonella cholaesuis* (primarily a pathogen in swine), and *Salmonella enteritidis* (a common cause of diarrheal infections in humans and animals) (Salyers and Whitt 1994). Today it is commonly accepted that there are only two species: *Salmonella enterica* and *Salmonella*

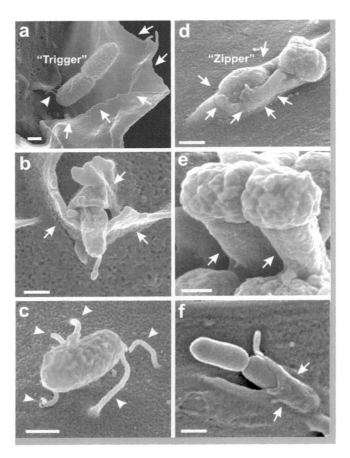

Fig. 2 Scanning electron micrographs of enteric bacterial pathogens interacting with epithelial cells in vitro. Selected examples include (**a**) *Salmonella enterica*, (**b**) *Campylobacter jejuni*, (**c**) *Shigella flexneri*, (**d**) *EHEC*, (**e**) *EPEC*, and (**f**) *Listeria monocytogenes*. The induction of membrane dynamics in cases of *Salmonella*, *Campylobacter*, and *Listeria* is indicated with arrows. *Salmonella* is a typical bacterium invading gastric epithelial cells by the trigger mechanism as indicated. The *arrows* for EHEC and *Listeria* indicate the tight engulfment of bacteria, which exhibit typical features of the zipper mechanism of invasion. EPEC induces classical actin-pedestal formation, as shown for two bacteria in panel (**e**) (*arrows*). *Arrowheads* indicate the presence of typical T3SS injection needles at the bacterial surface as observed for *Salmonella* and *Shigella*. Each *bar* represents 500 nm. For more details, see text

bongori (Boyd et al. 1996). *Salmonella enterica* was then classified into seven subspecies (I, II, IIIa, IIIb, IV, VI, and VII) containing more than 2,500 serovars according to the typing of different antigens. *Salmonella* spp. are able to infect numerous hosts and cause a broad spectrum of diseases in humans and animals, ranging from intestinal inflammation and gastroenteritis up to systemic infections and typhoid fever (Haraga et al. 2008; Tsolis et al. 2008). *Salmonella* spp. are the cause of sporadic food poisoning in developed countries but are especially prevalent in developing countries, where sanitation is poor and dairy and water supplies are contaminated with the bacterium. Animal food is also frequently contaminated with *Salmonella* spp. and may lead to infection in or colonization of domestic animals (Crump et al. 2002). Thus, most outbreaks in humans are associated with the consumption of contaminated eggs, egg products, poultry, and meat products (Mao et al. 2003). However, the pathogen is occasionally also detected in vegetables or fruits (Fang et al. 1991). The infective dose is moderate; approximately 10^2–10^3 ingested bacterial cells are sufficient to cause disease. Historically, the discussion about *Salmonella* spp. is divided into typhoid and nontyphoid species. Patients with typhoid (enteric) fever, usually caused by *S. typhi*, suffer from fever persisting over several days, abdominal pain, and headache (Lamps 2007). Abdominal rash ("rose spots"), delirium, hepatosplenomegaly, and leukopenia are also fairly common. In the second or third week after infection, watery diarrhea begins and may progress to severe GI bleeding and even perforation. In contrast, nontyphoid *Salmonella* (e.g., *S. enteritidis, S. typhimurium, S. muenchen, S. anatum, S. paratyphi,* and *S. give*) generally cause a milder, often self-limited gastroenteritis with vomiting, nausea, fever, and watery diarrhea (McGovern and Slavutin 1979; Boyd 1985; Pegues et al. 1995; Kelly and Owen 1997; Kraus et al. 1999). In the United States, about 1.5 million cases of nontyphoid *Salmonella* infection are reported each year, and 95% of those cases are related to food (Mead et al. 1999). This accounts for about 10% of foodborne enteric diseases in the United States. Although most *Salmonella* infections in developed countries resolve by antibiotic treatment and supportive care, enteric infections may progress to septicemia and death, particularly in the elderly, very young children, or patients who are debilitated. Delayed treatment is associated with higher mortality (Pegues et al. 1995). During infection, any level of the GI tract may be involved, but the ileum, appendix, and right colon are preferentially affected. The bowel wall is enlarged, with raised nodules corresponding to hyperplasic Peyer's patches. Apthoid ulcers overlying Peyer's patches, linear ulcers, discoid ulcers, or full-thickness ulceration and necrosis often occur when infection continues (McGovern and Slavutin 1979; Boyd 1985; Pegues et al. 1995; Kelly and Owen 1997; Kraus et al. 1999). In infections with nontyphoid *Salmonella*, the overall findings are rather uncomplicated. The lesions can be focal, and occasionally the mucosa is grossly normal or only mildly hyperemic and edematous. The pathological features are most often those of acute self-limited colitis, although severe cases may have significant crypt distortion.

Salmonella infections are dominated by their profound capabilities to invade host cells by a trigger mechanism and to proliferate intracellularly (Fig. 2a). All known *Salmonella* are highly invasive, facultative intracellular pathogens that preferentially enter the microfold cells (M cells) overlying small intestinal Peyer's patches (Jepson and Clark 2001) although they can also enter and pass through epithelial cells of the intestinal tract in vivo and in cultured polarized epithelial cells in vitro. In addition, *Salmonella* can penetrate the intestinal epithelial barrier by uptake into DCs that protrude into the intestinal lumen (Niess et al. 2005). Once the bacteria have crossed the epithelium, they either are present inside the DCs or are quickly taken up by those cells or macrophages within the lamina propria. Once internalized, macrophages then transport the bacteria from the GI tract to the bloodstream, ultimately leading to a systemic infection. Early studies have shown that neutrophils begin to accumulate underneath the epithelium within several hours of infection, indicating a rapid immune response to infection (Takeuchi 1967). The molecular basis of *Salmonella*'s virulence has been approached by screens for attenuated mutants, which lead to the identification and characterization of multiple genes involved in host cell entry and intracellular survival, replication, and spread. Many important virulence traits are clustered within the so-called *Salmonella* pathogenicity islands (SPIs) (Gal-Mor and Finlay 2006; Gerlach and Hensel 2007). In general, pathogenicity islands are large chromosomal regions that are present in pathogenic bacteria; they confer virulence properties and are absent in nonvirulent strains. They were first discovered in uropathogenic *E. coli* and since then have also been found in the chromosomes of many foodborne pathogens (Hacker et al. 2004; Gal-Mor and Finlay 2006). *Salmonella* strains can encode up to 17 SPIs (called SPI1–SPI17), and the availability of genome sequences of several *S. enterica* serotypes revealed serovar-specific SPIs (Gerlach and Hensel 2007). The best-characterized pathogenicity islands are SPI1 and SPI2. While SPI1 and SPI2 encode T3SSs with well-established roles in invasion (Patel and Galan 2005) and intracellular lifestyle (Kuhle and Hensel 2004), respectively, the role of many other putative SPIs still has to be elucidated. The events during the invasion of host cells have been studied in great detail both on the cellular and molecular levels. The two T3SSs (T3SS-1 and T3SS-2) can be described as multiprotein complexes spanning the inner and outer membranes of the Gram-negative bacterium to form needlelike injection devices for effector proteins. These effector proteins modify signaling events of the host cell, leading to multiple responses, most notably rearrangements in the actin network (Patel and Galan 2005; Schlumberger and Hardt 2006). The best-characterized T3SS-1 effector proteins are SopE, SopE2, and SopB, which act in concert to activate small Rho GTPases Cdc42 and Rac1 in the host cell, by mimicking the action of eukaryotic G-nucleotide exchange factors either directly (SopE/E2) or indirectly by the generation of phosphatidyl inositol phosphates (PIPs). The activated GTP-bound forms of Cdc42 and Rac1 act synergistically to stimulate the conversion of host monomeric actin (G-actin) into filamentous actin (F-actin) by an actin-polymerization machinery, the Arp2/3 (actin-related proteins 2 and 3) complex. In addition, the effectors SipA and SipC interact directly with actin, thus inducing de novo polymerization and stabilization of F-actin. Another effector, SptB, counteracts this signaling as a GTPase-activating protein (GAP), thus inactivating Cdc42 and Rac1 (Patel and Galan 2005; Schlumberger and Hardt 2006). It has been proposed that SptB may downregulate GTPase functions once the bacteria have successfully entered the host cell. A simplified model of this signaling is shown in Fig. 3a. As a result of this complex manipulation scenario, local membrane ruffles are formed that ultimately trigger the engulfment and internalization of the bacteria, a process called *macropinocytosis*. Recent studies demonstrated that this invasion phenotype is also closely linked to trafficking along microtubule networks and the induction of intestinal inflammation by *Salmonella* (Hapfelmeier and Hardt 2005; Gerlach and Hensel 2007).

3 *Campylobacter jejuni*

Campylobacter infections of the human GI tract are recognized as the leading causes of enteric bacterial infection (Nachamkin et al. 2008), which may be responsible for as many as 400–500 million bacterial gastroenteritis cases worldwide each year (Friedman et al. 2000). Statistical data show that *Campylobacter* infections of humans cause a considerable use of medication and health service burden. In the United States, it has been estimated that *Campylobacter*-associated illnesses cost up to $6.2 billion per year (Forsythe 2000). Remarkably, in many studies in the United States and other industrialized countries, *Campylobacters* were found to cause diarrheal disease more than two to seven times as frequently as *Salmonella* and *Shigella* species or pathogenic *E. coli* (Allos 2001; Tam 2001). The genus of this Gram-negative bacterium currently comprises 17 species; two of them, *C. jejuni* and *C. coli*, are most frequently isolated from infected humans. *Campylobacter jejuni* is a typical zoonotic pathogen, as it can be found as part of the normal GI flora in numerous mammals and birds. Thus, *C. jejuni* may contaminate poultry, beef, veal, pork, water, and milk during food processing, as mainly transmitted by the fecal–oral route (Potturi-Venkata et al. 2007). The infective dose is relatively moderate: As few as 500 ingested bacteria can cause symptomatic disease. *Campylobacter* remain highly motile in the intestinal mucus, and their microaerobic nature promotes survival in the mucus layer. As a consequence of infection, the bacteria colonize the ileum and colon, where they can interfere with normal secretory and absorptive functions in the GI tract. This may cause certain intestinal diseases typically associated with fever, malaise, abdominal pain, and

Fig. 3 Bacterial attachment, injection of toxins, and effector proteins involved in signal transduction and host cell invasion. Schematic representation of the initial interactions between selected pathogens and host cell leading to injection of bacterial virulence factors or triggering signaling pathways, which eventually result in bacterial uptake. The signaling of selected pathogens is shown in a simplified manner. Major bacterial and host cell factors of infections with (**a**) *Salmonella*, (**b**) *Campylobacter*, (**c**) *Shigella*, (**d**) *EHEC*, (**e**) *EPEC,* and (**f**) *Listeria* are summarized. For details, see text

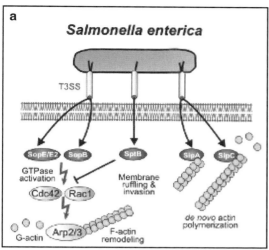
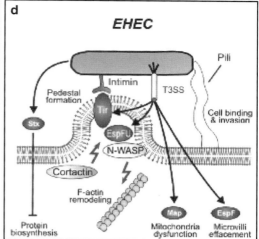
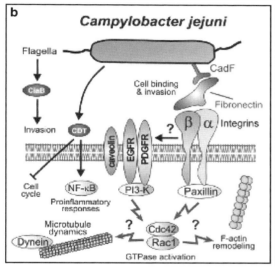
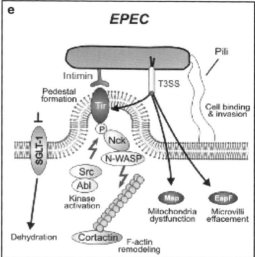
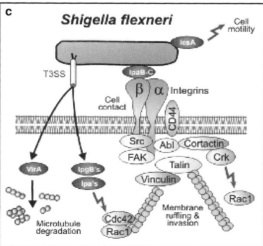
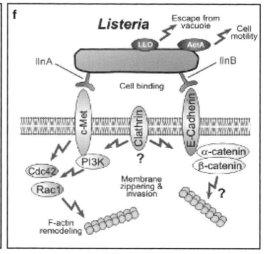

watery diarrhea that often contains blood and leukocytes (Wassenaar and Blaser 1999; Poly and Guerry 2008). Endoscopic findings are commonly nonspecific and include friable colonic mucosa with associated erythema and hemorrhage. Histological examination shows features of acute self-limited colitis, including a neutrophilic infiltrate in the lamina propria (Lamps 2007). Symptoms generally appear within 1–5 days of exposure and may last for 4–10 days. Most of these infections are self-limited as mentioned above, particularly in healthy persons, although relapse is common. In addition, individuals exposed to *C. jejuni* may develop postinfection sequelae, including Reiter's reactive arthritis or peripheral neuropathies such as Miller–Fisher and Guillain–Barré syndromes (Blaser and Engberg 2008). In contrast to the situation found in infected humans, *C. jejuni* is considered to be a commensal bacterium in chicken and other avian species. In particular, poultry is thought to be an important natural reservoir for *Campylobacters* that is supported by their perfect adaptive characteristics. For example, the optimal growth temperature of *C. jejuni* (42°C) is the same as that of the avian intestine. However, experimental infection of chicken with *C. jejuni* can also lead to diarrhea, but this is not typical. It appears that the human response to *C. jejuni* infection is more symptomatic than that of chicken. However, the molecular basis for the different outcomes of *C. jejuni* infection in humans versus chickens is not well understood and one of the pressing questions to be solved in the future.

Accumulating research work over the last few years has indicated that *C. jejuni* perturbs the normal absorptive capacity of the intestine by damaging epithelial cell function either directly by cell invasion and/or the production of toxin(s) or indirectly via the initiation of an inflammatory response (Ketley 1997; Wooldridge and Ketley 1997). The cytolethal distending toxin (CDT) is considered an important *C. jejuni* virulence factor that encodes a nuclease that damages DNA and causes cell-cycle arrest (Hickey et al. 2000; Lara-Tejero and Galan 2000). This toxin is essential for persistent infection of the GI tract and increases the severity of mucosal inflammation in susceptible mouse strains in vivo (Ge et al. 2008), and it was shown to play a role in *Campylobacter*-induced NF-κB activation and secretion of the proinflammatory chemokine interleukin-8 (IL-8) using polarized T84 human colonic epithelial cells as the in vitro model (Zheng et al. 2008). Early studies of intestinal biopsies from patients and in vitro infection of cultured human intestinal epithelial cells have demonstrated that *C. jejuni* is able to invade gut tissue cells (van Spreeuwel et al. 1985; Oelschlaeger et al. 1993; Wooldridge et al. 1996). Numerous studies demonstrated that *C. jejuni* encode a variety of adhesins, including CadF, JlpA, and PEB1 (Pei et al. 1998; Konkel et al. 2001; Poly and Guerry 2008). For example, CadF is a well-characterized bacterial outer membrane protein that binds fibronectin, an important extracellular matrix protein and bridging molecule, to integrin receptors (Moser et al. 1997). It has been described that host cell invasion of *C. jejuni* is one of the primary reasons for tissue damage in humans that involves a microtubule-dependent process (Oelschlaeger et al. 1993; Hu and Kopecko 1999). *C. jejuni* triggers membrane ruffling in cultured intestinal epithelial cells (INT-407) followed by invasion in a very specific manner, first with its tip followed by the end of the flagellum, and shows features of both the trigger and zipper mechanisms (Fig. 2b). Maximal adherence and invasion of INT-407 cells by *C. jejuni* require CadF and are accompanied with increased levels of tyrosine phosphorylation of host cell proteins (Biswas et al. 2004; Hu et al. 2006), such as the integrin-associated protein paxillin (Monteville et al. 2003). Interestingly, CadF is also involved in the activation of the small Rho GTPases Rac1 and Cdc42, which are required for the entry process (Krause-Gruszczynska et al. 2007). In addition, it has been shown that mutation of genes in the flagellar export system and ciaB (*Campylobacter* invasion antigen B), as well as deletion of the *kpsS* and *waaF* genes, which play a role in the biosynthesis of capsular polysaccharide and lipooligosaccharide, respectively, resulted in reduced bacterial adhesion and invasion in vitro, suggesting that these factors also play important roles in the host cell entry process (Karlyshev et al. 2000; Kanipes et al. 2004; Guerry 2007; Hu and Kopecko 2008; Larson et al. 2008). Host membrane caveolae, heterotrimeric G proteins, and certain protein kinases (EGF- and PDGF-receptor, phosphatidylinositol 3-kinase [PI3-K], and others) are also important for epithelial cell invasion of *C. jejuni* (Wooldridge et al. 1996; Hu et al. 2006; Watson and Galán 2008). A model of this signaling is shown in Fig. 3b. Once internalized in epithelial cells, *C. jejuni* colocalize specifically with microtubules and dynein (Hu and Kopecko 1999) and are able to survive for extended periods of time and ultimately induce a cytotoxic response in vitro (Konkel et al. 1992; Day et al. 2000). The *C. jejuni*-containing intracellular vacuole deviates from the canonical endocytic pathway, and thus may avoid delivery into lysosomes and subsequent bacterial killing (Watson and Galán 2008). The intracellular survival of *C. jejuni* may enhance its ability to evade the host immune system, cause relapse of the acute infection, and establish long-term persistent infections (Lastovica 1996; Day et al. 2000). However, the molecular mechanisms of *C. jejuni* host cell invasion as well as the complex interplay of different bacterial factors are still not clear and are currently under investigation in many research labs.

4 *Shigella* spp.

Shigellosis is an acute GI disorder caused by infections with species of the genus *Shigella*. *Shigella* spp. are human-adapted Gram-negative pathogens, capable of colonizing the gastric epithelium. *Shigella dysenteriae* is the most common species isolated, although cases of *S. sonnei* and *S. flexneri* infection are increasingly being reported in several countries. These pathogens are generally ingested from fecally contaminated water, but person-to-person transmission is also possible. The infective dose is very low, with as few as 10–100 bacteria (Acheson and Keusch 1995). The symptoms of shigellosis range from mild watery diarrhea to severe inflammatory bacillary dysentery, as characterized by strong abdominal cramps, fever, and stools containing blood and mucus (Jennison and Verma 2004; Niyogi 2005). Infants and small children, homosexuals, and malnourished patients are most commonly affected by *Shigella* infections (Lamps 2007). About 5–15% of all diarrheal episodes worldwide can be attributed to an infection with *Shigella* spp., including 1.1 million fatal cases (Kotloff et al. 1999). The disease is usually self-limiting but may become life-threatening if the infected person is immunocompromised or if adequate medical care is not available. Thus, two thirds of all episodes and deaths occur in children under 5 years old, especially in the developing world. Like salmonellosis, the gross and microscopic features of shigellosis may mimic chronic idiopathic inflammatory bowel disease (Lamps 2007). Besides the mentioned symptoms, perforation and hemolytic-uremic syndrome have also been described. The large bowel is typically affected (often the left colon most severely), but the ileum may also be involved. The mucosa is hemorrhagic, and variably ulcerated, sometimes with pseudomembranous exudates (Speelman et al. 1984). Early shigellosis has features of acute self-limited colitis with cryptitis, crypt abscesses (often superficial), and ulceration (Lamps 2007). Apthoid ulcers similar to Crohn's disease are variably present. As the infection and disease continue, mucosal destruction arises by the infiltration of neutrophils and other inflammatory cells into the lamina propria. Marked architectural distortion can be commonly observed, leading to diagnostic confusion with chronic idiopathic inflammatory bowel disease (Mathan and Mathan 1991). The differential diagnosis of early shigellosis is primarily that of other enteric infections, particularly enteroinvasive *E. coli* and *Clostridium difficile*. Later on in the course of the disease, it may be extremely difficult to distinguish shigellosis from Crohn's disease or ulcerative colitis both endoscopically and histologically (Lamps 2007). Although a simple combination of oral rehydration and antibiotics commonly leads to the rapid resolution of these infections, the emergence of multidrug-resistant *Shigella* strains and a continuous high disease incidence imply that shigellosis is still an unsolved global health problem (Sansonetti 2006).

Shigella spp. express neither classical adherence factors on their surface nor flagella. Following ingestion, the bacteria rapidly move through the small intestine to the colon and rectum, where they cross the epithelial barrier through M cells. In contrast to *Salmonella*, *Shigella* preferentially enters M cells of the colorectal mucosa rather than the distal small intestine. The specific receptors that account for this selectivity are unknown, but in vitro studies with cultured cells have demonstrated that β_1-integrins and CD44 receptor may play a role in the initial contact of *Shigella* with its host cell (Nhieu et al. 2005; Ogawa et al. 2008). In this way, *Shigella* breaches the epithelial barrier and immediately enters the macrophages that reside within the microfold-cell pocket. Once within the macrophages, the infecting bacteria disrupt the phagosomal membrane and disseminate from the phagosome into the macrophage cytoplasm, where they multiply and induce rapid apoptotic cell death in a caspase-1-dependent manner (Phalipon and Sansonetti 2007; Ogawa et al. 2008; Schroeder and Hilbi 2008). Bacteria released from dead macrophages can enter the surrounding enterocytes using a T3SS (Fig. 2c) that is encoded on a large virulence plasmid. T3SS-dependent injection of the Ipa effector proteins (IpaA-D) initiates actin-cytoskeletal and membrane remodeling processes that engulf the bacteria by macropinocytic ruffles (Nhieu et al. 2005). Exploring the process of how Ipa proteins alter the host cytoskeleton to induce the uptake process revealed that a complex of IpaB and IpaC binds the $\alpha_1\beta_5$-integrin receptor and the hyaluron receptor CD44 and induce actin rearrangements at the site of bacterial attachment (Watarai et al. 1996) (Fig. 3c). Earlier work demonstrated that IpaA binds to the focal adhesion protein vinculin and induces the recruitment of F-actin and the depolymerization of actin stress fibers (Bourdet-Sicard et al. 1999), but a recent study also showed that IpaA increases the activity of the small GTPase RhoA and decreases integrin affinity for extracellular matrix ligands by interfering with talin recruitment to the integrin's cytoplasmic tail (Demali et al. 2006). *Shigella* entry through the Ipa proteins further implicates the recruitment and activation of multiple other factors such as the tyrosine kinases FAK and Src, cortactin, Crk, Rac or Cdc42, which mediate massive actin polymerization in the vicinity of the original cup via the Arp2/3-complex (Burton et al. 2003; Bougneres et al. 2004). In addition, a new class of G-nucleotide exchange effector proteins has recently been discovered in *Shigella* that is also involved in invasion. Remarkably, IpgB1 functions to activate Rac1, and IpgB2 stimulates cellular responses activating RhoA (Huang et al. 2009). Using this multifactorial mechanism, *Shigella* invades the enterocytes. As soon as a bacterium is surrounded by a membrane vacuole within these cells, it disrupts the vacuolar membrane and escapes into the cytoplasm. *Shigella* movement is then triggered by the bacterial surface protein IcsA. IcsA has a high affinity to a major regulator of the actin-polymerization machinery, N-WASP (neuronal Wiskott–Aldrich syndrome protein), which recruits and activates the Arp2/3-complex (Phalipon and Sansonetti 2007; Ogawa et al. 2008; Schroeder and Hilbi 2008). The formation of actin tails pushes *S. flexneri* through the host cell cytoplasm, a process that is enhanced by secreting the T3SS-effector protein VirA, which is a protease of α-tubulin that destroys the surrounding microtubules (Yoshida et al. 2006). *Shigella* can multiply in the epithelial cell cytoplasm and move both intra- and intercellularly, and so infection of the intestinal epithelium by *Shigella* also elicits strong inflammatory and other responses (Phalipon and Sansonetti 2007; Ogawa et al. 2008; Schroeder and Hilbi 2008).

5 *Escherichia coli*

The famous laboratory strain K-12 and other nonpathogenic *Escherichia coli* isolates are common members in the normal GI microbial community. However, during evolution, some of these *E. coli* strains accumulated pathogenicity islands and other virulence elements in their genomes by horizontal DNA-transfer events (Hacker et al. 2004). Important examples of this category are enterohemorrhagic *E. coli* (EHEC) and enteropathogenic *E. coli* (EPEC), two emerging foodborne pathogens. One hallmark of EHEC and EPEC infections is their ability to colonize the gut mucosa and produce characteristic attaching and effacing lesions (A/E lesions), resulting in diarrheal and other diseases. A/E lesions are characterized by effacement of the intestinal brush border microvilli and close attachment of the bacterium to the enterocyte plasma membrane, leading to the formation of a characteristic pedestal-shaped, localized membrane protrusion that can extend up to 10 μm outwards from the cell periphery (Knutton et al. 1987). One of the most common strains of EHEC is O157:H7. This pathogen gained worldwide attention in 1993 when a massive outbreak in the United States was linked to contaminated hamburgers. EHEC O157:H7 is still the most relevant serotype in such foodborne outbreaks; however, an increased incidence of infections caused by non-O157:H7 was observed in other countries (Gerber et al. 2002). Since cattle have been shown to be a major reservoir of EHEC, raw food such as ground beef and milk are the most common sources of infection. A variety of other food types, such as fermented meat products, raw salad vegetable products, unpasteurized fresh fruit juice, and water, as well as person-to-person contact have also been linked to EHEC outbreaks (Olsen et al. 2002; Mao et al. 2003). Although pathogenic *E. coli* are not particularly resistant to harsh environmental conditions, some reports indicate that EHEC can tolerate a wide range of pH and water conditions as well as low temperature, indicating that there is considerable potential for these organisms to survive in and on food (Chikthimmah and Knabel 2001; Hancock et al. 2001). Because an infective dose of EHEC and EPEC is low (<100 bacteria), even the minimal contamination of food is of concern. When it comes to an infection, the bacteria can induce diarrhea, which is usually bloody, with severe abdominal cramps and mild or no fever. However, nonbloody, watery diarrhea may also occur in some cases. Only one third of infected persons have fecal leukocytes. Children and the elderly are at particular risk of serious illness, including the hemolytic-uremic syndrome or thrombotic thrombocytopenic purpura (Griffin et al. 1990; Kelly et al. 1990; Lamps 2007). Endoscopically, patients may have bowel edema, erosions, ulcers, and hemorrhage, and the right colon is usually more severely affected. The edema may be so marked as to cause obstruction, and surgical resection may be required to relieve this or to control bleeding. The lamina propria and submucosa contain marked edema and hemorrhage, with associated mucosal acute inflammation, cryptitis, crypt abscesses, ulceration, and necrosis. Crypt withering, such as that seen in other causes of ischemia, is often seen as well. Microthrombi may be observed within small vessels, and pseudomembranes are occasionally present (Griffin et al. 1990; Kelly et al. 1990; Lamps 2007).

The pathogenicity potential of EHEC is closely connected to the production of Shiga toxins (Stx1 and/or Stx2), which are related to the exotoxin of *Shigella dysenteriae* serotype-1 (Cleary 2004; Scheiring et al. 2008). These toxins act as inhibitors of protein biosynthesis and have profound effects on the signal transduction and immunological response in eukaryotic cells. In addition to the secretion of Shiga toxins, the production of EHEC hemolysin, serine protease, enterotoxin (EAST), catalase, pili, and other factors have also been implicated in the pathogenesis (Donnenberg and Whittam 2001; Vallance et al. 2002; Mao et al. 2003). In particular, EHEC O157:H7 produces long bundles of polar type-4 pili that mediate binding of the bacteria to epithelial cells and eventually cause bacterial invasion (Xicohtencatl-Cortes et al. 2009). Indeed, we could show that at least some EHEC bacteria can enter epithelial cells in vitro using a zipper-like mechanism (our unpublished data) (Fig. 2d). In addition, A/E lesions caused by both EHEC and EPEC in vivo are dependent on a T3SS that injects numerous effector proteins directly into host cells. The best-described effectors are encoded on the locus of enterocyte effacement (LEE) pathogenicity islands and display high levels of multifunctionality. The recent completion of the EPEC genome sequence suggests that there are at least 21 injected proteins (Dean and Kenny 2009). This T3SS acts together with the outer membrane adhesion molecule intimin to trigger actin-pedestal formation (Fig. 2e). Each of the two pathogens injects its own receptor called Tir (translocated intimin receptor), a T3SS effector molecule. Translocated Tir then inserts into the host cell plasma membranes, forming a hairpin loop, and interacts with intimin, both of which are required to trigger the actinpolymerization into focused pedestals just beneath attached bacteria. Despite similarities between the Tir molecules and the host components that associate with pedestals, EPEC-Tir and EHEC-Tir are not functionally interchangeable. Injected EPEC-Tir is tyrosine-phosphorylated by host cell kinases (mainly by members of the Src, Tec, and Abl kinase families) to mediate the binding of Nck, a host adaptor protein implicated in actin signaling. In contrast, EHEC-Tir cannot be phosphorylated, and pedestals are formed independently of Nck but require translocation of another bacterial factor (TccP/EspF[U]) in addition to Tir to trigger actin signaling (Campellone and Leong 2003; Backert and Selbach 2005; Hayward et al. 2006; Frankel and Phillips 2008; Dean and Kenny 2009). Otherwise, EPEC- and EHEC-induced pedestals are very similar. They are composed of F-actin and a variety of signaling factors, including actin-regulatory proteins such as N-WASP, Arp2/3, cortactin, and others, as well as numerous adaptor and focal adhesion proteins (Fig. 3d, e). However, the physiological significance of pedestal formation in vivo is unknown. One could predict that such a tight interaction between the bacterium and the host cell should severely impair the ingestion of bacteria by immune cells, which could be a possible strategy (albeit unique). Interestingly, EPEC directly inhibits phagocytosis, but the T3SS effectors triggering this antiphagocytic activity are unknown (Celli and Finlay 2002). In the light of these findings, it is remarkable that EPEC is able to invade nonphagocytic epithelial cells using Map and Tir effectors by synergistic mechanisms (Jepson et al. 2003). Particularly, the Map effector protein has two distinct functions within host cells: targeting mitochondria to elicit dysfunction and mediating Cdc42-dependent filopodia formation involved in host entry (Jepson et al. 2003). The promotion of EPEC invasion by Tir appears to involve interaction with intimin but is independent of pedestal formation. Finally, the phenomenon of effacement by EPEC has been shown to require the cooperative action of three injected effectors (Map, EspF, and Tir) as well as intimin and leads to the retention (not the release) of the detached microvilli structures (Dean et al. 2006). As a consequence of this, EPEC rapidly inactivates the sodium-d-glucose cotransporter (SGLT-1), which provides a plausible explanation for the rapid onset of severe watery diarrhea, given the crucial role of SGLT-1 in the daily uptake of approximately 6 L of fluids from the normal intestine (Dean et al. 2006). Given the multitude of EHEC and EPEC effectors, there are many more signaling pathways induced by these pathogens, and they need to be studied in more detail in future research (Campellone and Leong 2003; Hayward et al. 2006; Frankel and Phillips 2008; Dean and Kenny 2009).

6 *Listeria monocytogenes*

Listeriosis is an animal-borne and foodborne human disease that is caused by pathogenic bacteria of the genus *Listeria*. There are seven species within this genus, including *L. monocytogenes*, *L. ivanovii*, *L. innocua*, *L. seeligeri*, *L. welshimeri*, *L. grayii,* and *L. murrayi*. However, only two of them are pathogenic: *Listeria monocytogenes* can cause disease in both humans and animals, and *L. ivanovii* causes disease predominantly in sheep (Mead et al. 1999; Roberts and Wiedmann 2003; Mao et al. 2003). Although relatively uncommon, *L. monocytogenes* infections are almost exclusively foodborne (99%) and are mainly caused by the consumption of contaminated food products (Mead et al. 1999). *Listeria* species are Gram-positive, nonspore-forming rods that are commonly observed in the environment where they developed highly adaptive characteristics during their evolution. *Listeria* species can grow over a wide range of temperatures (1–45°C) and pHs (4.3–9.6), and even at salt concentrations of up to 10% (Seeliger and Jones 1986; Johnson et al. 1988). This ability to survive and multiply under conditions frequently used for food preservation makes *Listeria* particularly problematic to our food industry. Thus, *L. monocytogenes* is a common food contaminant and a major cause of food recalls due to bacterial contamination and outbreaks, particularly in developed countries and possibly worldwide (Mead et al. 1999; Farber and Peterkin 1991). These outbreaks are commonly linked to a wide variety of foods, including refrigerated foods, ready-to-eat foods (e.g., hot dogs, cold cuts), fresh vegetables, apple cider, and dairy products such as cheese (Hitchins and Whiting 2001; Asperger et al. 2001). The frequent occurrence of *L. monocytogenes* in food, coupled with a high mortality rate of 20–30% in those developing listeriosis, make these infections a serious public health concern (Mead et al. 1999; Farber and Peterkin 1991). *Listeria monocytogenes* causes *sepsis* and *meningitis*, usually affecting specific high-risk subgroups of the population such as the elderly, the immunocompromised, and fetuses. These diseases are due to *Listeria's* capacity to breach three host barriers during infection: the intestinal, the placental, and the blood–brain barriers. However, infection with *L. monocytogenes* in otherwise healthy individuals commonly causes self-limited gastroenteritis (Wing and Gregory 2002; Doganay 2003; Hof 2004). Listeriosis in animals, furthermore, represents not only a financial burden for the livestock industry but also a possible link between *Listeria* in the environment and human disease.

Listeria monocytogenes has evolved highly sophisticated strategies to infect its mammalian host and to survive as a facultative intracellular pathogen (Tilney and Portnoy 1989; Wing and Gregory 2002; Hamon et al. 2006; Dussurget 2008). At the cellular level, *Listeria* enters by a zipper mechanism characterized by a tight apposition of the plasma membrane around the entering bacteria (Fig. 2f). Two remarkable surface proteins, called internalin A and B (InlA and InlB), are crucial for mediating bacterial entry into mammalian cells. These adhesins interact with host cell transmembrane receptors, E-cadherin, and the hepatocyte growth factor receptor (c-Met), respectively (Fig. 3f). These interactions initiate a series of signaling events involving PI3-K, Cdc42, Rac, and possibly catenins and other factors, leading to actin polymerization, membrane invagination, and bacterial internalization. Investigations into InlA- and InlB-mediated entries have demonstrated that *Listeria* fully usurps the host cell cytoskeletal machinery (Rottner et al. 2005; Hamon et al. 2006; Bosse et al. 2007; Dussurget 2008). Moreover, recent studies have highlighted a role for the endocytic protein clathrin in *Listeria* InlB-mediated actin polymerization and entry, revealing a new role for this factor in bacteria-induced host entry. Furthermore, comparative studies have demonstrated that the clathrin-mediated endocytosis machinery is also used in the InlA–E-cadherin pathway and for the invasion of other bacteria that enter by the zipper mechanism (Cossart and Veiga 2008). In contrast, the clathrin-mediated endocytic machinery is not used by bacteria such as *Salmonella* that enter by the trigger mechanism (Cossart and Veiga 2008). However, the internalization process of *Listeria* results in the formation of intracellular vacuoles carrying the bacteria. A third bacterial protein, listeriolysin-O, rapidly lyses these vacuoles, releasing *Listeria* into the cytosol of the infected cell, where the bacterium can replicate. Certain phospholipases, which are secreted by *Listeria*, also play a role in this context (Hamon et al. 2006; Dussurget 2008). Similar to IcsA in *Shigella*, a fourth *Listeria* protein called ActA triggers a very efficient actin-polymerization process at the posterior pole of the bacterium that pushes the bacterium forward and allows active movement within the infected cell (Rottner et al. 2005). From time to time, intracellular bacteria may contact the membrane that allows *Listeria* to invade neighboring cells (Tilney and Portnoy 1989). This direct cell-to-cell spread allows bacteria to disseminate in the infected organism. Interestingly, most of the virulence proteins that have been identified in *L. monocytogenes* are under tight control of the transcriptional regulator PrfA, which is regulated by environmental conditions (Johansson et al. 2002). Finally, it has to be mentioned that *Listeria* infection induces a variety of other host cell signaling events and has also been established as a very useful model system to study host T cell responses (Pamer 2004).

7 Summary

Foodborne infections are a large health and economic problem worldwide. The WHO calculated that there were about 4.5 billion incidences of diarrhea that caused 1.8 million deaths in the year 2002. Approximately 99% of the cases occurred in developing countries, where poor hygiene and limited access to clean drinking water promote the spread of enteric diseases. Malnutrition and the lack of appropriate medical intervention contribute to the high mortality rate, especially for young children, the elderly, and immunocompromised persons. Infections with a large number of bacterial, viral, and parasitic pathogens have been implicated in these diseases. In this chapter we focused on foodborne bacterial pathogens and summarized important strategies and signaling mechanisms that result in colonization of the GI epithelium, where the bacteria can multiply and spread. We highlighted the strategies of important GI pathogens, with an emphasis on species such as *Salmonella, Campylobacter, Shigella, Escherichia,* and *Listeria* that represent paradigms of host–pathogen interactions. However, there are a variety of other foodborne bacteria, such as *Yersinia* or *Clostridium,* that are discussed elsewhere. Recognition of the genetic and functional bases of bacterial foodborne pathogenicity and analyses of cross-talks on the level of molecular signaling cascades between these pathogens and their mammalian target cells have illuminated the diversity but also common strategies of these interactions. Entire genome sequences are now available for many microbes that cause foodborne diseases, and the development of themed and whole-genome DNA microarrays as well as improved proteomics techniques might provide effective new tools for rapidly detecting and identifying such organisms, assessing their biological diversity, and understanding their ability to trigger certain diseases. However, since there are also substantial interactions between commensal and pathogenic bacteria, more information about individual members constituting the normal gut flora is also needed. Collective genomes (microbiome) of the human microbiota have now become important targets to be studied in both microbiology and human biology. The generated data will be accumulated and evaluated in future studies, including the International Human Microbiome Project. This information also provides fresh insights into the metabolic capacity and versatility of microbes, for example, specific metabolic pathways that might contribute to the growth and survival of pathogens in a range of niches, such as food-processing environments and the human host. Different concepts are emerging about how pathogens function, both within foods and in interactions with the host. The future should bring the practical benefits of genome sequencing and molecular infection research to the field of microbial food safety, including new strategies and tools for the identification and control of emerging foodborne pathogens.

References

Acheson, D.W.K., and G.T. Keusch. 1995. *Shigella* and enteroinvasive *E. coli*. In *Infections of the gastrointestinal tract*, ed. M.J. Blaser, P.D. Smith, and J.I. Ravdin, 763–784. New York: Raven.
Allos, B.M. 2001. *Campylobacter jejuni* infections: update on emerging issues and trends. *Clinical Infectious Diseases* 32: 1201–1206.
Alouf, J.E., and M.R. Popoff. 2005. *The comprehensive sourcebook of bacterial protein toxins*, 3rd ed. London: Academic.
Asperger, H., M. Wagner, and E. Brandl. 2001. An approach towards public health and food-borne human listeriosis—The Austrian *Listeria* monitoring. *Berliner und Münchener Tierärztliche Wochenschrift* 114: 446–452.
Backert, S., and W. Koenig. 2005. Interplay of bacterial toxins with host defence: molecular mechanisms of immunomodulatory signalling. *International Journal of Medical Microbiology* 295: 519–530.
Backert, S., and M. Selbach. 2005. Tyrosine-phosphorylated bacterial effector proteins: the enemies within. *Trends in Microbiology* 13: 476–484.
Beutler, B., Z. Jiang, P. Georgel, K. Crozat, B. Croker, S. Rutschmann, X. Du, and K. Hoebe. 2006. Genetic analysis of host resistance: toll-like receptor signaling and immunity at large. *Annual Review of Immunology* 24: 353–389.
Biswas, D., H. Niwa, and K. Itoh. 2004. Infection with *Campylobacter jejuni* induces tyrosine-phosphorylated proteins into INT-407 cells. *Microbiology and Immunology* 48: 221–228.
Blaser, M.J., and J. Engberg. 2008. Clinical sspects of *Campylobacter jejuni* and *Campylobacter coli* infections. In *Campylobacter,* ed. I. Nachamkin, C.M. Szymanski, and M.J. Blaser, 99–121. Washington, DC: ASM Press.
Boquet, P., and E. Lemichez. 2003. Bacterial virulence factors targeting Rho GTPases: parasitism or symbiosis? *Trends in Cell Biology* 13: 238–246.
Bosse, T., J. Ehinger, A. Czuchra, S. Benesch, A. Steffen, X. Wu, K. Schloen, H.H. Niemann, G. Scita, T.E. Stradal, C. Brakebusch, and K. Rottner. 2007. Cdc42 and phosphoinositide 3-kinase drive Rac-mediated actin polymerization downstream of c-Met in distinct and common pathways. *Molecular and Cellular Biology* 27: 6615–6628.

Bougneres, L., S.E. Girardin, S.A. Weed, A.V. Karginov, J.C. Olivo-Marin, J.T. Parsons, P.J. Sansonetti, and G.T. Van Nhieu. 2004. Cortactin and Crk cooperate to trigger actin polymerization during *Shigella* invasion of epithelial cells. *The Journal of Cell Biology* 166: 225–235.

Bourdet-Sicard, R., M. Rudiger, B.M. Jockusch, P. Gounon, P.J. Sansonetti, and G.T. Nhieu. 1999. Binding of the *Shigella* protein IpaA to vinculin induces F-actin depolymerization. *The EMBO Journal* 18: 5853–5862.

Boyd, J.F. 1985. Pathology of the alimentary tract in *Salmonella typhimurium* food poisoning. *Gut* 26: 935–944.

Boyd, E.F., F.S. Wang, T.S. Whittam, and R.K. Selander. 1996. Molecular genetic relationships of the *salmonellae*. *Applied and Environmental Microbiology* 62: 804–808.

Burns, D., J.T. Barbieri, B.H. Iglewski, and R. Rappuoli. 2003. *Bacterial protein toxins*. Washington, DC: ASM Press.

Burton, E.A., R. Plattner, and A.M. Pendergast. 2003. Abl tyrosine kinases are required for infection by *Shigella flexneri*. *The EMBO Journal* 22: 5471–5479.

Campellone, K.G., and J.M. Leong. 2003. Tails of two Tirs: actin pedestal formation by enteropathogenic *E. coli* and enterohemorrhagic *E. coli* O157:H7. *Current Opinion in Microbiology* 6: 82–90.

Celli, J., and B.B. Finlay. 2002. Bacterial avoidance of phagocytosis. *Trends in Microbiology* 10: 232–237.

Chen, L.F., and W.C. Greene. 2004. Shaping the nuclear action of NF-κB. *Nature Reviews Molecular Cell Biology* 5: 392–401.

Chikthimmah, N., and S.J. Knabel. 2001. Survival of *Escherichia coli* O157:H7, *Salmonella typhimurium* and *Listeria monocytogenes* in and on vacuum packaged Lebanon bologna stored at 3.6 and 13.0 degrees C. *Journal of Food Protection* 64: 958–963.

Cleary, T.G. 2004. The role of Shiga-toxin-producing *Escherichia coli* in hemorrhagic colitis and hemolytic uremic syndrome. *Seminars in Pediatric Infectious Diseases* 15: 260–265.

Cossart, P., and P.J. Sansonetti. 2004. Bacterial invasion: the paradigms of enteroinvasive pathogens. *Science* 304: 242–248.

Cossart, P., and E. Veiga. 2008. Non-classical use of clathrin during bacterial infections. *Journal of Microscopy* 231: 524–528.

Crump, J.A., P.M. Griffin, and F.J. Angulo. 2002. Bacterial contamination of animal feed and its relationship to human food-borne illness. *Clinical Infectious Diseases* 35: 859–865.

Day Jr., W.A., J.L. Sajecki, T.M. Pitts, and L.A. Joens. 2000. Role of catalase in *Campylobacter jejuni* intracellular survival. *Infection and Immunity* 68: 6337–6345.

Dean, P., and B. Kenny. 2009. The effector repertoire of enteropathogenic *E. coli*: ganging up on the host cell. *Current Opinion in Microbiology* 12: 101–109.

Dean, P., M. Maresca, S. Schüller, A.D. Phillips, and B. Kenny. 2006. Potent diarrheagenic mechanism mediated by the cooperative action of three enteropathogenic *Escherichia coli*-injected effector proteins. *Proceedings of the National Academy of Sciences of the United States of America* 103: 1876–1881.

Demali, K.A., A.L. Jue, and K. Burridge. 2006. IpaA targets beta1 integrins and rho to promote actin cytoskeleton rearrangements necessary for *Shigella* entry. *Journal of Biological Chemistry* 281: 39534–39541.

Doganay, M. 2003. Listeriosis: clinical presentation. *FEMS Immunology and Medical Microbiology* 35: 173–175.

Donnenberg, M.S., and T.S. Whittam. 2001. Pathogenesis and evolution of virulence in enteropathogenic and enterohemorrhagic *Escherichia coli*. *The Journal of Clinical Investigation* 107: 539–548.

Dussurget, O. 2008. New insights into determinants of *Listeria monocytogenes* virulence. *International Review of Cell and Molecular Biology* 270: 1–38.

Eckburg, P.B., E.M. Bik, C.N. Bernstein, E. Purdom, L. Dethlefsen, M. Sargent, S.R. Gill, K.E. Nelson, and D.A. Relman. 2005. Diversity of the human intestinal microbial flora. *Science* 308: 1635–1638.

Eckmann, L., and M.F. Kagnoff. 2005. Intestinal mucosal responses to microbial infection. *Springer Seminars in Immunopathology* 27: 181–196.

Fang, G., V. Araujo, and R.L. Guerrant. 1991. Enteric infections associated with exposure to animals or animal products. *Infectious Disease Clinics of North America* 5: 681–701.

Farber, J.M., and P.I. Peterkin. 1991. *Listeria monocytogenes*, a food-borne pathogen. *Microbiological Reviews* 55: 476–511.

Forsythe, S.J. 2000. Food poisoning microorganisms. In *The microbiology of safe food*, ed. S.J. Forsythe, 87–148. Abingdon: Blackwell Science.

Frankel, G., and A.D. Phillips. 2008. Attaching effacing *Escherichia coli* and paradigms of Tir-triggered actin polymerization: getting off the pedestal. *Cellular Microbiology* 10: 549–556.

Friedman, C.R., J. Neimann, H.C. Wegener, and R.V. Tauxe. 2000. Epidemiology of *Campylobacter jejuni* infections in the United States and other industrialized nations. In *Campylobacter*, ed. I. Nachamkin and M.J. Blaser, 121–138. Washington, DC: ASM Press.

Gal-Mor, O., and B.B. Finlay. 2006. Pathogenicity islands: a molecular toolbox for bacterial virulence. *Cellular Microbiology* 8: 1707–1719.

Ge, Z., D.B. Schauer, and J.G. Fox. 2008. *In vivo* virulence properties of bacterial cytolethal-distending toxin. *Cellular Microbiology* 10: 1599–1607.

Gerber, A., H. Karch, F. Allerberger, H.M. Verweyen, and L.B. Zimmerhackl. 2002. Clinical course and the role of shiga toxin-producing Escherichia coli infection in the hemolytic-uremic syndrome in pediatric patients, 1997–2000, in Germany and Austria: a prospective study. *The Journal of Infectious Diseases* 186: 493–500.

Gerlach, R.G., and M. Hensel. 2007. *Salmonella* pathogenicity islands in host specificity, host pathogen-interactions and antibiotics resistance of *Salmonella enterica*. *Berliner und Münchener Tierärztliche Wochenschrift* 120: 317–327.

Ghosh, S., and M.S. Hayden. 2008. New regulators of NF-kappaB in inflammation. *Nature Reviews Immunology* 8: 837–848.

Griffin, P.M., L.C. Olmstead, and R.E. Petras. 1990. *Escherichia coli* 0157:H7-associated colitis: a clinical and histological study of 11 cases. *Gastroenterology* 99: 142–149.

Guerry, P. 2007. *Campylobacter* flagella: not just for motility. *Trends in Microbiology* 15: 456–461.

Hacker, J., B. Hochhut, B. Middendorf, G. Schneider, C. Buchrieser, G. Gottschalk, and U. Dobrindt. 2004. Pathogenomics of mobile genetic elements of toxigenic bacteria. *International Journal of Medical Microbiology* 293: 453–461.

Hamon, M., H. Bierne, and P. Cossart. 2006. *Listeria monocytogenes*: a multifaceted model. *Nature Reviews Microbiology* 4: 423–434.

Hancock, D., T. Besser, J. Lejeune, M. Davis, and D. Rice. 2001. The control of VTEC in the animal reservoir. *International Journal of Food Microbiology* 66: 71–78.

Hapfelmeier, S., and W.D. Hardt. 2005. A mouse model for *S. typhimurium*-induced enterocolitis. *Trends in Microbiology* 13: 497–503.

Haraga, A., M.B. Ohlson, and S.I. Miller. 2008. *Salmonellae* interplay with host cells. *Nature Reviews Microbiology* 6: 53–66.

Hayward, R.D., J.M. Leong, V. Koronakis, and K.G. Campellone. 2006. Exploiting pathogenic *Escherichia coli* to model transmembrane receptor signalling. *Nature Reviews Microbiology* 4: 358–370.

Hickey, T.E., A.L. McVeigh, D.A. Scott, R.E. Michietutti, A. Bixby, S.A. Carroll, A.L. Bourgeois, and P. Guerry. 2000. *Campylobacter jejuni* cytolethal distending toxin mediates release of interleukin-8 from intestinal epithelial cells. *Infection and Immunity* 68: 6535–6541.

Hitchins, A.D., and R.C. Whiting. 2001. Food-borne *Listeria monocytogenes* risk assessment. *Food Additives and Contaminants* 18: 1108–1117.

Hof, H. 2004. An update on the medical management of listeriosis. *Expert Opinion on Pharmacotherapy* 5: 1727–1735.

Hooper, L.V., and J.I. Gordon. 2001. Commensal host-bacterial relationships in the gut. *Science* 292: 1115–1118.

Hornef, M.W., M.J. Wick, M. Rhen, and S. Normark. 2002. Bacterial strategies for overcoming host innate and adaptive immune responses. *Nature Immunology* 3: 1033–1040.

Hu, L., and D.J. Kopecko. 1999. *Campylobacter jejuni* 81–176 associates with microtubules and dynein during invasion of human intestinal cells. *Infection and Immunity* 67: 4171–4182.

Hu, L., and D.J. Kopecko. 2008. Cell biology of human host cell entry by *Campylobacter jejuni*. In *Campylobacter*, ed. I. Nachamkin, C.M. Szymanski, and M.J. Blaser, 297–313. Washington, DC: ASM Press.

Hu, L., J.P. McDaniel, and D.J. Kopecko. 2006. Signal transduction events involved in human epithelial cell invasion by *Campylobacter jejuni* 81–176. *Microbial Pathogenesis* 40: 91–100.

Huang, Z., S.E. Sutton, A.J. Wallenfang, R.C. Orchard, X. Wu, Y. Feng, J. Chai, and N.M. Alto. 2009. Structural insights into host GTPase isoform selection by a family of bacterial GEF mimics. *Nature Structural and Molecular Biology* 16: 853–860.

Jennison, A.V., and N.K. Verma. 2004. *Shigella flexneri* infection: pathogenesis and vaccine development. *FEMS Microbiology Reviews* 28: 43–58.

Jepson, M.A., and M.A. Clark. 2001. The role of M cells in *Salmonella* infection. *Microbes and Infection* 3: 1183–1190.

Jepson, M.A., S. Pellegrin, L. Peto, D.N. Banbury, A.D. Leard, H. Mellor, and B. Kenny. 2003. Synergistic roles for the Map and Tir effector molecules in mediating uptake of enteropathogenic *Escherichia coli* (EPEC) into non-phagocytic cells. *Cellular Microbiology* 5: 773–783.

Johansson, J., P. Mandin, A. Renzoni, C. Chiaruttini, M. Springer, and P. Cossart. 2002. An RNA thermosensor controls expression of virulence genes in *Listeria monocytogenes*. *Cell* 110: 551–561.

Johnson, J.L., M.P. Doyle, R.G. Cassens, and J.L. Schoeni. 1998. Fate of *Listeria monocytogenes* in tissues of experimentally infected cattle and in hard salami. *Applied and Environmental Microbiology* 54: 497–501.

Kanipes, M.I., L.C. Holder, A.T. Corcoran, A.P. Moran, and P. Guerry. 2004. A deep-rough mutant of *Campylobacter jejuni* 81–176 is noninvasive for intestinal epithelial cells. *Infection and Immunity* 72: 2452–2455.

Karlyshev, A.V., D. Linton, N.A. Gregson, A.J. Lastovica, and B.W. Wren. 2000. Genetic and biochemical evidence of a *Campylobacter jejuni* capsular polysaccharide that accounts for Penner serotype specificity. *Molecular Microbiology* 35: 529–541.

Kelly, J.K., and D.A. Owen. 1997. Bacterial diarrheas and dysenteries. In *Pathology of infectious diseases*, ed. D.H. Connor, F.W. Chandler, and D.A. Schwartz, 421–429. Stamford: Appleton & Lange.

Kelly, J., A. Oryshak, M. Wenetsek, J. Grabiec, and S. Handy. 1990. The colonic pathology of *E. coli* 0157:H7 infection. *The American Journal of Surgical Pathology* 14: 87–92.

Ketley, J.M. 1997. Pathogenesis of enteric infection by *Campylobacter*. *Microbiology* 143: 5–21.

Knutton, S., D.R. Lloyd, and A.S. McNeish. 1987. Adhesion of enteropathogenic *Escherichia coli* to human intestinal enterocytes and cultured human intestinal mucosa. *Infection and Immunity* 55: 69–77.

Konkel, M.E., S.F. Hayes, L.A. Joens, and W. Cieplak Jr. 1992. Characteristics of the internalization and intracellular survival of *Campylobacter jejuni* in human epithelial cell cultures. *Microbial Pathogenesis* 13: 357–370.

Konkel, M.E., M.R. Monteville, V. Rivera-Amill, and L.A. Joens. 2001. The pathogenesis of *Campylobacter jejuni*-mediated enteritis. *Current Issues in Intestinal Microbiology* 2: 55–71.

Kotloff, K.L., J.P. Winickoff, B. Ivanoff, J.D. Clemens, D.L. Swerdlow, P.J. Sansonetti, G.K. Adak, and M.M. Levine. 1999. Global burden of *Shigella* infections: implications for vaccine development and implementation of control strategies. *Bulletin of the World Health Organization* 77: 651–666.

Kraus, M.D., B. Amatya, and Y. Kimula. 1999. Histopathology of typhoid enteritis: morphologic and immunophenotypic findings. *Modern Pathology* 12: 949–955.

Krause-Gruszczynska, M., M. Rohde, R. Hartig, H. Genth, G. Schmidt, T. Keo, W. Koenig, W.G. Miller, M.E. Konkel, and S. Backert. 2007. Role of the small Rho GTPases Rac1 and Cdc42 in host cell invasion of *Campylobacter jejuni*. *Cellular Microbiology* 9: 2431–2444.

Kuhle, V., and M. Hensel. 2004. Cellular microbiology of intracellular *Salmonella enterica*: functions of the type III secretion system encoded by *Salmonella* pathogenicity island 2. *Cellular and Molecular Life Sciences* 61: 2812–2826.

Lamps, L.W. 2007. Infective disorders of the gastrointestinal tract. *Histopathology* 50: 55–63.

Lara-Tejero, M., and J.E. Galan. 2000. A bacterial toxin that controls cell cycle progression as a deoxyribonuclease-1 like protein. *Science* 290: 354–357.

Larson, C.L., J.E. Christensen, S.A. Pacheco, S.A. Minnich, and M.E. Konkel. 2008. *Campylobacter jejuni* secretes proteins via the flagellar type III secretion system that contribute to host cell invasion and gastroenteritis. In *Campylobacter*, ed. I. Nachamkin, C.M. Szymanski, and M.J. Blaser, 315–332. Washington, DC: ASM Press.

Lastovica, A.J. 1996. *Campylobacter/Helicobacter* bacteraemia in Cape Town, South Africa 1977–95. In *Campylobacters, helicobacters and related organisms*, ed. D.G. Newell, J.M. Ketley, and R.A. Feldman, 475–479. New York: Plenum Press.

Liston, A., and S. McColl. 2003. Subversion of the chemokine world by microbial pathogens. *BioEssays* 25: 478–488.

Mao, Y., C. Zhu, and E.C. Boedeker. 2003. Food-borne enteric infections. *Current Opinion in Gastroenterology* 19: 11–22.

Mathan, M.M., and V.I. Mathan. 1991. Morphology of rectal mucosa of patients with shigellosis. *Reviews of Infectious Diseases* 13(Suppl. 4): 314–318.

McGovern, V.J., and L.J. Slavutin. 1979. Pathology of *Salmonella* colitis. *The American Journal of Surgical Pathology* 3: 483–490.

Mead, P.S., L. Slutsker, V. Dietz, L.F. McCaig, J.S. Bresee, C. Shapiro, P.M. Griffin, and R.V. Tauxe. 1999. Food-related illness and death in the United States. *Emerging Infectious Diseases* 5: 607–625.

Monack, D.M., A. Mueller, and S. Falkow. 2004. Persistent bacterial infections: the interface of the pathogen and the host immune system. *Nature Reviews Microbiology* 2: 747–765.

Monteville, M.R., J.E. Yoon, and M.E. Konkel. 2003. Maximal adherence and invasion of INT 407 cells by *Campylobacter jejuni* requires the CadF outer-membrane protein and microfilament reorganization. *Microbiology* 149: 153–165.

Moser, I., W. Schroeder, and J. Salnikow. 1997. *Campylobacter jejuni* major outer membrane protein and a 59-kDa protein are involved in binding to fibronectin and INT-407 cell membranes. *FEMS Microbiology Letters* 157: 233–238.

Nachamkin, I., C.M. Szymanski, and M.J. Blaser. 2008. *Campylobacter*. Washington, DC: ASM Press.

Nhieu, G.T., J. Enninga, P. Sansonetti, and G. Grompone. 2005. Tyrosine kinase signaling and type III effectors orchestrating *Shigella* invasion. *Current Opinion in Microbiology* 8: 16–20.

Niess, J.H., S. Brand, X. Gu, L. Landsman, S. Jung, B.A. McCormick, J.M. Vyas, M. Boes, H.L. Ploegh, J.G. Fox, D.R. Littman, and H.C. Reinecker. 2005. CX3CR1-mediated dendritic cell access to the intestinal lumen and bacterial clearance. *Science* 307: 254–258.

Niyogi, S.K. 2005. Shigellosis. *Journal of Microbiology* 43: 133–143.

Oelschlaeger, T.A., P. Guerry, and D.J. Kopecko. 1993. Unusual microtubule-dependent endocytosis mechanisms triggered by *Campylobacter jejuni* and *Citrobacter freundii*. *Proceedings of the National Academy of Sciences of the United States of America* 90: 6884–6888.

Ogawa, M., Y. Handa, H. Ashida, M. Suzuki, and C. Sasakawa. 2008. The versatility of *Shigella* effectors. *Nature Reviews Microbiology* 6: 11–16.

Olsen, S.J., G. Miller, T. Breuer, M. Kennedy, C. Higgins, J. Walford, G. McKee, K. Fox, W. Bibb, and P. Mead. 2002. A waterborne outbreak of *Escherichia coli* O157:H7 infections and hemolytic uremic syndrome: implications for rural water systems. *Emerging Infectious Diseases* 8: 370–375.

Palm, N.W., and R. Medzhitov. 2009. Pattern recognition receptors and control of adaptive immunity. *Immunological Reviews* 227: 221–233.

Pamer, E.G. 2004. Immune responses to *Listeria monocytogenes*. *Nature Reviews Immunology* 4: 812–823.

Patel, J.C., and J.E. Galan. 2005. Manipulation of the host actin cytoskeleton by *Salmonella* – all in the name of entry. *Current Opinion in Microbiology* 8: 10–15.

Pédron, T., and P. Sansonetti. 2008. Commensals, bacterial pathogens and intestinal inflammation: an intriguing ménage à trois. *Cell Host and Microbe* 3: 344–347.

Pegues, D.A., E.L. Hohmann, and S.I. Miller. 1995. *Salmonella* including *S. typhi*. In *Infections of the gastrointestinal tract*, ed. M.J. Blaser, P.D. Smith, and J.I. Ravdin, 785–809. New York: Raven.

Pei, Z., C. Burucoa, B. Grignon, S. Baqar, X.Z. Huang, D.J. Kopecko, A.L. Bourgeois, J.L. Fauchere, and M.J. Blaser. 1998. Mutation in the peb1A locus of *Campylobacter jejuni* reduces interactions with epithelial cells and intestinal colonization of mice. *Infection and Immunity* 66: 938–943.

Phalipon, A., and P.J. Sansonetti. 2007. Shigella's ways of manipulating the host intestinal innate and adaptive immune system: a tool box for survival? *Immunology and Cell Biology* 85: 119–129.

Poly, F., and P. Guerry. 2008. Pathogenesis of *Campylobacter*. *Current Opinion in Gastroenterology* 24: 27–31.

Potturi-Venkata, L.P., S. Backert, S.L. Vieira, and O.A. Oyarzabal. 2007. Evaluation of logistic processing to reduce cross-contamination of commercial broiler carcasses with *Campylobacter* spp. *Journal of Food Protection* 70: 2549–2554.

Roberts, A.J., and M. Wiedmann. 2003. Pathogen, host and environmental factors contributing to the pathogenesis of listeriosis. *Cellular and Molecular Life Sciences* 60: 904–918.

Rottner, K., T.E. Stradal, and J. Wehland. 2005. Bacteria-host-cell interactions at the plasma membrane: stories on actin cytoskeleton subversion. *Developmental Cell* 9: 3–17.

Salyers, A.A., and D.D. Whitt. 1994. *Bacterial pathogenesis*. Washington, DC: ASM Press.

Sansonetti, P.J. 2004. War and peace at mucosal surfaces. *Nature Reviews Immunology* 4: 953–964.

Sansonetti, P.J. 2006. Shigellosis: an old disease in new clothes? *PLoS Medicine* 3: e354.

Scheiring, J., S.P. Andreoli, and L.B. Zimmerhackl. 2008. Treatment and outcome of Shiga-toxin-associated hemolytic uremic syndrome (HUS). *Pediatric Nephrology* 23: 1749–1760.

Schlumberger, M.C., and W.D. Hardt. 2006. *Salmonella* type III secretion effectors: pulling the host cell's strings. *Current Opinion in Microbiology* 9: 46–54.

Schroeder, G.N., and H. Hilbi. 2008. Molecular pathogenesis of *Shigella* spp.: controlling host cell signaling, invasion, and death by type III secretion. *Clinical Microbiology Reviews* 21: 134–156.

Seeliger, H.P.R., and D. Jones. 1986. Listeria. In *Bergey's manual of systematic bacteriology*, ed. J. Butler, 1235–1245. Baltimore: Williams and Wilkins.

Sougioultzis, S., and C. Pothoulakis. 2003. Bacterial infections: small intestine and colon. *Current Opinion in Gastroenterology* 19: 23–30.

Speelman, P., I. Kabir, and M. Islam. 1984. Distribution and spread of colonic lesions in shigellosis: a colonoscopic study. *Journal of Infectious Diseases* 150: 899–903.

Stecher, B., and W.D. Hardt. 2008. The role of microbiota in infectious disease. *Trends in Microbiology* 16: 107–114.

Takeuchi, A. 1967. Electron microscope studies of experimental *Salmonella* infection. I. Penetration into the intestinal epithelium by Salmonella typhimurium. *The American Journal of Pathology* 50: 109–136.

Tam, C.C. 2001. *Campylobacter* reporting at its peak year of 1998: don't count your chickens yet. *Communicable Disease and Public Health* 4: 194–199.

Tato, C.M., and C.A. Hunter. 2002. Host-pathogen interactions: subversion and utilization of the NF-κB pathway during infection. *Infection and Immunity* 70: 3311–3317.

Thanassi, D.G., and S.J. Hultgren. 2000. Multiple pathways allow protein secretion across the bacterial outer membrane. *Current Opinion in Cell Biology* 12: 420–430.

Tilney, L.G., and D.A. Portnoy. 1989. Actin filaments and the growth, movement, and spread of the intracellular bacterial parasite. Listeria monocytogenes. *The Journal of Cell Biology* 109: 1597–1608.

Tsolis, R.M., G.M. Young, J.V. Solnick, and A.J. Bäumler. 2008. From bench to bedside: stealth of enteroinvasive pathogens. *Nature Reviews Microbiology* 6: 883–892.

Vallance, B.A., C. Chan, M.L. Robertson, and B.B. Finlay. 2002. Enteropathogenic and enterohemorrhagic *Escherichia coli* infections: emerging themes in pathogenesis and prevention. *Canadian Journal of Gastroenterology* 16: 771–778.

van Spreeuwel, J.P., G.C. Duursma, C.J. Meijer, R. Bax, P.C. Rosekrans, and J. Lindeman. 1985. *Campylobacter colitis*: histological immunohistochemical and ultrastructural findings. *Gut* 26: 945–951.

Wassenaar, T.M., and M.J. Blaser. 1999. Pathophysiology of *Campylobacter jejuni* infections of humans. *Microbes and Infection* 1: 1023–1033.

Watarai, M., S. Funato, and C. Sasakawa. 1996. Interaction of Ipa proteins of *Shigella flexneri* with alpha5beta1 integrin promotes entry of the bacteria into mammalian cells. *The Journal of Experimental Medicine* 183: 991–999.

Watson, R.O., and J.E. Galán. 2008. *Campylobacter jejuni* survives within epithelial cells by avoiding delivery to lysosomes. *PLoS Pathogens* 4: e14.

Wing, E.J., and S.H. Gregory. 2002. *Listeria monocytogenes*: clinical and experimental update. *Journal of Infectious Diseases* 185: 18–24.

Wooldridge, K.G., and J.M. Ketley. 1997. *Campylobacter* – host cell interactions. *Trends in Microbiology* 5: 96–102.

Wooldridge, K.G., P.H. Williams, and J.M. Ketley. 1996. Host signal transduction and endocytosis of *Campylobacter jejuni*. *Microbial Pathogenesis* 21: 299–305.

World Health Organization. 2004. Global burden of disease (GBD) 2002 estimates. WHO, Geneva, Switzerland. http://www.who.int/topics/global_burden_of_disease/en/.

Xicohtencatl-Cortes, J., V. Monteiro-Neto, Z. Saldaña, M.A. Ledesma, J.L. Puente, and J.A. Girón. 2009. The type 4 pili of enterohemorrhagic *Escherichia coli* O157:H7 are multipurpose structures with pathogenic attributes. *Journal of Bacteriology* 191: 411–421.

Yoshida, S., Y. Handa, T. Suzuki, M. Ogawa, M. Suzuki, A. Tamai, A. Abe, E. Katayama, and C. Sasakawa. 2006. Microtubule-severing activity of *Shigella* is pivotal for intercellular spreading. *Science* 314: 985–989.

Zheng, J., J. Meng, S. Zhao, R. Singh, and W. Song. 2008. *Campylobacter*-induced interleukin-8 secretion in polarized human intestinal epithelial cells requires *Campylobacter*-secreted cytolethal distending toxin- and Toll-like receptor-mediated activation of NF-kappaB. *Infection and Immunity* 76: 4498–4508.

Microbiology Terms Applied to Food Safety

Anup Kollanoor-Johny, Sangeetha Ananda Baskaran, and Kumar Venkitanarayanan

1 Introduction

A cell is the structural and functional unit of every living organism. It is the smallest unit that performs vital physiological functions of life and maintains homoeostasis in the organisms. Living organisms are broadly classified as prokaryotes and eukaryotes based on the differences in cellular complexity. All bacteria are unicellular and are prokaryotes ("pro" meaning before and "karyon" meaning nut or kernel). These cells can form clusters, chains, and tetrads and can be either motile or nonmotile. On the other hand, eukaryotes ("eu" meaning good, true) can be either unicellular (photosynthesis-active algae and nonphotosynthetic protozoa) or multicellular (plants, animals, and fungi).

Prokaryotes and eukaryotes differ in various aspects (Edwards 2000). Generally, prokaryotes are much smaller (1–10 μm) than eukaryotes (10–100 μm) and have a linear/circular molecule of DNA as their genetic material but lack a nuclear envelope. Eukaryotes, on the other hand, have their linear genetic material contained within the nucleus, which is surrounded by a double-walled "nuclear membrane." The prokaryotic cell wall (5–80 nm) contains murein, which is composed of a large number of polysaccharide molecules tethered together by short chains of amino acids. Muramic acid, one of the major molecules of murein, is absent in eukaryotes. The prokaryotic cell membrane, unlike its eukaryotic counterpart, lacks cholesterol and other steroids and includes energy-generating components. Excluding ribosomes, which are the compartments of protein synthesis in both cell types, all the other cellular structures – the organelles such as mitochondria, Golgi bodies, endoplasmic reticulum, mitotic spindles, centrioles, and vacuoles – are absent in prokaryotic cells. In addition, the latter do not contain basic protein structures such as histones that tightly bind to the DNA. Although prokaryotes lack several organelles, they are equipped with flagella, pili, fimbriae, and special surface proteins for movement, attachment, colonization, invasion, and reproduction (Jonson et al. 2004; Ray 2001). Prokaryotes divide mostly by binary fission although gene transfer and genetic recombination can occur without gamete or zygote formation. Eukaryotes, on the other hand, divide by mitotic divisions (Sadava et al. 2009). These differences between prokaryotes and eukaryotes are summarized in Table 1.

A. Kollanoor-Johny • S.A. Baskaran • K. Venkitanarayanan (✉)
Department of Animal Science, University of Connecticut, Storrs, CT, USA
e-mail: kumar.venkitanarayanan@uconn.edu

Table 1 Characteristics of prokaryotes and eukaryotes

Prokaryotes	Eukaryotes	Common features
No nuclear envelope	Nuclear envelope	DNA
Small and simple	Complex	Ribosomes
No organelles	All organelles	Cytoplasm
Unicellular	Unicellular or multicellular	Cell membrane
Naked DNA	DNA/RNA	
Reproduction – binary fission	Reproduction via mitosis	

2 Bacterial Cell Wall and Gram's Reaction

Outside the plasma membrane that surrounds the cytoplasm, all bacterial cells have a cell envelope that forms the outer portion. The cell envelope of Gram-positive bacteria includes the cell wall and the periplasm, whereas that of a Gram-negative cell consists of an outer membrane (OM), a cell wall, and a periplasmic space above the inner membrane. The OM is composed of the lipopolysaccharides (LPSs), lipoproteins, and phospholipids. LPSs and lipoproteins are embedded in the phospholipid bilayer, with the hydrophobic fatty acid part facing in and the hydrophilic part facing out. The OM is a barrier against many extracellular factors, including enzymes, denaturing agents, or certain antibiotics, and it elicits immunogenic reactions. Underneath the OM is the layer that contains a thin film of peptidoglycan (1–7 nm). The space between the inner and outer membranes is called the periplasmic space, where there are enzymes to break down large molecules. The OM and the peptidoglycan are linked to each other by lipoprotein bridges. On the other hand, Gram-positive bacteria have a thick cell wall composed of mucopeptides, teichoic acids, and lipoteichoic acids. Teichoic acids are immunogenic and can induce tumor necrosis factor α and interleukin 1. These bacteria have multiple layers (20–80 nm) of peptidoglycan in a large polymer network in the bacterial cell wall formed by repeating disaccharides interconnected by polypeptides (Ray 2001).

Gram-negatives and Gram-positives are distinguished from each other using Hans Christian Gram's staining technique based on their cell wall's capacity to retain crystal violet dye (Cabeen and Jacobs-Wagner 2005). Gram-positive bacteria retain the dye, whereas Gram-negatives do not. Gram-positives have a thicker, multilayered cell wall (Bhavsar et al. 2004) and multiple layers of peptidoglycan. However, Gram-negatives have a thinner cell wall and peptidoglycan layer. Upon staining with crystal violet dye, both types of bacteria take up the stain and appear blue. In the following step, Gram's iodine solution enters the cells and forms a water-insoluble complex with the dye. After decolorization with alcohol or acetone solvent, it is theorized that the Gram-positive cell wall is dehydrated by the solvent, making it nonpermeable. The Gram-negative cell wall, on the other hand, becomes more permeable due to the loss of lipid content on the cell wall. The crystal violet – iodine complex is solubilized during the decolorization step and escapes a Gram-negative cell, whereas the thick, dehydrated, nonpermeable cell wall prevents it from getting out of a Gram-positive cell. Consequently, Gram-negative bacteria will take up the counterstain, Safranin, and appear pink (Murray 1995).

3 Multiplicity of Food Organisms

From a food microbiologist's perspective, there are several microorganisms such as bacteria, viruses, fungi, protozoans, and parasites for which foods serve as vehicles of transmission (Table 2). Among these agents, several bacteria are most commonly implicated in foodborne outbreak

Table 2 Important foodborne bacteria and associated foods

Bacteria	Gram +/−	Shape	Biochemical reaction	Foods
Acinetobacter	−	Rods	Oxidase negative, strict aerobes	Refrigerated fresh produce
Aeromonas	−	Rods	Sugar fermentation producing gas	Fish products
Alcaligenes	−	Rods	Alkaline reaction in litmus milk	Raw milk, poultry products
Alteromonas	−	Rods	Strict aerobes	Seafood
Arcobacter	−	Curved/s-shaped	Aerotolerant, oxidase- and catalase-positive	Poultry, raw milk, cattle, and swine products
Bacillus	+	Rods	Aerobes, mesophilic	Vegetables
Brochothrix	+	Rods	Rhamnose and hippurate reactions	Processed meats
Campylobacter	−	Spirally curved rods	Microaerobic	Poultry, fish, and pork
Carnobacterium	+	Rods	Catalase-negative, heterofermentative	Vacuum-packaged meats and related products, fish, and poultry meats
Citrobacter	−	Rods	Slow lactose-fermenting, citrate-fermenting	Vegetables and fresh meats
Clostridium	+	Rods	Anaerobic, mesotrophic, psychrotrophic, thermotrophic	Canned foods
Corynebacterium	+	Rods	Mesotrophs	Vegetable and meat products
Enterobacter	−	Rods	Facultatively anaerobic	Infant foods
Enterococcus	+	Coccus	Facultatively anaerobic	Fermented foods
Erwinia	−	Rods	Facultatively anaerobic, catalase-negative, oxidase-positive	All fruits, especially apples, pears, and carrots
Escherichia	−	Rods	Facultatively anaerobic, oxidase-negative, catalase-positive	Ground beef, fresh produce
Flavobacterium	−	Rods	Yellow to red pigment producer	Refrigerated fruits and vegetables, fish
Hafnia	−	Rods	Lysine- and ornithine-positive	Refrigerated meat and vegetable products
Kocuria	+		Oxidase-negative Catalase-positive	Fermented foods
Lactobacillus	+	Rods	Catalase-negative, Microaerophilic	Dairy products, apple and pear mashes; refrigerated, stored, and vacuum-packaged meat
Lactococcus	+	Spherical or ovoid	Catalase-negative	Unpasteurized milk
Leuconostoc	+	Cocci	Catalase-negative, heterofermentative	Cheese, dairy products
Listeria	+	Cocco-bacillary	Catalase-positive and oxidase-negative	Unpasteurized cheese, milk, ready-to-eat foods
Micrococcus	+	Cocci	Catalase-positive	High-salt-containing products
Moraxella	−	Rods	Oxidase-positive	Meats, poultry, offals, fish, seafood, ready meals
Paenibacillus	+	Rods	Spore-forming aerobes	Cooked chilled foods

(continued)

Table 2 (continued)

Bacteria	Gram +/−	Shape	Biochemical reaction	Foods
Pantoea	−	Rods	Motile	Plants, water
Pediococcus	+	Coccus	Homofermentative, catalase-negative, motile, facultative anaerobe	Meat starter culture
Proteus	−	Rods	Motile, swarming	Vegetable and meat products
Pseudomonas	−	Rods	Blue-green pigments	Fresh foods, especially meat, vegetables, poultry and seafood products
Psychrobacter	−	Cocco-bacilli	Nonmotile, catalase- and oxidase-positive, non-glucose fermenter, egg yolk positive, and Tween-80 hydrolyzing	High-salt products, meats, poultry, fish
Salmonella	−	Rods	Catalase-positive, oxidase-negative	Poultry, meat, vegetables
Serratia	−	Rods	Aerobic, proteolytic, red pigments	Refrigerated meat and vegetables
Shewanella	−	Straight or curved rods	Nonpigmented, motile, oxidase positive	Aquatic foods
Shigella	−	Rods	Facultatively anaerobic, nonmotile, utilize glucose-producing acid	Salads, raw vegetables, poultry, milk and other dairy products
Staphylococcus	+		Catalase-positive	High-salt foods
Vagococcus	+	Cocci	Motile, catalase-negative, hydrogen sulfide production	Fish and water
Vibrio	−	Straight or curved rods	Comma-shaped, oxidase-positive, catalase-variable	Fish products, seafood
Weisella	+	Cocco-bacillary, rarely appearing as rods	Gas from carbohydrates	Traditional fermented foods
Yersinia	−	Straight rods	Facultatively anaerobic	Pork, milk

episodes. Foodborne diseases in human beings are caused either by direct contact with infected food animals/animal products (zoonotic) or humans, such as a food handler, or by direct ingestion of contaminated foods. The Centers for Disease Control and Prevention reported an estimated 76 million cases of foodborne diseases annually in the United States, with approximately 300,000 hospitalizations and 5,000 deaths (Mead et al. 1999). There are three important terms with regard to foodborne diseases: foodborne infections, foodborne intoxications, and foodborne toxicoinfections. Foodborne infection is the condition caused by the ingestion of viable cells of a pathogen. For example, *Escherichia coli* O157:H7 and *Salmonella* Enteritidis infections are brought about by the ingestion of food contaminated with living cells of these pathogens. Foodborne intoxication is the condition in which preformed toxins in the food produced by a toxigenic pathogen act as the underlying cause of disease. For example, *Clostridium botulinum* produces a neurotoxin on foods that, when consumed, bring about a serious condition called botulism. Finally, foodborne toxicoinfection is that in which the ingestion of viable pathogenic cells causes the production of toxins inside the human body, leading to infection episodes. For example, *Vibrio cholerae* produces cholera toxin inside the body after getting ingested by the host (Bhunia 2007). The morphology, Gram's reaction, biochemical properties, and associated foods with important foodborne bacteria are given in Table 2.

4 Bacterial Growth

Foodborne bacteria follow the classical sigmoidal growth curve that involves lag phase, log phase, stationary phase, and the death phase, albeit the Gompertz, Baranyi, and the three-phase models do not recognize the death phase (Buchanan et al. 1997). In the lag phase, bacteria will sense the environment and prepare for an active multiplicative period. Bacterial will synthesize cell components such as ribosomes, ATP, and cofactors with no increase in cell numbers. Then, in the log phase, bacteria will grow and divide at a maximal rate in an exponential pattern. The time that is required for bacteria to double the number at a specific temperature – known as the generation time – remains constant during this phase. The stationary phase starts when the cell density reaches approximately 10^8–10^9 colony-forming units (CFU). This follows with the death phase, in which there will be a gradual decline in the number of living cells because of the nutrient-deprived environment and the accumulation of toxic biochemical wastes.

Normally in foods, there will be a mixed population of microorganisms including bacteria, fungi, yeasts, and molds. Since these organisms have different optimum growth conditions, the counts of some microorganisms at a given temperature at a certain time may be higher than the others. The temperature of 35°C allows the growth of many bacteria, whereas during storage at 4°C (refrigeration), some bacteria will outcompete those that will grow better at 35°C. There is another situation wherein at the same temperature two or more organisms can grow, but because of the differences in their generation times, one overcomes the other(s) (Ray 2001; Sinell 1992). This is true about bacteria predominating yeasts and molds, resulting in spoilage of foods. There are certain foods that exploit the qualities of symbiosis in which two or more bacteria grow together in a given food for their benefit, helping in the development of a desired quality in those foods. For example, two lactic acid bacteria, *Streptococcus thermophilus* and *Lactobacillus delbrueckii* subsp. *bulgaricus,* utilize their biochemical machineries to help each other's growth in foods such as yogurt (Tamime 2002). Individually, they produce fewer flavor compounds such as acetaldehyde, whereas when put together, they produce more than the additive amount of acetaldehyde produced by them individually, thereby bringing about the desired flavor in yogurt, an interesting example of synergism or proto-cooperation in foods. On the contrary, there are conditions where one bacterium inhibits or reduces the growth of the other, known as antagonistic growth (Sinell 1992).

5 Indicators of Contamination

Certain bacteria are often indicators of possible food contamination hazards. Their presence in food or water indicates the potential presence of pathogens. They reflect the microbiological quality of foods relative to product shelf life or their safety from foodborne pathogens. They are generally the enteric indicator organisms such as *E. coli* and other coliforms, and fecal streptococci (Ray 2001).

6 Important Factors Influencing Microbial Growth

Several factors determine the growth of microorganisms in foods. They can be classified as intrinsic and extrinsic to foods (Jay 2000). The former is characterized by the factors inherently present in foods, including pH, moisture content, nutrient availability, oxidation-reduction potential, biological

Table 3 Range of certain important parameters for growth

Bacteria	pH range	Growth temperature, °C	Water activity, a_w
Bacillus cereus	4.9–9.2	0–5	0.95
Campylobacter spp.	5.8–9.0	30–47	0.98
Clostridium botulinum	4.3–8.2	0–15	0.97
Clostridium perfringens	5.0–8.5	10–15	0.94
Escherichia coli	4.2–9.0	5–10	0.96
Listeria monocytogenes	4.2–9.8	0–5	0.92
Salmonella	3.8–9.2	5–10	0.90
Staphylococcus aureus	4.0–9.5	5–10	0.84
Vibrio parahaemolyticus	4.8–11.0	5–10	0.94
Yersinia enterocolitica	4.2–9.0	0–5	0.96

structures, and presence of antimicrobial compounds. Extrinsic factors are the properties of the storage environment of foods that affect both the foods and associated microorganisms, and include temperature of storage, relative humidity, presence and activities of other organisms, and gaseous environment. The optimum temperature, pH range, and water activity of some of the important foodborne bacteria are given in Table 3.

7 Classification of Foodborne Bacteria

7.1 Based on Temperature Tolerance

Bacteria can be classified based on their temperature for growth. Psychrotrophs ("cold-loving") are those bacteria that are capable of growing at low temperatures, ranging from 0°C to 5°C. They can also live in an optimum temperature range of 20–30°C and are mostly associated with spoilage of foods. *Pseudomonas* is the most commonly found one (Kraft 1992). Other major psychrotrophic foodborne pathogens include *Listeria monocytogenes* and *Yersinia enterocolitica*. Mesophiles are those bacteria that like to grow at a temperature range between 20°C and 45°C, with their optimum temperature for growth being between 30°C and 40°C. Most of the enteric pathogens such as *E. coli* O157:H7 and *Salmonella* Enteritidis do grow very well at this temperature range. However, unlike the psychrotrophs, they do not grow at lower temperatures. As their name indicates, thermophilic ("temperature-loving") bacteria like to grow at high temperatures, especially between 45°C and 65°C. The genera *Clostridium* and *Bacillus* come under this category (Jay 2000), which are of concern to the canning industry. Additionally, there are other bacteria that are capable of surviving even pasteurization temperatures (but below boiling temperature) and are called thermodurics. Micrococcus, *Lactobacillus,* and *Bacillus* (spores) are examples of thermodurics.

7.2 Based on Oxygen Requirements

Based on the oxygen requirements for growth, bacteria are classified as *aerobic*, anaerobic, microaerobic, and *facultatively anaerobic*. Aerobic bacteria require oxygen for growth. They have no fermentative pathways and generally produce superoxide dismutase. Examples are *Pseudomonas*, *Brevibacterium,* and *Bacillus*. On the other side, anaerobic bacteria dislike oxygen and the presence of oxygen will inhibit their growth. They lack superoxide dismutase and catalase. They are fermenters but cannot use oxygen as a terminal oxygen acceptor. *Clostridium botulinum* and *Propionibacterium* are examples. Facultative anaerobes are fundamentally anaerobes, but can grow under aerobic conditions also if conditions insist. Members of Enterobacteriaceae are typical examples. Organisms such as *Campylobacter* that require low but not full oxygen tension are called microaerobic.

7.3 Based on Physiological Tolerance Levels

Physiological tolerance levels are very important for some bacteria to adapt under extreme conditions. Bacteria often upregulate certain genes under stress conditions for survival purposes. *Staphylococcus aureus*, *Pediococcus,* and *Vibrio* are able to tolerate high-salt concentrations and are called *halotolerant* ("halo" meaning salt). Microorganisms growing at low water activity are termed *xerophiles*. These include osmophiles and halophiles, wherein halophiles are microorganisms that live and grow in salty environments. They are classified into moderate and extreme halophiles, of which the former require 0.2–0.5 M NaCl for their growth and play a significant role in the spoilage of salted food products (*Vibrio parahaemolyticus*), whereas extreme halophiles require 3 M NaCl for growth (archaebacteria members). A similar term, *osmophiles,* is applied to a group of bacteria that grow at a relatively higher osmotic environment. *S. aureus* and *Leuconostoc* are examples. *Saccharophiles* are osmophiles that tolerate high sugar, and bacteria capable of survival at very low pH, such as *Lactobacillus* and *Pediococcus*, are called *acidurics* (Ray and Daeschel 1992).

7.4 Based on Substrate Breakdown

Bacteria differ in their capacity to lyse substrates. The proteolytic bacteria (*Clostridium* and *Pseudomonas*) are capable of hydrolyzing proteins. The lipolytic bacteria like *Pseudomonas* and *Flavobacterium* are capable of hydrolyzing triglycerides. *Bacillus* and *Aeromonas* can break down complex carbohydrates and are called *saccharolytic*. In terms of their capacities to produce acids from carbohydrates, certain bacteria are lactic acid producers (*Lactococcus*, *Pediococcus*), acetic acid producers (*Acetobacter*), propionic acid producers (*Propionibacterium*), and butyric acid producers (*Clostridium butyricum*) (Ray and Daeschel 1992). Some of them produce gases such as carbon dioxide, hydrogen, and hydrogen sulfide (*Leuconostoc, Enterobacter, Clostridium, Lactobacillus,* etc.), and some others produce slime (*Xanthomonas, Alcaligenes, Enterobacter, Lactococcus,* and *Lactobacillus*).

7.5 Based on Extensiveness of Hazard

Based on the degree of hazard impact on health, foodborne pathogens are divided into those that cause severe direct health hazards, such as *C. botulinum*, *L. monocytogenes*, *E. coli O157:H7,* and *Shigella dysenteriae*; those that cause moderate hazards with a potentially extensive spread, such as *Salmonella* spp. and enterox igenic *E. coli*; and those that cause moderate hazard with limited spread. *S. aureus*, *C. perfringens*, *B. cereus*, *Campylobacter,* and *Yersinia* are examples.

8 Microorganisms Involved in Food Safety

8.1 Gram-Positive Foodborne Pathogens

Among the various species of Gram-positive bacteria responsible for foodborne outbreaks, *L. monocytogenes* causes fatal conditions in pregnant women, children, the elderly, and the immunocompromised. *Staphylococcus aureus* causes serious outcomes due to the heat-stable enterotoxin in foods, resulting in food poisoning. Certain foodborne pathogens are difficult to inactivate due to their capacity to produce spores. Spores are survival structures that help bacteria to tide over unfavorable environmental conditions. Many spores are resistant to cooking temperatures and can survive freezing and drying. Bacteria execute resistance in spores by synthesizing new enzymes such as dipicolinic acid synthetase and heat-resistant catalase, by increasing or decreasing other enzymes, by synthesizing calcium dipicolinate in the spore core, and by producing a keratin spore coat. *Bacillus cereus, Clostridium perfringens,* and *C. botulinum* are examples of bacteria that produce spores. Spores of *C. botulinum* type A and B are heat-resistant, whereas spores of type E are heat-labile. *C. botulinum* produces a neurotoxin (botulinum toxin) resulting in a condition called botulism (Latin, *botulus*, meaning "sausage"), a rare but serious paralytic illness caused by botulinum toxin. There are three types of botulism associated with foods, namely, (1) foodborne botulism caused by ingestion of foods contaminated with botulinum toxin, (2) infant botulism due to intestinal colonization and toxin production in infants, and (3) adult intestinal toxemia botulism, an uncommon form of intestinal colonization and toxin production by *C. botulinum* in adults (Sobel 2005). In infant botulism, the toxin will be present in the stool excreted by the infected infant. Poorly canned alkaline vegetables and smoked fish are implied in adult botulism. Honey is commonly associated with infant botulism. Another *Clostridium* species, *C. perfringens,* causes enteritis in humans.

8.2 Gram-Negative Foodborne Pathogens

Many of the Gram-negative bacteria causing foodborne contaminations are pathogens. In general, the Gram-negative bacterial infections result in manifestations such as diarrhea, abdominal cramps, and intestinal disorders. Although diarrhea is difficult to define quantitatively, it is the passage of stool in excess than normal with a changed consistency observed with a recent change in the bowel movements. The syndrome of dysentery is characterized by blood and pus in the stools, abdominal cramps, tenesmus, and fever. Gross blood in the stools is the most reliable sign (Ghai et al. 2004; Winn et al. 2005). Many Gram-negative foodborne pathogens such as *Campylobacter*, *Salmonella*, *Shigella*, *E. coli* O157:H7, enteroinvasive *E. coli* (EIEC), and *Y. enterocolitica* produce bloody diarrhea, whereas enterotoxigenic *E. coli* (ETEC), *Vibrio cholerae,* and *C. perfringens* cause watery diarrhea. Information on some of the important foodborne diarrheal pathogens is given in Table 4.

8.3 Foodborne Prions

Bovine spongiform encephalopathy (BSE), or *"mad cow disease,"* is a chronic degenerative disease affecting the central nervous system of cattle (USFDA 2005). The condition is caused by prions, a modified form of a normal eukaryotic protein capable of causing cells to produce more abnormal proteins. The disease in cattle is characterized by a prolonged incubation period (months to years) and a progressive, debilitating, neurological illness with characteristic histological lesions such as fibrils, vacuoles, and amyloid deposits in the brain resulting in death (Scott et al. 1989, 1990). No inflammatory or immune response will be present (Scott et al. 1990). The U.S. Department of

Table 4 Microbial diarrhea of foodborne importance

Organism	Symptoms	Duration	Incubation period
Campylobacter (inflammatory diarrhea)	Abdominal pain, enteritis with diarrhea (may be bloody),[a] fever	1–2 days mild, <1 week normal, self-limiting	2–5 days
Salmonella (inflammatory diarrhea)	Diarrhea (occasionally blood), abdominal cramps, fever, nausea	4–7 days, mostly recovery without treatment[a]	8–48 h
Shigella (inflammatory diarrhea)	Watery diarrhea first, lower abdominal cramps, pus and blood in the stools (dysentery)	5–7 days, rarely requires hospitalization	1–7 days
Yersinia enterocolitica (inflammatory diarrhea)	Fever, diarrhea, frequently with blood in stools, noninflammatory gastroenteritis	1 day to 3 weeks	4–7 days
Enterohemorrhagic *E. coli* (noninflammatory/inflammatory diarrhea)[b]	Abdominal cramps with blood in stools	1–2 weeks	3–5 days
Vibrio parahaemolyticus (noninflammatory)	Watery diarrhea with headache, nausea, vomiting, fever and chills	3–10 days, self-limiting[a]	5–92 h

[a]CDC (2009)
[b]Miller et al. (1994)

Agriculture cautioned that there might be a possibility of contracting a type of transmissible spongiform encephalopathy called as the variant Creutzfeldt – Jakob disease by humans upon eating contaminated beef. The agency reported that the epidemic ended by banning beef byproducts from being incorporated in feeds supplied to cattle and sheep (USDA 2009).

8.4 Molds of Foodborne Importance

Foodborne molds can be of spoilage or pathogenic significance. Pathogenic molds produce mycotoxins, which are the secondary metabolites produced at the end of the exponential (log) phase of growth. Among the spoilage causing fungi, *Aspergillus* species are involved in spoilage of foods such as jams, cured ham, fruits, and vegetables. Other fungi such as *Alternaria, Geotrichum*, and *Mucor* cause spoilage of vegetables. *Penicillium expansum*, a pathogen in fresh fruits, is psychrotrophic and can grow at temperatures as low as −2 to −3°C. They produce a toxin called *patulin,* which is potentially carcinogenic and produces immunological, neurological, and gastrointestinal toxic effects in animals (Davis and Diener 1987). Other species such as *Penicillium verrucosum* and *P. commune* are also capable of producing toxins. *Aspergillus flavus* and *A. parasiticus* produce toxins called aflatoxins (Kurtzman et al. 1987), and the common ones are B1, B2, G1, G2, M1, and M2 (Ayres et al. 1980). In humans, aflatoxins are reported to cause liver cancer. It results in stunted growth in children if they are exposed to toxic levels during the neonatal period.

8.5 Foodborne Parasites and Protozoa

Foods can play an important role in the transmission of a variety of parasites of eukaryotic origin, including nematodes, cestodes, and protozoans. Parasites such as *Taenia saginata, T. solium*, and other *Taenia* spp. can be transmitted to humans by the consumption of raw or undercooked meat or meat products. In addition, raw sea and freshwater foods including fish, mollusks, and frogs can serve as vehicles of some parasites (Pozio 2003). Important foodborne protozoans are *Cryptosporidium, Giardia, Entamoeba* spp., and *Cyclospora,* which are commonly associated with fresh produce. The use of untreated water for food preparation also plays a role in their transmission to humans. Like foodborne viruses, parasites do not multiply in foods but survive in moist foods for months (Dawson 2005; Dawson et al. 2005).

9 Bacteria Involved in the Spoilage of Foods

Among the various bacteria in foods, some cause spoilage of foods and food products. Spoilage is a condition resulting from microbial activity that is detected by changes in odor, flavor, and appearance. In foods such as meat, bacteria use the low-molecular-weight substances such as lactate, glucose-1-phosphate, and glucose-6-phosphate followed by the use of creatine, and peptides such as carnosine and anserine. Once these compounds are used up, amino acids are potentially utilized as substrates. Finally, proteins are catabolized, resulting in the production of ammonia, cadaverine, putrescine, and hydrogen sulfide; the signs of spoilage start to appear. In general, bacterial spoilage occurs when the bacterial population of food reaches 10^7–10^8 per g or cm^2 or mL (ICMSF 2005).

Among several factors favoring spoilage, water activity (a_w) is critical in meat spoilage. Water activity refers to the water available for the biochemical reactions. It is the ratio of the vapor pressure of water in the food to the vapor pressure of pure water at the same temperature. If meat is stored anaerobically at higher water activity ($a_w > 0.95$), *Pseudomonas, Flavobacterium, Alcaligenes, Moraxella,* and *Bacillus* constitute the predominant spoilage flora. These bacteria cause proteolysis and lipolysis and result in pigment and slime production. If meat is stored anaerobically at a relatively high water activity ($a_w > 0.95$) at a lower temperature, lactic acid bacteria dominate, resulting in acid production. Slime and green color development follow. If meat is stored anaerobically at a higher water activity at a high temperature, *Clostridium* spp. take the upper hand, resulting in noxious odors due to anaerobic catabolism of proteins. When spoilage occurs in eggs, *Pseudomonas, Flavobacterium, Proteus,* and *Salmonella* grow exuberantly in the decaying menstruum. In case of spoilage of fish, apart from *Pseudomonas* and *Flavobacterium, Vibrio* and *Micrococcus* may be involved (Vanderzant and Splittstoesser 1992). Spoilage of fruits and vegetables is generally encountered with *Pseudomonas, Bacillus, Clostridium,* and *Erwinia* (Kraft 1992; Vanderzant and Splittstoesser 1992). Although not commonly seen, there are certain spoilage bacteria that produce spores. For example, *Sporolactobacillus*, a Gram-positive organism, and *Desulfotomaculum*, a Gram-negative spore former, cause the spoilage of canned food.

10 Bacteria in Foods Beneficial to Humans

Although a number of bacterial species encountered in foods are either spoilage or pathogenic organisms, there are several others that are beneficial to the food. The lactic acid bacteria (LAB) are important in the fermentation of several food products. They include important Gram-positive genera such as *Enterococcus, Lactococcus, Lactobacillus, Leuconostoc, Pediococcus,* and *Streptococcus*. They obtain energy by substrate-level phosphorylation while oxidizing carbohydrates. There are two groups of LAB, based on the end products of glucose metabolism. Homofermentative LAB are those that produce lactic acid as a result of glucose fermentation. *Lactobacillus acidophilus, L. bulgaricus, L. delbrueckii, L. lactis, Pediococcus acidilactici, Lactococcus lactis,* and *L. cremoris* are examples. *Pediococcus cerevisiae* is used as a starter culture in fermented sausage, whereas *P. acidilactici* and *P. pentosauces* are used in making summer sausages. Heterofermentative LAB are those that utilize glucose and produce equal volumes of lactate, ethanol, and carbon dioxide. They produce flavor and aroma compounds such as acetaldehyde and diacetyl. *Lactobacillus brevis, L. buchneri*, and all members of *Leuconostoc* are the important ones (Jay 2000). *Leuconostoc dextranicum* and *L. cremoris* are required for dairy fermentations, wherein they ferment citric acid in milk, producing diacetyl. Another species, *L. mesenteroides,* is used in sauerkraut fermentation. Another common species of LAB, *Streptococcus thermophilis*, is used in the manufacture of Swiss cheese and yogurt (Sieuwerts et al. 2008; Tamime 2002). *Lactococcus lactis* is responsible for the natural souring of milk. It produces nisin, an antibiotic peptide that inhibits *Bacillus* and *Clostridium*. *Brevibacterium linens* is another bacterium important in the ripening of cheese surfaces, especially of Brick cheese and Limburger cheese. *Propionibacterium shermanii* and *Propionibacterium freundenreichii* are useful in making Swiss cheese. Another beneficial bacterium, *Micrococcus*, a nonmotile bacterium that produces yellow, orange, and pink pigments, is very salt-tolerant and therefore used in cured meats, brines, and curing tanks.

11 Summary

The biology of unicellular organisms has always provided important insides on the host-pathogen interactions and in the way bacteria, viruses, and other organisms survive in and contaminate different environments. Prokaryotes are still grouped based on the staining of their cell walls, although new insights into their cellular composition reflect a vast diversity within this group of microorganisms. The survival, and in some instances the growth, of bacteria in foods is the response to factors associated with the microorganisms themselves or the foods. Many microorganisms from foods have been used beneficially to impart flavors, to preserve foods, or to create new food products, as in the case of yeast used to make bread. Yet some microorganisms can be harmful to humans. Foodborne diseases are classified as infections, intoxications, and toxicoinfections based on how bacteria produce disease. Some bacteria are often used as indicators of possible food contamination.

References

Ayres, J.C., J.O. Mundt, and W.E. Sandine. 1980. *Microbiology of foods*, 658–683. San Francisco: W.H. Freeman.
Bhavsar, A.P., L.K. Erdman, J.W. Schertzer, and E.D. Brown. 2004. Teichoic acid is an essential polymer in *Bacillus subtilis* that is functionally distinct from teichuronic acid. *Journal of Bacteriology* 186: 7865–7873.
Bhunia, A.K. 2007. *General mechanism of pathogenesis in foodborne pathogens*, 93–103. New York: Springer.
Buchanan, R.L., R.C. Whiting, and W.C. Damert. 1997. When is simple good enough: a comparison of Gompertz, Baranyi, and the three-phase linear models for fitting bacterial growth curves. *Food Microbiology* 14: 313–326.
Cabeen, M.T., and T. Jacobs-Wagner. 2005. Bacterial cell shape. *Nature Reviews Microbiology* 3: 601–610.
CDC. 2009. Centers for Disease Control and Prevention. General information on various bacteria. http://www.cdc.gov/nczved/divisions/dfbmd/diseases/. Accessed 3 Dec 2010.
Davis, N.D., and U.L. Diener. 1987. Mycotoxins. In *Food and beverage mycology*, 2nd ed, ed. L.R. Beuchat, 517–570. New York: Van Nostrand Reinhold.
Dawson, D. 2005. Foodborne protozoan parasites. *International Journal of Food Microbiology* 103: 207–227.
Dawson, D.J., A. Paish, L.M. Staffell, I.J. Seymour, and H. Appleton. 2005. Survival of viruses on fresh produce, using MS2 as a surrogate for norovirus. *Journal of Applied Microbiology* 98: 203–209.
Edwards, G.I. 2000. *Biology the easy way*, 3rd ed, 370. Hauppauge: Barron's Educational Series.
Ghai, O.P., P. Gupta, and V.K. Paul. 2004. Diseases of gastrointestinal system and liver. In *Essential pediatrics*, 6th ed., 269. New Delhi: CBS Publishers and Distributers.
ICMSF (International Commission on Microbiological Specifications for Foods). 2005. *Microorganisms in foods: microbial ecology of food commodities*, 2nd ed, 31–32. New York: Kluwer Academic/Plenum.
Jay, J.M. 2000. *Modern food microbiology*, 6th ed, 38–52. Frederick: Aspen.

Jonson, A.-B., S. Normark, and M. Rhen. 2004. Fimbriae, pili, flagella and bacterial virulence. In *Concepts in bacterial virulence*, ed. W. Russel and H. Herwald, 66–89. Basel: Karger.

Kraft, A.A. 1992. *Psychrotrophic bacteria in foods. Disease and spoilage*. Boca Raton: CRC Press.

Kurtzman, C.P., B.W. Horn, and C.W. Hesseltine. 1987. *Aspergillus nominus*, a new aflatoxin producing species related to *Aspergillus flavus* and *Aspergillus tamari*. *Antonie Van Leeuwenhoek* 53: 147–158.

Mead, P.S., L. Slutsker, V. Dietz, L.F. McCaig, J.S. Bresee, C. Shapiro, P.M. Griffin, and R.V. Tauxe. 1999. Food-related illness and death in the United States. *Emerging Infectious Diseases* 5: 607–625.

Miller, J.R., L.J. Barrett, and K. Kotloff. 1994. A rapid test for infectious and inflammatory enteritis. *Archives of Internal Medicine* 154: 2660.

Murray, P.R. 1995. *Manual of clinical microbiology*, 6th ed. Washington, DC: American Society of Microbiology.

Pozio, E. 2003. Foodborne and waterborne parasites. *Acta Microbiologica Polonica* 52: 83–96.

Ray, B. 2001. *Fundamental food microbiology*, 2nd ed, 15–17. Boca Raton: CRC Press.

Ray, B., and M.A. Daeschel. 1992. *Food biopreservatives of microbial origin*. Boca Raton: CRC Press, pp. 81, 102, 137, 155, 177.

Sadava, D., D.M. Hillis, H.C. Heller, and M. Berenbaum. 2009. Genes and heredity – the cell cycle and cell division. In: *Life: the science of biology*, 210–220. New York: Macmillan.

Scott, P.R., B.M. Aldridge, M. Clarke, and R. Will. 1989. Bovine spongiform encephalopathy in a cow in the United Kingdom. *Journal of the American Veterinary Medical Association* 195: 1745–1747.

Scott, A.C., G.A.H. Wells, M.J. Stack, H. White, and M. Dawson. 1990. Bovine spongiform encephalopathy: detection and quantification of fibrils, fibril protein (PrP) and vacuolation in brain. *Veterinary Microbiology* 23: 295–304.

Sieuwerts, S., A.M. de Bok, J. Hugenholtz, and E.T.J. van Hylckama Vlieg. 2008. Unraveling microbial interactions in food fermentations: from classical to genomic approaches. *Applied and Environmental Microbiology* 74: 4997–5007.

Sinell, H.J. 1992. Interacting factors affecting mixed populations. In *Microbial ecology of foods*, vol. I, ed. J.H. Silliker, 215. New York: Academic.

Sobel, J. 2005. *Botulism Clinical Infectious Disease* 41: 1167–1173.

Tamime, A.Y. 2002. Fermented milks: a historical food with modern applications – a review. *European Journal of Clinical Nutrition* 56(Suppl. 4): S2–S15.

USDA. 2009. Agricultural research service. BSE ("Mad cow disease") and other TSE diseases. http://www.ars.usda.gov/Main/docs.htm?docid=13677. Accessed on 3 Dec 2010.

USFDA. 2005. Commonly asked questions about BSE in products regulated by FDA's Center for Food Safety and Applied Nutrition (CFSAN). http://catalogue.nla.gov.au/Record/4592529. Accessed 3 Dec 2010.

Vanderzant, C., and D.F. Splittstoesser. 1992. *Compendium of methods for the microbiological examination of foods*, 3rd ed. Washington, DC: American Public Health Association.

Winn Jr., W., S. Allen, W. Janda, E. Koneman, G. Procop, P. Schreckenberger, and G. Woods. 2005. *Koneman's color atlas and textbook of diagnostic microbiology*, 6th ed. Baltimore: Lippincott Williams and Wilkins.

Methods for Identification of Bacterial Foodborne Pathogens

Ramakrishna Nannapaneni

1 Introduction

The Centers for Disease Control and Prevention (CDC) estimates that 76 million people contract various foodborne illnesses annually, leading to 325,000 hospitalizations and 5,000 deaths in the United States (Mead et al. 1999). Over the last 20 years, joint efforts by food safety regulation and food industries have led to a significant drop in the incidence of foodborne illnesses, from approximately seven to three cases per million people (CDC 2007). Bacteria are still the leading cause of foodborne illnesses in most countries. Among the bacterial pathogens, 10 species are most commonly associated with foodborne outbreaks: *Listeria monocytogenes*, *Escherichia coli* O157:H7, *Salmonella* spp., *Campylobacter jejuni*, *Bacillus*, *Staphylococcus*, *Yersinia*, *Shigella*, *Clostridium*, and *Vibrio* (CDC 2009a; Foley et al., 2009; Gerner-Smidt et al., 2006). The infectious dose of these foodborne bacterial pathogens may vary from 10 to 1,000 bacterial cells. Such infective cells of these harmful pathogenic bacteria surviving in soil, water, and tissues of plant or animal origin may contaminate various food products during food production and processing. Pathogenic bacteria may then multiply under suitable conditions in raw products or finished ready-to-eat food products, causing potential food safety risks. Recently, many food products of both animal and plant origins have been implicated in foodborne illnesses. These products include meat and poultry products, dairy products, seafood products, as well as fresh fruits and vegetables (CDC 2009b; Todd 2006; Scallan 2007).

For the routine implementation of the Hazard Analysis and Critical Control Points program, simple, reliable, and user-friendly detection methods are essential for ensuring the safety of large quantities of food products processed by the food industry. All food processing industries and regulatory agencies continuously depend on evolving new technologies for the routine monitoring of food products for the presence of major bacterial pathogens. Due to the large-scale globalization of food ingredients and processed food products, approved pathogen-detection methods are routinely used by many suppliers, processors, and regulatory agencies from "farm to fork." Also, to distinguish closely related pathogenic strains to confirm the sources during outbreaks, the accurate identification and detection of diverse pathogenic strains are critical (Swaminathan and Feng 1994; de Boer and Beumer 1999).

R. Nannapaneni (✉)
Department of Food Science, Nutrition and Health Promotion,
Mississippi State University, Mississippi State, MS 39762, USA
e-mail: nannapaneni@fsnhp.msstate.edu

Research indicates that a combination of detection technologies is essential for accurately monitoring low numbers of target pathogens in the processing environment and in food products. Highly sensitive, reliable, easy-to-use, low-cost, and specific assays for detecting pathogens are continuously evolving due to a strong demand by both regulatory agencies and food processing industries. The detection technologies used for monitoring the presence or destruction of harmful pathogens in food products or in the processing environment are classified into three major groups: (1) culture-dependent assays; (2) rapid immunological or antibody-based assays; and (3) rapid molecular or nucleic acid-based assays. Many factors influence the sensitivity and accuracy of detection of the target pathogenic bacteria in a specific food product, and these detection technologies are therefore continuously being refined by the diagnostic industries.

2 Culture-Based Methods for Isolation and Identification

Culture-based assays are well established and routinely used in all microbiological laboratories. These methods are generally the least expensive but are laborious and time-consuming, and it can take days or weeks to complete the full isolation of the target pathogen. In most cases, these presence and absence assays do not provide any information about the virulence of the strains isolated. Thus, further subcharacterization by biochemical or molecular approaches is required. One critical advantage of the culture methods is the ability to isolate the live cells of the suspected pathogen from any target food product or environmental sample to trace back the potential sources of the foodborne bacterial pathogen during an outbreak. The presumptive colonies of the target bacterial pathogen cells isolated from suspected samples are used for follow-up confirmative assays. The suspected bacterial cells, if culturable, can be easily isolated in laboratory growth media conditions to confirm their identity by closely matching with the existing biochemical and physiological patterns of known strains of the same species. If the isolated strains of the target pathogen contain unique biochemical characteristics that do not fit into the patterns of the characteristic species, their reliable identification by biochemical assay becomes a challenge. Therefore, cultural methods are frequently integrated with molecular assays for species typing or subtyping of strains after their initial isolation and preservation.

The first step in the *standard cultural methods* for isolating and detecting the presence or absence of target pathogens is the successful preenrichment of the suspected samples in either nonselective or selective broth. Such enrichment techniques have been well established for all major bacterial pathogens suitable for various food products. For example, of all the *Listeria* species, only one species is pathogenic. Cultural methods for presumptive *L. monocytogenes* require isolation of esculin-positive cells using Oxford or PALCAM agar medium with *Listeria* selective supplements to eliminate competing species. These methods have been standardized for the enumaration of *L. monocytogenes* from various food products (Anonymous 2004). This involves primary enrichment in half-strength Fraser broth and then spread-plating on selective PALCAM agar for presumptive isolation of *L. monocytogenes,* followed by biochemical API test for confirmation (Lin et al. 2006). Also, a low number of *L. monocytogenes* cells surviving in environmental samples from processing plants can easily be detected by highly sensitive primary enrichment broths such as University of Vermont broth, *Listeria* repair broth, and Oxoid Listeria enrichment broth (D'Amico and Donnelly 2009). Table 1 shows the most commonly used commercial biochemical kits for the identification of foodborne bacterial pathogens after their isolation from a suspected sample. In these tests, a disposable strip containing small quantities of media or substrates is inoculated with the pure culture of bacterial cells and incubated for 18–24 h to provide identification based on the stored database.

Table 1 Commercial kits for identification of bacterial foodborne pathogens by biochemical assays

Biochemical kit	Assay basis	Time	Assay specificity	Manufacturer
MicroLog microStation identification system	Interpreting patterns of 95 sole-carbon utilization by bacteria	24–48 h	Presumptive genus & species identification for over 150 bacteria	Biolog Inc.
Microgen *Listeria* ID kit	10 sugar fermentation biochemical tests	18–24 h	*Listeria* genus identification	Microgen Bioproducts Ltd.
API 20E system for *Listeria, Enterobacteriaceae, Staphylococcus, Campylobacter*	20–30 min biochemical tests	24 h	Genus identification	bioMerieux Inc.
MICRO-ID system for *Listeria, Enterobacteriaceae*	20 sugar fermentation biochemical tests	24 h	Genus identification	REMEL

Culture-dependent assays are also invaluable for enumerating the specific pathogen load in suspected samples. The classical most probable number (MPN) method is routinely used for the quantification of *L. monocytogenes* in foods. In the MPN method, replicated dilution series of suspected sample extracts are first performed in selective enrichment broth, followed by plating on selective agar plates for quantitative isolation of presumptive colonies for species identification (De Martinis et al. 2007). The advantage of the conventional MPN method is that it allows accurate quantification of low numbers of *L. monocytogenes* (<100 CFU/g) in a food matrix containing a large number of background flora. However, MPN is laborious and requires several days for the confirmation of results (Gasanov et al. 2005).

Simple, novel enumeration methods by direct plating have now been developed for all major foodborne bacterial pathogens by adding fluorogenic or chromogenic enzyme substrates into a nonselective or selective agar medium that can easily distinguish pathogenic species from nonpathogenic species. Such differential chromogenic agars eliminate the need for further biochemical tests to identify pathogenic species. Chromogenic agar-based protocols allow the specific enumeration of *L. monocytogenes* within 48 h compared to other previously approved USDA, FDA, or AOAC methods, which take 4–7 days (Hegde et al. 2007; Lin et al. 2006; Ritter et al. 2009). Some of the most commonly used selective and differential media are listed in Table 2. Chromogenic agar media are available for the routine isolation and identification of *E. coli* O157:H7, *Salmonella*, *Clostridium perfringens*, *Bacillus* spp., and *S. aureus* from food products (Manafi 2000). Oxoid Chromogenic Listeria Agar (OCLA) allows the isolation, enumeration, and presumptive identification of *Listeria* spp. and *L. monocytogenes* from food. This is accomplished by the detection of b-glucosidase activity common to all *Listeria* species, resulting in distinct blue colonies. Phosphatidylinositol phospholipase C (PIPLC) or phosphatidylcholine phospholipase C (PCPLC) activity forms a clearly visible, opaque white halo around pathogenic *Listeria* colonies. Both PIPLC and PCPLC are associated with the virulent *L. monocytogenes*, but some strains of *L. ivanovii* may also possess these enzymes and have been shown to be pathogenic in animals and humans. A new, highly selective chromogenic agar has been developed for the enumeration of *Campylobacter* in which *C. jejuni* and *C. coli* turn dark red in 48 h.

Also, cultural methods are essential for recovering the injured or sublethally stressed bacterial cells from samples that are subjected to various food processing steps, such as heat, cold, acid, alkali, and osmotic shock. Sublethally injured *Salmonella* and *L. monocytogenes* from milk, shell

Table 2 Commonly used selective and differential plating medium for the isolation, detection, and identification of bacterial foodborne pathogens

Culture medium	Microcolonies observed	Time	Assay specificity	Manufacturer
Modified Oxford agar for *Listeria*	*Listeria* forms black colonies surrounded by black halos	24 h	Presumptive, genus-specific	Oxoid
PALCAM agar for *Listeria*	*Listeria* forms gray-green colonies with a black sunken center and a black halo	24 h	Presumptive, genus-specific	Oxoid
Sorbitol-MacConkey (SMAC) for *E. coli* O157:H7	Wild *E. coli* ferment sorbitol forming pink colonies while most *E. coli* O157:H7 are sorbitol-negative, forming colorless colonies	24 h	Presumptive isolation of sorbitol-negative *E. coli* O157:H7	PML Micro-biologics
Rapid L. mono plates	*Listeria* forms yellow to white colonies, while *L. monocytogenes* forms blue colonies without a yellow halo	24 h	Specific for *L. monocytogenes*, which can be distinguished from *L. ivanovii* and other *Listeria* species	Bio-Rad Laboratories
Chromogenic *Listeria* Agar (OCLA)	*Listeria* colonies are blue, while *L. monocytogenes* and *L. ivanovii* colonies have an opaque white halo	24 h	Detects b-glucosidase activity common to *Listeria* and phosphatidylinositol phospholipase C or phosphatidylcholine phospholipase C activity specific for *L. monocytogenes* and *L. ivanovii*	Oxoid
BBL CHROMagar O157 (CHROM) or *Salmonella*	*E. coli* O157:H7 or *Salmonella* produces rose to purple color colonies	24 h	Presumptive identification for *E. coli* O157:H7 or *Salmonella*	Becton Dickinson
Salmonella Chromogenic agar	*Salmonella* produces magenta colonies	24 h	Specific for *Salmonella* species and *Shigella dysenteriae*	Oxoid
Campyfood ID Agar	*Campylobacter* forms burgundy-red to orange-red colonies on the light beige agar	48 h	Genus-specific for *Campylobacter*; will not distinguish *C. jejuni* from *C. coli*	bioMérieux
Brilliance™ CampyCount Agar	*C. jejuni* and *C. coli* turn dark red	48 h	Genus-specific for *Campylobacter*; will not distinguish *C. jejuni* from *C. coli*	Oxoid

eggs, and ready-to-eat meats were detected by double enrichment in nonselective modified universal broth and then in selective enrichment broth (Peng and Shelef 2001). Resuscitation and growth of low numbers of sublethally injured *L. monocytogenes* were recovered in various enrichment broths by an overnight enrichment method (Jasson et al. 2009). Culture methods, however, are currently limited in their ability to detect viable but nonculturable cells (Gracias and McKillip 2004).

3 Immunological or Antibody-Based Assays

Antibodies are powerful tools for detecting the target pathogen cells or toxins even from crude food extracts. Specific and sensitive antibodies must first be produced by immunizing laboratory animals using a specific antigenic fraction of the target pathogen, such as cell surface lipopolysaccharides, capsular polysaccharides, membrane proteins, or flagella. The production of genus-specific and species-specific polyclonal and monoclonal antibodies is required for developing specific foodborne pathogen diagnostic assays. Polyclonal antibodies can bind to multiple antigenic sites on the target pathogen and can be produced by affinity purification of the crude serum, which removes the cross-reacting antibodies against closely related antigens. Polyclonal antibodies are produced in limited quantities from each batch and differ in their affinity, specificity, and sensitivity against the target antigen from batch to batch. Monoclonal antibodies recognize a unique site on the surface of the antigen and are highly specific for the target pathogen. Monoclonal antibodies can be produced in unlimited quantities through continuous culturing of immortal hybridoma clones.

Specific antisera are used in traditional agglutination methods for species and serotype confirmation. For example, *L. monocytogenes* serotypes can be detected by variation in the somatic (O) and flagellar (H) antigens, allowing the differentiation of 13 different serotypes. However, these tests depend on the availability of quality antisera, which can differentiate the four main serotypes 1/2a, 1/2b, 1/2c, and 4b (Wiedmann et al. 1997) that are epidemiologically relevant for humans.

Compared to cultural methods, immunoassays can quickly screen a large number of samples in a relatively shorter time for the presence of the suspected pathogen or toxins in complex food matrices. Various types of immunoassays are widely used for the detection of bacterial cells and toxins using specific polyclonal and monoclonal antibodies. Enzyme-linked immunosorbent assays (ELISAs) are highly specific antigen–antibody interactions conducted in a microtiter plate for rapidly detecting and quantifying target pathogen cells or toxins. In an ELISA, the antigen or the antibody is immobilized onto a solid surface and then treated with sample extracts. When the immobilized antigen–antibody complex is formed in microtiter plates, it is detected by a specific antibody probe linked to an enzyme that yields a measurable color reaction when exposed to a substrate. Of all the immunoassay formats, indirect ELISA and sandwich ELISA are the most commonly used. Indirect ELISA is very efficient in simultaneously screening many primary antibodies against a target antigen using a secondary antibody conjugated to an enzyme. Once a target-specific primary antibody is identified by indirect ELISA, it is then used for conjugating to an enzyme for use in a sandwich ELISA format. Such ELISA techniques are now routinely developed for various foodborne bacterial pathogens. Rapid immunodiagnostic assays are now available that allow for the monitoring of the presence of multiple toxins in a single assay. For example, VIDAS SET2 enables the detection of staphylococcal enterotoxins A, B, C, and D using highly specific and sensitive immunoassays formats (Bennett 2005). Alternatively, a variety of immunoassay formats, including immunofluorescence or immunogold-labeled antibodies, are extensively used for the detection of *L. monocytogenes* and listeriolysin O toxin in food (Churchill et al. 2006). Some commonly used immunodiagnostics assays for bacterial foodborne pathogens are listed in Table 3.

Of all the immunological assays, one of the simplest formats is the lateral flow immunoassay, which can detect the presence or absence of a target pathogen in a large number of samples within 15 min. However, this type of assay requires as high as 10^7 or 10^9 CFU/mL (colony-forming units/milliliter) of target cells (Bohaychuk et al. 2005). In other assay formats, antibodies are used to speed up the capture of target bacterial cells or their toxins in complex sample matrices by using antibody-coated beads. Recently, specific antibodies conjugated to immunomagnetic beads have been widely used for the rapid separation of the target pathogen cells from crude sample extracts (Benoit and Donahue 2003; Stevens and Jaykus 2004). Captured immunomagnetic beads can be plated out on agar to isolate the live cells of the target pathogen (Ueda et al. 2006). Achieving detection limits of 10^4–10^6 is currently possible by the application of antibody-coated magnetic beads for

Table 3 Immunodiagnostics and DNA-based identification assays for bacterial foodborne pathogens

Target agent	Principle	Time	Assay specificity	Manufacturer
Listeria	Latex agglutination	24 h	Presumptive, genus-specific	Remel
Listeria and E. coli O157:H7	Immunoprecipitation	24 h	Presumptive, genus-specific	Biocontrol
E. coli O157: H7	Latex agglutination	24 h	Presumptive detection of E. coli O157:H7	Oxoid
Listeria	Lateral flow immunoassay	48 h	Specific to L. monocytogenes flagellar B antigen	Oxoid
Listeria	Sandwich ELISA	48 h	Genus-specific	Biocontrol
Listeria	Sandwich ELISA	48 h	Genus-specific	Neogen
L. monocytogenes	Semiautomated immunoassay	48 h	Species-specific	bioMérieux
L. monocytogenes, E. coli O157:H7, Salmonella and Campylobacter	Recirculating immunomagnetic separation technology	48 h	Presumptive, species-specific	Matrix MicroScience
L. monocytogenes, Campylobacter	Use of a single-stranded DNA probe with a chemiluminescent label that is complementary to the ribosomal RNA of the target organism; labeled DNA: RNA hybrids measured by luminometer	30 min[a]	Highly specific for L. monocytogenes or C. jejuni	Gen-probe
L. monocytogenes, E. coli O157:H7, Salmonella and Campylobacter	Polymerase chain reaction using a specific genetic sequence unique for each target pathogen	48 h	Species-specific	Qualicon
L. monocytogenes, Campylobacter, E. coli O157:H7, Salmonella	Nucleic acid hybridization using a specific 16 s ribosomal RNA sequence unique for target pathogen	48 h	Species-specific	Neogen

[a] Accuprobe assay time for detecting L. monocytogenes or C. jejuni without including the time it takes to grow and isolate their microcolonies from a suspect sample

L. monocytogenes detection (Duffy et al. 1997). The rapid detection of *L. monocytogenes* can be accomplished with the immunoassay such as *Listeria* Rapid Test (Oxoid Ltd., Basingstoke, UK). A large number of sample aliquots can be easily screened by rapid immunoassay to confirm the presence of *L. monocytogenes* after the preenrichment step. Pathatrix uses a recirculating immunomagnetic separation technology to selectively bind and purify the target pathogen cells in a high-volume food sample rinse. Small magnetic beads coated with antibodies specific to a target pathogen are immobilized inside a sample tube using a magnet, and then the food sample is pumped through the tube for 30 min, where immobilized paramagnetic antibody beads will capture or bind to specific target bacterial cells from the complex food matrix. Captured bacterial cells are further tested by direct plating, ELISA, or polymerase chain reaction (PCR). The Pathatrix systems are available for *Listeria, Salmonella, Campylobacter,* and *E. coli* O157, and detection limits are around 1–10 CFU per 25-g sample after preenrichment steps. Pathatrix systems can also pool up to five subsamples for cost savings.

Biosensor assay formats have been developed for the rapid detection of foodborne bacterial pathogens and their toxins. In a majority of the biosensor assay formats, biologically active antibodies are integrated to a transducer and a signal is generated when the target pathogen cells are captured. Recently, nanoparticles of silica (60 nm long) treated with antibodies were used to detect single *E. coli* O157:H7 in ground beef in 30 min without amplification (Zhao et al. 2004). Surface plasmon resonance (SPR) is another example of a biosensor assay that has been developed for various foodborne bacterial pathogens. A prototype detection of *E. coli* O157:H7 in pure cultures by a hybrid microfluidic SPR has been developed, although it has not been used for the detection of *E. coli* O157:H7 in actual food samples (Zordan et al., 2009). A simultaneous detection of *E. coli* O157:H7, *Salmonella* Typhimurium, *L. monocytogenes*, and *Campylobacter jejuni* was demonstrated using an eight-channel SPR sensor (Taylor et al. 2006). However, these assays have not been validated in routine food microbiology laboratories. The cells of two pathogens, *E. coli* O157:H7 and *Salmonella* Typhimurium, were detected by fluorescent-labeled antibody probes specific for each pathogen in a single test with minimum detection limits of 10^6 CFU/ml in 2.5 h (Gehring et al. 2008). However, many technological difficulties have to be resolved in removing the interference from complex food matrices in biosensor assays in order to achieve real-time detection of target pathogen cells.

4 Nucleic Acid-Based Assays

Molecular methods are now increasingly being used for the detection and identification of foodborne bacterial pathogens after single or double preenrichment steps in selective broth. The presence of unique DNA or RNA signatures of a pathogen makes it possible to distinguish a pathogenic strain from a closely related nonpathogenic strain (Peters 2009). With this technology, it is now possible to overcome many limitations of cultural or antibody-based techniques. Using DNA sequencing technology, a rapid and efficient sequencing of complete microbial genomes is now paving the way for the thorough understanding of the genomes of different microorganisms, and it is leading to the development of new molecular detection technologies as well as gene expression assays. Some DNA-based assays for foodborne bacterial pathogen detection include PCR assays, nucleic acid sequence-based amplification (NASBA) assays, and microarrays (Hyytiä-Trees et al. 2007; Kerouantona et al., 2009; McKillip and Drake 2004). Some commonly used molecular methods for bacterial foodborne pathogens are listed in Table 3.

In PCR, a specific nucleic acid sequence is amplified for the identification of the genus and species. The PCR can amplify a target sequence within a gene or in repetitive areas or an arbitrary sequence within the genome. PCR targeting specific genes or highly conserved genes or a high degree of specificity to all major foodborne bacterial pathogens has been developed. However, PCR is affected by the complexity of food matrices, interference, and inhibitors in food and therefore requires clean sample extracts compared to immunoassays. There are now many sensitive methods for the detection of *L. monocytogenes* by PCR, using primers targeted to virulence and nonvirulence factors, including hemolysin, invasion-associated protein, and 16 S rRNA genes (Gasanov et al. 2005; Niederhauser et al. 1992; Wang et al. 1992; Wiedmann et al., 1997). Also, amplification of the *prf*A gene is commonly used in PCR assays to check for *L. monocytogenes;* it has been validated and suggested as an international standard PCR method for identifying *L. monocytogenes* in food (Rossmanith et al. 2006). PCR assays have been developed for the rapid, sensitive, and specific detection of all major foodborne pathogens. PCR-based BAX systems are now widely used by the food industries and regulatory research laboratories to screen food samples for the presence of several foodborne bacteria in food samples (Stewart and Gendel 1998; Hoffman and Wiedmann 2001). The PCR method based on 16 S rRNA genes for the detection of *L. monocytogenes* presents better sensitivity due to the presence of multiple copies of the genes in the cell to maintain an adequate concentration of ribosomes (De Martinis 2007).

In a quantitative real-time PCR (qrt-PCR) assay, target bacterial nucleic acids can be monitored during the amplification process. Many improvements in qrt-PCR assays have been achieved by eliminating PCR inhibition to achieve an adequate concentration of target nucleic acids to avoid false-positives and false-negatives. qrt-PCR assays are being developed for the accurate quantification of *L. monocytogenes* or *Campylobacter* in environmental and food samples (Liu 2006; Mafu et al. 2009). In a recent study, a conventional MPN method was combined with qrt-PCR based on 16 S rRNA genes for the enumeration of *L. monocytogenes* in naturally and artificially contaminated, minimally processed leafy vegetables. Based on experimentally inoculated product with MPN quantification, the qrt-PCR assay was capable of detecting 1–5 CFU per 50 g with no interference from the microflora present in these food samples (Aparecida de Oliveira 2010). *Campylobacter* in various food products was detected up to a sensitivity of 4.2×10^3 CFU/mL.

Conventional PCR will not determine if the bacteria are viable or dead; therefore, reverse transcription PCR (RT-PCR) can be used for the detection of viable cells (Flekna et al. 2007). A filtration-based PCR assay has been developed for *L. monocytogenes* and *Salmonella* that can detect only viable cells within 30 min (D'Urso et al. 2009). Another specific way to detect viable cells of target pathogens is by designing an mRNA-based assay. In bacteria, mRNA is unstable and easily degradable, and it is only produced when the bacteria are in the multiplication phase. Messenger RNA serves as a template for the synthesis of a single-stranded DNA in the 5′ to 3′ direction. In these assays, RNA is extracted from food products. It is transformed into DNA by means of reverse transcription, which is detected by a complementary marked probe with a fluorescent compound (Nocker and Camper 2009). The separation of whole bacterial cells or the quality of the extraction of DNA, RNA, and proteins for molecular assays depends on efficient extraction protocols. Current extraction using stomaching or pulsifying allows the preparation of homogenates for direct plating but does not allow an efficient release or concentration of foodborne pathogen cells. Further improvement in highly sensitive and faster PCR assays for the quantification of *L. monocytogenes* or *Campylobacter* will depend on improvements in recovery and extraction protocols.

Nucleic acid sequence-based amplification (NASBA) is an alternative method to PCR that also involves the amplification of target nucleic acid sequences. NASBA involves the simultaneous enzyme activity of reverse transcriptase, T7 RNA polymerase enzyme and RNAase in combination with two oligonucleotides for the cyclical and exponential amplification of the target sequences. Such transcription-based amplification results in a 10^{14}-fold selective amplification of RNA. This method allows the amplification of RNA sequences in the presence of DNA background since DNA strands are not melted out. Since RNAs synthesized in any cell are unique to that cell's type, NASBA can be used to distinguish live cells of different foodborne bacterial pathogens present in a mixture, and such methods have been developed for *L. monocytogenes* (Cook 2003; Gasanov et al. 2005; Nadal et al. 2007). NASBA has yet to evolve as a rapid, robust, sensitive, and specific semiquantitative assay for the foodborne bacterial pathogens, and no commercial assays are currently available using NASBA.

With the development of microarray technology, it is now possible to work with a large number of primer sequences for pathogen identification. There is also a potential for the detection of multiple target pathogens by microarray methods. DNA array technology allows the detection of unlimited amounts of unknown DNA sequences in a single assay. Microarray-based detection combines powerful nucleic acid amplification strategies with the massive screening capability for a high level of sensitivity, specificity, and throughput. In microarrays, specific detector oligonucleotides are immobilized onto a solid support for hybridization with homologous-labeled target amplicons that can be detected. Based on the full genome sequences now available for many foodborne bacterial pathogens, it is now possible to identify nucleic acid sequences associated with specific virulence factors or serotype-specific markers. O-antigen gene clusters of four different serogroups of *E. coli* (O7, O111, O104, O157) are now detected rapidly using DNA microarray technology (Guo et al. 2004). It is now possible to analyze the expression of many genes in a single reaction and in an efficient manner in order to detect multiple pathogens in a mixed sample to understand the fundamental aspects of physiology at a molecular level. Microarrays have been developed for the identification of various foodborne bacterial pathogens, including *L. monocytogenes* and *E. coli* O157:H7 (Al-Khaldi et al. 2004; Cebula et al. 2005). A high level of sensitivity, specificity, and throughout can be achieved by microarrays, but this technology is currently extremely expensive.

5 Summary

Foodborne pathogenic bacteria are often present in low numbers or may be injured or damaged during processing treatments; therefore, appropriate enrichment and resuscitation steps are essential for their isolation and detection in the presence of other competing microflora. Due to their uneven distribution, many technological hurdles have yet to be overcome in the rapid isolation of these pathogenic bacterial cells from complex matrices. As a result of potential interferences from food matrices or due to their being present in low numbers, a combination of cultural and immunological or cultural and molecular methods are essential for the effective isolation, detection, and identification of major foodborne bacterial pathogens. Commercial PCR methods are now routinely being used to screening for target pathogens in combination with preenrichment cultural methods. There is a great need for efficient sample preparation methods that are less expensive and are compatible with immunological or molecular assays to allow for the near-real-time detection of foodborne pathogens. More advances still have to be made to realize the simultaneous detection and identification of multiple target bacterial pathogens from complex food extracts through an integration of the cultural, immunological, and molecular approaches.

References

Al-Khaldi, S.F., M.M. Mossoba, A.A. Ismail, and F.S. Fry. 2004. Accelerating bacterial identification by infrared spectroscopy by employing microarray deposition of microorganisms. *Foodborne Pathogens and Disease* 1: 172–177.
Anonymous. 2004. Microbiology of food and animal feeding stuffs – Horizontal method for detection and enumeration of *Listeria monocytogenes*. Part 2. Enumeration method, amendment 1: 2004. Modification of the enumeration medium. International Standard ISO 11290–2. Geneva: International Organization for Standardization
Aparecida de Oliver, M., E.G. Abeid Ribeiro, A.M. Morato Bergamini, and E.C. Pereira De Martinis. 2010. Quantification of *Listeria monocytogenes* in minimally processed leafy vegetables using a combined method based on enrichment and 16S rRNA real-time PCR. *Food Microbiology* 2010 27:19–23

Bennett, R.W. 2005. *Staphylococcal* enterotoxin and its rapid identification in foods by enzyme-linked immunosorbent assay-based methodology. *Journal of Food Protection* 68: 1264–1270.

Benoit, P.W., and D.W. Donahue. 2003. Methods for rapid separation and concentration of bacteria in food that bypass time-consuming cultural enrichment. *Journal of Food Protection* 66: 1935–1948.

Bohaychuk, V.M., G.E. Gensler, R.K. King, J.T. Wu, and L.M. McMullen. 2005. Evaluation of detection methods for screening meat and poultry products for the presence of foodborne pathogens. *Journal of Food Protection* 68: 2637–2647.

CDC (Centers for Disease Control and Prevention). 2007. Preliminary FoodNet data on the incidence of infection with pathogens transmitted commonly through food – 10 states, 2006. *Morbidity and Mortality Weekly Report* 56: 336–339.

CDC (Centers for Disease Control and Prevention). 2009a. Preliminary FoodNet data on the incidence of infection with pathogens transmitted commonly through food – 10 States, 2008. *Morbidity and Mortality Weekly Report* 58: 333–337.

CDC (Centers for Disease Control and Prevention). 2009b. Surveillance for foodborne disease outbreaks – United States, 2006. *Morbidity and Mortality Weekly Report* 58(22): 609–615.

Cebula, T.A., E.W. Brown, S.A. Jackson, M.K. Mammel, A. Mukherjee, and J.E. LeClerc. 2005. Molecular applications for identifying microbial pathogens in the post-9/11 era. *Expert Review of Molecular Diagnostics* 5: 431–445.

Churchill, R.L., H. Lee, and J.C. Hall. 2006. Detection of *Listeria monocytogenes* and the toxin listeriolysin O in food. *Journal of Microbiological Methods* 64: 141–170.

Cook, N. 2003. The use of NASBA for the detection of microbial pathogens in food and environmental samples. *Journal of Microbiological Methods* 53: 165–174.

D'Amico, D.J., and C.W. Donnelly. 2009. Detection, isolation, and incidence of *Listeria* spp. in small-scale artisan cheese processing facilities: a methods comparison. *Journal of Food Protection* 72: 2499–2507.

D'Urso, O.F., P. Poltronieri, S. Marsigliante, C. Storelli, M. Hernández, and D. Rodríguez-Lázaro. 2009. A filtration-based real-time PCR method for the quantitative detection of viable *Salmonella enterica* and *Listeria monocytogenes* in food samples. *Food Microbiology* 26: 311–316.

de Boer, E., and R.R. Beumer. 1999. Methodology for detection and typing of foodborne microorganisms. *International Journal of Food Microbiology* 50: 119–130.

De Martinis, E.C., R.E. Duvall, and A.D. Hitchins. 2007. Real-time PCR detection of 16S rRNA genes speeds most-probable-number enumeration of foodborne Listeria monocytogenes. Journal of Food Protection 70:1650–1655.

Duffy, G., J.J. Sheridan, H. Hofstra, D.A. McDowell, and I.S. Blair. 1997. A comparison of immunomagnetic and surface adhesion immunofluorescent techniques for the rapid detection of *Listeria monocytogenes* and *Listeria innocua* in meat. *Letters in Applied Microbiology* 24: 445–450.

Flekna, G., P. Stefanic, M. Wagner, F.J. Smulders, S.S. Mozina, and I. Hein. 2007. Insufficient differentiation of live and dead *Campylobacter jejuni* and *Listeria monocytogenes* cells by ethidium monoazide (EMA) compromises EMA/real-time PCR. *Research in Microbiology* 158: 405–412.

Foley, S.L., A.M. Lynne, and R. Nayak. 2009. Molecular typing methodologies for microbial source tracking and epidemiological investigations of Gram-negative bacterial foodborne pathogens. *Infection, Genetics and Evolution* 9: 430–440.

Gasanov, U., D. Hughes, and P.M. Hansbro. 2005. Methods for the isolation and identification of *Listeria* spp. and *Listeria monocytogenes*: a review. *FEMS Microbiology Reviews* 29: 851–875.

Gehring, A.G., D.M. Albin, S.A. Reed, S.I. Tu, and J.D. Brewster. 2008. An antibody microarray, in multiwell plate format, for multiplex screening of foodborne pathogenic bacteria and biomolecules. *Analytical and Bioanalytical Chemistry* 391:497–506.

Gerner-Smidt, P., K. Hise, J. Kincaid, S. Hunter, S. Rolando, E. Hyytiä-Trees, E.M. Ribot, B. Swaminathan, and P. Taskforce. 2006. PulseNet USA: a five-year update. *Foodborne Pathogens and Disease* 3: 9–19.

Gracias, K.S., and J.L. McKillip. 2004. A review of conventional detection and enumeration methods for pathogenic bacteria in food. *Canadian Journal of Microbiology* 50: 883–890.

Guo, H., L. Feng, J. Tao, C. Zhang, and L. Wang. 2004. Identification of *Escherichia coli* O172 O-antigen gene cluster and development of a serogroup-specific PCR assay. *Journal of Applied Microbiology* 97: 181–190.

Hegde, V., C.G. Leon-Velarde, C.M. Stam, L.A. Jaykus, and J.A. Odumeru. 2007. Evaluation of BBL CHROMagar Listeria agar for the isolation and identification of *Listeria monocytogenes* from food and environmental samples. *Journal of Microbiological Methods* 68: 82–87.

Hoffman, A.D., and M. Wiedmann. 2001. Comparative evaluation of culture- and BAX polymerase chain reaction-based detection methods for *Listeria* spp. and *Listeria monocytogenes* in environmental and raw fish samples. *J Food Protection* 64: 1521–1526.

Hyytiä-Trees, E.K., K. Cooper, E.M. Ribot, and P. Gerner-Smidt. 2007. Recent developments and future prospects in subtyping of foodborne bacterial pathogens. *Future Microbiology* 2: 175–185.

Jasson, V., A. Rajkovic, J. Debevere, and M. Uyttendaele. 2009. Kinetics of resuscitation and growth of *Listeria monocytogenes* as a tool to select appropriate enrichment conditions as a prior step to rapid detection methods. *Food Microbiology* 26: 88–93.

Kérouantona, A., M. Maraulta, L. Petita, J. Grouta, T.T. Daoa, and A. Brisabois. 2009. Evaluation of a multiplex PCR assay as an alternative method for *Listeria monocytogenes* serotyping. *Journal of Microbiological Methods* 80: 134–137.

Lin, C.M., L. Zhang, M.P. Doyle, and B. Swaminathan. 2006. Comparison of media and sampling locations for isolation of *Listeria monocytogenes* in Queso fresco cheese. *Journal of Food Protection* 69: 2151–2156.

Liu, D. 2006. Identification, subtyping and virulence determination of *Listeria monocytogenes*, an important foodborne pathogen. *Journal of Medical Microbiology* 55: 645–659.

Mafu, A.A., M. Pitre, and S. Sirois. 2009. Real-time PCR as a tool for detection of pathogenic bacteria on contaminated food contact surfaces by using a single enrichment medium. *Journal of Food Protection* 72: 1310–1314.

Manafi, M. 2000. New developments in chromogenic and fluorogenic culture media. *International Journal of Food Microbiology* 60:205–218.

McKillip, J.L., and M. Drake. 2004. Real-time nucleic acid-based detection methods for pathogenic bacteria in food. *Journal of Food Protection* 67: 823–632.

Mead, P.S., L. Slutsker, V. Dietz, L.F. McCaig, J.S. Bresee, C. Shapiro, P.M. Griffin, and R.V. Tauxe. 1999. Food-related illness and death in the United States. *Emerging Infectious Diseases* 5: 607–625.

Nadal, A., A. Coll, N. Cook, and M. Pla. 2007. A molecular beacon-based real time NASBA assay for detection of *Listeria monocytogenes* in food products: Role of target mRNA secondary structure on NASBA design. *Journal of Microbiological Methods* 68: 623–632.

Niederhauser, C., U. Candrian, C. Hofelein, M.H. Jermini, P. Buhler, and J. Luthy. 1992. Use of polymerase chain reaction for detection of *Listeria monocytogenes* in food. *Applied and Environmental Microbiology* 58: 1564–1568.

Nocker, A., and A.K. Camper. 2009. Novel approaches toward preferential detection of viable cells using nucleic acid amplification techniques. *FEMS Microbiology Letters* 291: 137–142.

Peng, H., and L.A. Shelef. 2001. Automated simultaneous detection of low levels of listeriae and salmonellae in foods. *International Journal of Food Microbiology* 63: 225–233.

Peters, T.M. 2009. Pulsed-field gel electrophoresis for molecular epidemiology of food pathogens. *Methods in Molecular Biology* 551: 59–70.

Ritter, V., S. Kircher, K. Sturm, P. Warns, and N. Dick. 2009. USDA FSIS, FDA BAM, AOAC, and ISO culture methods BD BBL CHROMagar *Listeria* Media. *Journal of AOAC International* 92(4): 1105–1117.

Rossmanith, P., M. Krassnig, M. Wagner, and I. Hein. 2006. Detection of *Listeria monocytogenes* in food using a combined enrichment/real-time PCR method targeting the prfA gene. *Research in Microbiology* 157: 763–771.

Scallan, E. 2007. Activities, achievements, and lessons learned during the first 10 years of the foodborne diseases active surveillance network: 1996–2005. *Clinical Infectious Diseases* 44: 718–725.

Stevens, K.A., and L.A. Jaykus. 2004. Bacterial separation and concentration from complex sample matrices: a review. *Critical Reviews in Microbiology* 30: 7–24.

Stewart, D., and S.M. Gendel. 1998. Specificity of the BAX polymerase chain reaction system for detection of the foodborne pathogen *Listeria monocytogenes*. *Journal of AOAC International* 81: 817–822.

Swaminathan, B., and P. Feng. 1994. Rapid detection of food-borne pathogenic bacteria. *Annual Review of Microbiology* 48: 401–426.

Taylor, A.D., J. Ladd, Q. Yu, S. Chen, J. Homola, and S. Jiang. 2006. Quantitative and simultaneous detection of four foodborne bacterial pathogens with a multi-channel SPR sensor. *Biosens Bioelectron*. 22:752–758.

Todd, E.C. 2006. Challenges to global surveillance of disease patterns. *Marine Pollution Bulletin* 53: 569–578.

Ueda, S., T. Maruyama, and Y. Kuwabara. 2006. Detection of *Listeria monocytogenes* from food samples by PCR after IMS-plating. *Biocontrol Science* 11: 129–134.

Wang, R.-F., W.W. Cao, and M.G. Johnson. 1992. 16 S rRNA-Based probes and polymerase chain reaction method to detect *Listeria monocytogenes* cells added to foods. *Applied and Environmental Microbiology* 58: 2827–2831.

Wiedmann, M. 2002. Molecular subtyping methods for *Listeria monocytogenes*. *Journal of AOAC International* 85: 524–531.

Wiedmann, M., J.L. Bruce, C. Keating, A.E. Johnson, P.L. McDonough, and C.A. Batt. 1997. Ribotypes and virulence gene polymorphisms suggest three distinct *Listeria monocytogenes* lineages with differences in pathogenic potential. *Infection and Immunity* 65: 2707–2716.

Zhao, X., L.R. Hilliard, S.J. Mechery, Y. Wang, R.P. Bagwe, S. Jin, and W. Tan. 2004. A rapid bioassay for single bacterial cell quantitation using bioconjugated nanoparticles. *Proceedings of the National Academy of Sciences USA* 101: 15027–15032.

Zordan, M.D., M.M. Grafton, G. Acharya, L.M. Reece, C.L. Cooper, A.I. Aronson, K. Park, and J.F. Leary. 2009. Detection of pathogenic *E. coli* O157:H7 by a hybrid microfluidic SPR and molecular imaging cytometry device. *Cytometry A* 75: 155–162.

Methods for Epidemiological Studies of Foodborne Pathogens

Omar A. Oyarzabal

1 Introduction

The surveillance of foodborne agents relies heavily on techniques that can provide quick genetic fingerprinting profiles of the agents implicated in cases or outbreaks of foodborne illnesses. These fingerprinting profiles can be of great utility when linking food products to specific outbreaks or when carrying out other epidemiological studies to understand the sources of contamination for specific foodborne pathogens.

There are several molecules that can be used as markers to create unique profiles or types to differentiate strains within a bacterial pathogen. Some methods are based on the phenotypic characteristics of the pathogen under study, and in some cases, these phenotypic tests can provide useful information about the pathogens, as in the case of antibiotic susceptibility profiles or the serotyping of *Salmonella* or *E. coli* O157:H7 isolates. However, phenotypic tests have major drawbacks: They require a long time to generate the results, and these results can still be subject to interpretation. In addition, the incorporation of well-defined controls is indispensable when interpreting the results. Antibiotics susceptibility tests and serotyping are examples of techniques that require the use of several controls as well as trained personnel to interpret the results. For these reasons, molecular techniques have many advantages over phenotypic tests and are the preferred techniques used to type foodborne bacteria.

Within the molecular methods, the study of DNA has resulted in a variety of techniques that can be incorporated in different microbiology laboratories to type bacterial pathogens (Foley et al. 2009). Although lipid and protein studies can also generate unique fingerprinting profiles, these methods are still too complex to perform and interpret in regular laboratories, and there have not been many technological advances to make these techniques easier. In addition, lipid profiles are more conserved among bacteria and may not generate the necessary variability needed to differentiate isolates within a species. Furthermore, because proteins are sensitive to changes in the environment (denaturing, etc.), the sample preparation process is cumbersome, but it is crucial for generating reliable fingerprinting profiles. The DNA molecule does not have these limitations and has much more resiliency, allowing for less stringent sample collection, handling, and storage. For some DNA techniques, however, where the integrity of the DNA is vital for the reproducibility of the results, sample preparation does play an important role. Yet even in these cases, the protocol for the collection and handling of DNA samples is much less stringent than the collection and handling of samples for protein analysis.

O.A. Oyarzabal (✉)
Department of Biological Sciences, Alabama State University, Montgomery, AL, USA
e-mail: oaoyarzabal@gmail.com

Typing methods are not used to identify bacterial species. Although some methods could be used for this purpose, it would be expensive and cumbersome to do so. Therefore, typing methods are used to study the intraspecies patterns on different isolates. This chapter presents an overview of the DNA-based methodology used to type foodborne bacteria. Several definitions are presented in the text to help understand the terminology used in molecular epidemiology, and although several classifications can be used to group DNA-based techniques, I have restricted this chapter to the following: (1) amplification methods, which are based on the amplification of particular DNA segments using polymerase chain reaction (PCR) assays; (2) amplification and restriction analysis, in which PCR products are restricted using specific restriction enzymes before analysis; (3) restriction of the whole chromosome, which includes only pulsed-field gel electrophoresis (PFGE); and (4) sequence-based methods, primarily multilocus sequence typing (MLST), which identifies the polymorphisms of specific loci in the genome. Several terms are defined throughout the text, with an emphasis on the advantages and disadvantages of the most common methods to type foodborne pathogens for epidemiological studies. A complete description of the variables that affect the adoption of any particular technique is beyond the objectives of this chapter. However, the references cited are a good resource for further reading in this area.

2 Methods Based on PCR Amplification

The methodology known as PCR has opened up many opportunities for molecular studies in biology. The principle is based on the enzymatic amplification of DNA using primers that are specific to the DNA segments to be amplified (Mullis et al. 1986). The degree of complementarity of the primers and the variable temperatures for annealing that can be used make PCR a simple and relatively inexpensive methodology for the identification of nearly any DNA segment. PCR assays can target specific segments within a bacterial species and therefore can be used for the specific identification of unknown isolates. This chapter does not review the PCR technique, but for those interested in the details of this methodology, we suggest reading some of the recent books and reviews (Kennedy and Oswald 2011; Weissensteiner et al. 2004).

2.1 Repetitive Polymerase Chain Reaction (REP-PCR)

The genome of some bacterial species has repetitive sequences that are specific to each organism. These sequences are commonly found in noncoding regions of bacterial genomes (Hulton et al. 1991; Stern et al. 1984; Versalovic et al. 1991). The amplification of segments from these repetitive sequences with the use of standard primers through PCR assays is a simple technique that can assign molecular fingerprints to different isolates within a bacterial species. There is already a commercially available system (DiversiLab, bioMérieux, Hazelwood, MO) that types bacterial isolates based on the separation of different amplified fragments from repetitive sequences. The system separates the fragments after PCR amplification using a microfluidic chip (Agilent Technologies, Palo Alto, CA), creating densitometric curves of each segment and therefore generating a fingerprint profile for each isolate. The data are transferred to the DiversiLab Web Interface, and the results are generated using the Pearson correlation coefficient to create cluster analyses. The dendrogram representing DNA relatedness is created by software with an Internet interface. This software uses a cutoff of 95% to analyze different REP-PCR profiles.

This automated system appears to have a niche for the typing of bacterial isolates with genomes larger than two megabases, such as *Salmonella* and *E. coli*. In the case of *Salmonella enterica*, one of the features sought in any typing technique is a strong correlation with serotyping, which is the most important technique to classify *Salmonella* but is, at the same time, an expensive, time-consuming method that only few laboratories can afford. Therefore, if the typing method can also determine the serotype with a high correlation, the method will be more accepted in food microbiology and diagnostics laboratories. Although some work suggests that REP-PCR can be used to predict *Salmonella* serotypes (Wise et al. 2009), a large database is necessary to include isolates that do not show a strong correlation between REP-PCR and serotype (Rasschaert et al. 2005).

One of the limitations of REP-PCR is the reproducibility of the technique, meaning that repeating the assay on the same isolates may generate different fingerprinting profiles. Several adjustments to the PCR assay during amplification may improve the results of REP-PCR typing. For instance, avoiding DNA contamination when using randomly designed primers, or increasing the annealing temperature during the PCR amplification step may help improve the reproducibility of REP-PCR (Johnson and Clabots 2000; Riley 2004a). However, for foodborne pathogens with compact, reduced genomes of less than two megabases (as in the case of *Campylobacter* spp.), REP-PCR, although with good discrimination, may lack epidemiologic concordance (defined as the agreement between the grouping of strains by a given method and the available epidemiological information about those strains), and can even produce identical patterns from isolates of different species (Behringer et al. 2011).

3 Methods Based on PCR Amplification and Restriction

Another discovery that revolutionized molecular biology and allowed for the manipulation of genes and the initiation of biotechnology as we currently know them is the identification of DNA restriction enzymes (Danna and Nathans 1971). Restriction fragment length polymorphism (RFLP) is one of the first and basic methods used in molecular laboratories to study the variability of DNA sequences by combining the use of restriction enzymes to generate DNA fragments of a defined length followed by standard electrophoresis in agarose gels to separate the different fragments according to their molecular size. Originally, RFLP was developed to study specific DNA sequences from complex populations of DNA fragments using hybridization methods (Southern 1975) or by transferring the DNA from an agarose gel to nitrocellulose membranes for further hybridization with radioactive probes (Botstein et al. 1980; Jeffreys and Flavell 1977). The further development of PCR allowed for the development of simplified protocols for RFLP, and today PCR-RFLP is one of the simplest typing methods because it does not require complex and expensive equipment. One of the first approaches using RFLP to type bacterial foodborne pathogens was applied for the typing of the *Salmonella* serotypes Typhimurium, Dublin, and Enteritidis (Tompkins et al. 1986). The technique was called chromosomal probe fingerprinting because the whole chromosome was cut with *Eco*Rl, *Hind*lll, and *Pst*l restriction enzymes, and sequences of cloned chromosomal DNA were used as probes to highlight restriction site heterogeneity. Nowadays, most of the PCR-RFLP simplified protocols amplify a segment of DNA, usually a given gene or part of an operon, of approximately 1,400–2,500 base pairs (bp). Then the amplified DNA product is cleaved using restriction enzymes and the fragment are visualized using gel electrophoresis. PCR also allowed for the design of a technique called amplified fragment length polymorphism (AFLP), based on the selective PCR amplification of restriction fragments from genomic DNA that was previously cleaved with restriction enzymes. Typically, 50–100 restriction fragments can be detected on polyacrylamide gels with AFLP (Vos et al. 1995).

One of the first techniques used to type bacterial pathogens was *ribotyping*, a method based on the RFLP analysis associated with the ribosomal operon (Grimont and Grimont 1986). *Operons* are clusters of genes that have related functions and most of the time are regulated in coordination.

The ribosomal operon is a polycistronic operon that includes the 16S and 23S rRNA genes and at least one tRNA gene. Conventional ribotyping makes use of restriction enzymes, for example, *Eco*Rl, to restrict the total genomic DNA, the separation of the different fragments by gel electrophoresis, the transfer of the fragments with Southern blot, and the hybridization of the transferred fragments with a labeled ribosomal operon probe.

In the 1990s, PCR became a common technique in microbiology laboratories worldwide, and PCR ribotyping emerged as a simplified technique for molecular epidemiology. Even an automated method (RiboPrinter, Dupont, Qualicon) is now available that compares newly acquired RFLPs to existing databases. The basics of the technique include the amplification of a segment of the operon, including the internal spacer region (ISR), by using a 3′ end primer complementary to the 16S rRNA gene sequence and a 5′ end primer complementary to the 23S rRNA gene (Bouchet et al. 2008; Kostman et al.1992). But the consistent length and degree of conservation of the ISR have limited the use of ribotyping for some bacterial species, such as *Listeria* and *Streptococcus*. However, the technique has found to be a reliable tool for species identification, and in the case of a few *Salmonella* enteric serotypes, it can even generate predictable identification of these serotypes in unknown isolates. Variations of the PCR ribotyping method, including the whole amplification of the ribosomal operons, have also been developed.

Recent studies of the molecular basis of ribotyping have revealed that the term "ribotyping" may be a misnomer and that the resolved RFLPs are due to polymorphisms in housekeeping genes that are flanking the ribosomal operons. Therefore, it is now possible to design RFLP schemes that take into consideration the housekeeping genes flanking the ribosomal operons to perform evolutionary studies (Bouchet et al. 2008). In the future, newly developed RFLP schemes may provide additional tools for molecular epidemiology to study foodborne pathogens.

Another example of a PCR-RFLP method for typing bacterial foodborne pathogens is the amplification of a 1,400-bp segment of the *flaA* gene for the typing of *Campylobacter jejuni* (Nachamkin et al. 1993). The use of a PCR-amplified product restricted with *Dde*I generates DNA segments that can be separated by agarose gel electrophoresis and easily visualized for interpretation. The overall procedure just requires standard equipment for a PCR assay, and if only a few strains are to be tested, it can be easily implemented in most food microbiology laboratories. This method has been proven to have a good reproducibility in an interlaboratory study (Harrington et al. 2003). The protocol can be improved by increasing the number of restriction bands generated with the addition of a restriction step using the enzyme *Hin*fI. Yet, although the standardization to generate reproducible results is relatively simple, evidence of intergenomic recombination, that is, recombination between *flaA* genes of different strains, and intragenomic recombination between the *flaA* and *flaB* genes of the same strain support the concept that this methodology may not be stable for population studies of *C. jejuni*. The extent of recombination is not clear, but recombination in these genes allows *C. jejuni* to diversify its antigenic characteristics and therefore evade the immunological responses of the host (Harrington et al. 1997). In addition, the same PCR-RFLP patterns can be found in *C. jejuni* and *C. coli* isolates, which means that strains that may be epidemiologically nonrelated may appear similar with this method (Behringer et al. 2011).

4 Methods Based on Restriction of the Whole Genome

Pulsed-field gel electrophoresis (PFGE) originated from the need to map large genomic DNA fragments that cannot be separated by conventional agarose gel electrophoresis. Until 1984, most of the measurements of large DNA molecules were done by electron microscopy, a slow and not very accurate technique (Southern and Elder 1995). But Schwartz and Cantor (1984) demonstrated that large DNA segments could be separated using a technique that was originally called *pulsed-field*

Fig. 1 PFGE profiles of strains of *Campylobacter jejuni*. The bacterial DNA was restricted using *Sma*I and the bands were separated using a CHEF machine from Bio-Rad Laboratories (Hercules, CA)

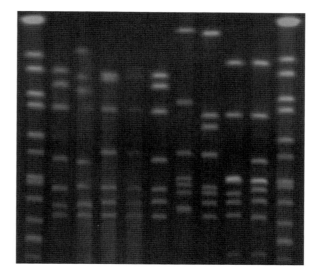

gradient gel electrophoresis. This technique later developed into what we now know as PFGE and was first used to type a foodborne bacterial pathogen by Böhm and Karch (1992).

PFGE is based on the macrorestriction of the bacterial chromosome using restriction enzymes that recognize six to eight bases and produce "rare" cuts. These enzymes, such as *Kpn*I, *Sma*I, and *Not*I, generate a few large DNA segments from the whole genome (Fig. 1). These segments are then separated using a system that switches the voltage periodically in three directions, one running through the central axis of the gel and the other two running at 120° angles on both sides. Adjustments in the switching, or pulse, time allows for the discrimination of smaller or larger molecules. The PFGE system most commonly used is the contour-clamped homogeneous electric field, known as CHEF.

Although PFGE is a relatively time-consuming technique, there are several 24-h protocols that can be applied for the typing of most foodborne bacterial pathogens. By 1996, when the Centers for Disease Control and Prevention started PulseNet (the national molecular subtyping network for foodborne disease surveillance), PFGE was the technique of choice (Swaminathan et al. 2001).

For many bacterial pathogens, PFGE is regarded as *the* reference typing method (Maslow et al. 1993a, b). The advantage of PFGE is that the whole genome is used to determine the fingerprinting profile. In most of the evaluation studies, PFGE has performed similarly to, or better than, most of the other typing methods. However, the time required to properly incorporate the technique is more than that needed with the PCR-based techniques, and adherence to the protocols is very important to ensure repeatability. Another drawback of PFGE is that restriction-based methods are more subjective in the interpretation of results than sequence-based methods.

5 Methods Based on DNA Sequencing

Multilocus sequence investigation utilizes nucleotide polymorphisms in relatively conserved housekeeping genes and is a powerful technique for bacterial population studies (Maiden et al. 1998).

The methodology for sequencing DNA has greatly improved in the last 15 years, with new high-throughput DNA methods that can sequence the whole genome of a bacterium in 2–3 days. Although these methods are still expensive for regular application in food and clinical microbiology laboratories, the increased availability of benchtop instruments, coupled with a reduced cost of materials and supply, will expand the sequencing capabilities of smaller laboratories in the next decade. The expected explosion of sequence data will bring several challenges to the biology community, especially in data storage, retrieval, and analysis. However, there will be a myriad of benefits from using sequencing techniques for typing purposes, and in the long term, these sequencing methods may replace the current typing methods based on PCR and restriction profiling.

5.1 Multilocus Sequence Typing

The last two decades have seen an explosion in the amount of DNA sequence information generated from different biological systems, including the sequencing of the whole genome of most of the bacterial foodborne pathogens. From the two methods for DNA sequencing described in 1977, the enzymatic chain termination method has proven to be more resilient and more widely used for DNA sequencing. Over the years, this method has been automated and radioactive compounds have been substituted by fluorescence compounds for labeling purposes. The sequencing protocols have also been expedited by the incorporation of PCR reactions and the use of robotic systems to enhance the processing of the samples and sequencing (Griffin and Griffin 1993; Messing 2001). However, this method has faced increased scrutiny in the last 10 years as scientists have looked for alternatives to cope with the large amount of DNA sequencing needed in biological and biomedical studies. From the newer technologies, the so-called next-generation or massive parallel sequencing technologies are bringing swift and massive sequencing output to exponentially increase the generation of DNA sequences in biology (Mardis 2008).

This section will review multilocus sequence typing (MLST) and a variation of this protocol called multilocus virulence sequence typing (MLVST). The sequence data for these methods is usually collected using traditional sequencing protocols based on capillary systems; therefore, this section will not review the different sequencing protocols. It is important to note, however, that as the high-throughput sequencing protocols become simpler and more affordable in cost, the time will come when the whole genome of an organism will be generated in hours to allow for a full comparison.

In general, MLST is based on the sequencing of housekeeping gene fragments, and the number of genes that are targeted varies according to different microorganisms. For instance, in the case of *Campylobacter jejuni*, MLST amplifies a segment of seven housekeeping genes: a 477-bp segment of the *aspA* gene (aspartate ammonia-lyase); a 489-bp segment of the *atpA* gene (ATP synthase F1 sector, alpha subunit, or *uncA*); a 402-bp segment of the *glnA* gene (glutamine synthetase); a 402-bp segment of the *gltA* gene (citrate synthase); a 507-bp segment of the *glyA* gene (serine hydroxymethyltransferase); a 498-bp segment of the *pgm* gene (phosphoglucomutase); and a 459-bp segment of the *tkt* gene (transketolase), for a total of 3,309 bp. For each locus, alleles are given arbitrary numbers and the clonality of the sequence types is assigned based on allele variations from a number of given genes in a bacterial species.

A variation of the traditional MLST method, which targets housekeeping genes, is the development of mutilocus virulence sequence typing (MLVST), in which segments of genes associated with virulence or pathogenicity are sequenced to determine their allelic profiles. Some MLVST protocols have been designed already for some foodborne pathogens, such as *Listeria monocytogenes*. These protocols may provide information on the potential pathogenicity of the strains under study.

MLST can be applied to almost all bacterial species and other haploid organisms, including those that are difficult to cultivate. Because MLST analyzes nucleotide polymorphisms in relatively conserved housekeeping genes, this technique has an important application for population studies (Maiden et al. 1998).

The availability of reliable DNA sequence data allowed for the development of MLST as a portable method that relies on the electronic storage of DNA data in an Internet database for other scientists to use. These features are important characteristics in a typing method to complement band sequence protocols, which have little portability and are less robust for comparison among laboratories (Maiden et al. 1998). MLST has allowed for the expansion of a global database per species and the exchange of molecular typing data for global epidemiology via the Internet. The MLST website (http://www.mlst.net/) provides complete information on how to start MLST analysis, collect the sequences, and submit the sequences to the database for free access by other scientists. Thus, a scientist can compare the newly acquired sequence data from one of the target genes and immediately know if an identical ST has already been listed on the MLST site. If not, the scientist can submit the new sequence, which will be annotated and uploaded within a few days of submission. The website also offers the opportunity to download all the known STs for that specific organism and even the actual sequences for each allele.

One of the limitations of MLST is the fact that this technique requires high-quality sequences and therefore is the most expensive of all the current typing techniques. As the sequencing methods become more robust and accessible, it is expected that MLST profiles will be generated more rapidly and at a lower cost.

6 Analysis and Interpretation of Results

The goal of any typing method is to determine the DNA relatedness among strains from the same species. However, the simple question of "are two strains similar?" is much more difficult to answer because different methods provide different information. Therefore, two strains may be similar with one method but different with another method. Most of the mathematical algorithms used to analyze fingerprinting profiles have been borrowed from other disciplines, such as systematics and molecular biology, and their application in molecular epidemiology is aimed at responding to the epidemiological questions that we may have. Therefore, it is important that clearly defined epidemiological questions are stated before deciding on the implementation of a given method.

The most appropriate way to analyze the typing methods used in epidemiology is to determine the epidemiological concordance between the results from a given method and the epidemiological data associated with the tested strains. A method that generates results with high epidemiological concordance will guarantee that these results will be a meaningful response to the epidemiological question that one is trying to resolve. For instance, if a method cannot group strains based on their origin, source, or geographical area, the results may not generate any useful information to the scientists.

Although systematics and molecular epidemiology are examples of disciplines that incorporate fingerprinting methodologies, in the latter the relatedness of the strains is not known a priori and the questions to answer are more temporally constrained. For instance, associating a specific strain to a food source during an outbreak study may involve the collection of strains for a few weeks up to a few months. In the case of phylogenetics studies of a given bacterial species, such as *Listeria monocytogenes*, scientists are interested in a large collection of isolates from different sources and from different times over a long period of years to evaluate the diversity within this species.

Fingerprinting techniques compare different characters, and these techniques can be used to study different taxa (singular=taxon). In general, fingerprinting techniques are used to study operational taxonomic units (OTUs), which include anything that reproduces with modification, from kingdoms to single nucleotides (Mushegian 2007). The goal of fingerprinting methods is to identify similarities or differences among a group of OTUs. The analysis of the result is done using different mathematical models. For the purpose of interpreting data for epidemiology studies, a simple approach is to classify or group the strains under study using phenetic methods, which do not take into account any evolutionary assumption. In the phenetic approach, also called *numerical taxonomy*, organisms are grouped or classified based on mathematical models applied to the characteristic(s) under study. Thus, numerical codes are created that reveal similarities or differences among organisms, and although these groupings do not reflect any phylogenetic relationships per se, it is assumed that similarity signifies genetic relationships based on the methodology used. Another approach to categorize or group strains could be using cladistic methods, where the final goal is to reveal evolutionary relationships among taxa (Riley 2004b). Most of the mathematical models and computer algorithms used to interpret the different fingerprinting patterns are derived from the numerical taxonomy, and the comparison of fingerprinting data is done by comparing densitometric curves or bands/peaks positions.

6.1 Comparison and Analysis of Band Patterns

PCR methods and most of the restriction methods generate DNA segments that can be visualized in gel electrophoresis or can be interpreted as densitometric curves. In all these band migration methods, the presence of a band or the surface created by the densitometric curve is the character under study. The idea is to determine if the presence of one specific band, or the surface covered by a densitometric curve, is unique to a given strain or is a trait shared with other strains. Because fingerprinting data can be analyzed as binary characters, the comparison of different fingerprinting profiles is based on the distance or similarity of the different OTUs or characters under analysis. In the case of band comparison methods, a comparison can be made between a band and its presence in two fingerprinting profiles, or OTUs, by just drawing a simple 2×2 table. The formulas used to calculate the similarity coefficient, sometimes called the *similarity index*, are based on how much importance is given to characters that are present in the two fingerprinting profiles (concordant characters) or are present in one but not the other fingerprinting profile (discordant characters). From the 2×2 tables, the concordant characters are "a" and "d" and the discordant characters are "b" and "c" (Fig. 2). If the index is based only on the presence of a given characteristic (a band, for instance), the formulas of Jaccard and Dice can be used, but if the index does not place any weight on the concordance or discordance of the characters, then a simplified index can be used (Riley 2004b):

$$\text{Dice index} = \frac{2a}{2a+b+c}$$

$$\text{Simple index} = \frac{a+d}{a+b+c+d}$$

Other indexes place different weight on the concordance or discordance of characters. For instance, the Sneath and Sokal's index gives more weight to concordant characters than discordant characters:

$$\text{Sneath and Sokal index} = \frac{2(a+d)}{2(a+d)+(b+c)}$$

The indices by Jaccard and Dice are most commonly used to compare gel electrophoresis band patterns. It is standard to create digital images of the restriction profiles, for instance, from RFLP or

		Presence of band A in FP1		
		Yes	No	
Presence of band A in FP2	Yes	Present in FP1 and 2 (a)	Present in FP2 but not in FP1 (b)	a+b
	No	Present in FP1 but not in FP2 (c)	Absent in both FPs (d)	c+d
		a+c	b+d	a+b+c+d

Fig. 2 The presence or absence (matching/nonmatching) of a band in two fingerprinting profiles (FP). The true positives are represented in cell *a*, while the true negatives are represented in cell *d*

PFGE analyses, and convert them into TIFF format. These images are then uploaded into the software for analysis. Several software programs have been developed for these studies, but the different modules of BioNumerics and GelCompar II (Applied Maths, Sint-Martens-Latem, Belgium) have become the standard for the analysis of this kind of fingerprinting data. In general, techniques that use restriction endonucleases as the last step before electrophoresis can be analyzed by band-matching coefficients, while others are better analyzed with densitometric curves (Vauterin and Vauterin 2006).

6.2 Comparison and Analysis of Densitometric Curves

Densitometric curves are created as arrays of *n* values. A correlation coefficient is then calculated between two arrays of values. This correlation measures the goodness of fit of these values with a linear regression. The most common index calculated with densitometric curve values is the Pearson product-moment correlation, which is sensitive to the intensity of individual bands (Vauterin and Vauterin 2006). Densitometric curves are a better fit for the analysis of the results generated by techniques that produce peaks that are overlapping or have different intensities. These results are usually generated by repetitive PCR and RAPD methods, in which the annealing temperature is usually low, and the band intensity can vary considerably.

6.3 Analysis of DNA Sequence Data

The results from sequence data can be analyzed with many different programs available on the Internet. Basically, the analysis is performed by sequence comparison algorithms, which are the basis of programs such as ClustalW and BLAST, and several other sequence comparison programs available for free online. In addition, the alignment of sequence data is the first step for further clustering and phylogenetic or epidemiological analysis.

The algorithms used in sequence alignment search for the homologies present in two sequences and create gaps in one sequence to make the homology segment match in the final consensus sequence. To avoid fragmentation of segments if gaps are freely introduced, gap penalties are introduced by these algorithms. Although there are several methods to assign gap penalties, all of the methods take into consideration how frequently deletions and insertions occur in a given DNA segment. In general terms, the probability of having a gap is inversely proportional to the size of the gap, and if a gap is created, the open gap penalty is introduced. If gaps are increased by single positions, then the unit gap penalty is introduced (Li and Graur 1991; Vauterin and Vauterin 2006). The choices of gap penalties influence the results of sequence alignment; therefore, the assumptions taken into consideration when incorporating the gap penalties have a major impact on the final sequence alignment results.

The dot-matrix method is one of the original methods used for sequence alignment studies, and the algorithm developed by Needleman and Wunsch in 1970 is still the basis of many current programs, although with many modifications. With faster and more powerful computers, the programs for alignment and DNA studies are becoming extremely sophisticated, but the premises of most of these algorithms and programs are still related to the concept that we currently have on the evolution of DNA.

6.4 Cluster Analysis

In most molecular epidemiology studies, many strains are part of the analysis. Therefore, comparing multiple strains can provide a series of clusters, or groups, of the strains with meaningful epidemiological values. The results from the pairwise comparisons described in Sects. 6.1 and 6.2 are usually analyzed using cluster algorithms to create trees based on similarities or distances, which is a strong visual method to determine relatedness among strains.

Cluster analysis can be defined as a set of analytical techniques used to organize data into groups based on their similarity or difference indices. The hierarchical structures, or trees, created with these methods are not a reflection of any evolutionary or phylogenetic relationship, and therefore "phenograms" or "dendograms" are better terms to describe these trees (Riley 2004b). Several algorithms used for cluster analysis have been borrowed from evolutionary models. The two primary clustering methods are the distance method and the maximum parsimony method. Two of the distance methods, the unweighted pair group method using arithmetic averages (UPGMA) and the neighbor joining methods, are most commonly used to cluster fingerprinting data based on their similarities or distances (Fig. 3). The UPGMA method resorts to a sequential cluster algorithm in which relationships are identified based on similarity. By doing so, the algorithm creates a composite out, which is now compared to another OTU, and so forth. Finally, a resemblance matrix is created that assumes that all OTUs or taxa are equally distant from a root. In the neighbor joining method, the distance between two OTUs is the shortest or most direct distance that could be calculated. Because several trees are constructed in this way, the algorithm selects the tree with the shortest branch length, which may or may not be the most appropriate tree to describe a set of OTUs.

7 Understanding Relatedness in Molecular Epidemiology

The ability of a molecular technique to assign types to unknown strains is based on the probability that this technique will make an assignment that reflects the true epidemiological relationship of these strains. In many instances, it is difficult to determine the exact epidemiological relationship of two strains with just one fingerprinting technique, but just adding more techniques may not provide more discrimination and may only increase the cost of the analysis. A term that is commonly used to assess the robustness of fingerprinting techniques is called *discriminatory power* which is

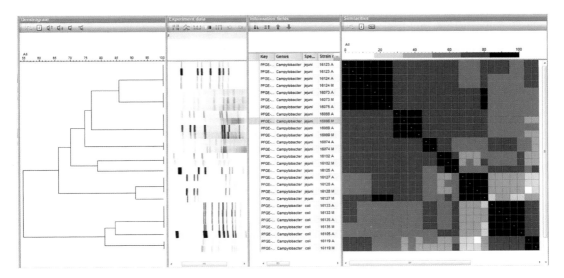

Fig. 3 Screen shot of BioNumerics showing the PFGE profiles of strains of *Campylobacter jejuni* and *Campylobacter coli*. The *left side* of the picture shows the dendogram created using the Dice coefficient for pairwise comparisons and the UPGMA clustering algorithm. The rest of the screen shows the experimental data and the similarity matrix calculated for these strains

calculated by determining Simpson's index of diversity (SID) for the method (Hunter and Gaston 1988). The SID describes a method's ability to assign a different type to two unrelated strains sampled randomly and taken from the population of a given species (van Belkum et al. 2007). The formula used to calculate the SID is

$$D = 1 - \frac{1}{N(N-1)} \sum_{J=1}^{N} aj$$

In this formula, aj is the number of strains in the population that are indistinguishable from the *j*th strain, and N is the number of strains in the population (Hunter 1990). The SID varies from 0 to 1, with 0 meaning that all the strains have different types, and 1 meaning that all the strains have the same type. Typing methods that generate SID values above 0.9 are considered appropriate for typing bacterial pathogens. However, the selection of the method depends on many variables, some of which include the cost and the degree of difficulty in performing the technique and interpreting the results. A more important consideration is the congruence of the results with epidemiological data. For instance, it is not difficult, expensive, or complex to perform PFGE analysis on a small number of bacterial isolates, but a large number of PFGE profiles can only be interpreted with computer programs, and thus the cost of performing such an analysis may increase significantly. For a method to produce valuable results, it is more important to have a strong concordance with epidemiological data, and to be validated this way, than to just have high discriminatory power. In other words, the random generation of profiles that do not correlate with epidemiological data may not provide valuable fingerprinting data.

More recently, other statistical methods have been used to calculate the concordance among different methods as well as the concordance of the types obtained based on the epidemiological data available for a given set of strains. For instance, bootstrap methods with a large number of iterations can now be used to calculate the approximate 95% confidence interval from an SID value. The adjusted Rand and Wallace coefficients are also examples of new statistical methods applied to the interpretation of the results obtained from typing techniques. The adjusted Rand coefficient is used to evaluate the extent of agreement between two typing methods. This coefficient incorporates the inter-cluster distances and is a powerful methodology to evaluate cases of OTU pairs grouped in the same cluster by one method but separated by another. The Wallace coefficient calculates the probability that OTUs grouped in the same cluster under a method are also in the same cluster under another method, and vice versa. This coefficient basically estimates how much new information is obtained from a typing method in comparison with another method. A high value of Wallace's coefficient indicates that grouping, as defined by one method, could have been predicted by another method. These two methods can be found in the online tools for quantitative assessment of classification agreement at the following URL: http://darwin.phyloviz.net/ComparingPartitions/index.php?link=Tool (Carrico et al. 2006; Pinto et al. 2007, 2008).

To understand the relatedness of two strains, it is important to understand first some basic terms that can be easily confused. A bacterial *isolate* is a group of bacterial cells in pure culture derived from a singly colony, while a *strain* is an isolate or group of isolates with similar phenotypic and/or genotypic characteristics. A term commonly used when dealing with molecular techniques for fingerprinting is *clone*, which means an isolate or group of isolates descending from a common ancestral strain. In most cases, the results generated by fingerprinting techniques do not tell us if two strains are clonally related. We may infer some clonality, for instance, if these two strains are very similar or identical. But if we do not have other information about those strains, it is better not to use the term "clones" and refer to the strains as being "similar to" based on the techniques used. The best methodology for determining clonality among bacterial strains is DNA sequencing, but if the DNA genes or segments sequenced are conserved, we may believe that two strains may belong to the same clone when, in reality, they may be different in many other aspects. Conversely, if the target sequence is quite variable among strains from the same species, as is the case with some pathogenicity genes, we may believe that two strains may not share any clonality when, in reality, they may be descended from a common ancestor only a few generations back in their ancestry. It is important not to use the term "clonality" when analyzing results from techniques that target a small DNA region, like RFLP-PCR based on a gene, or REP-PCR, etc.

There is a series of terms also used to analyze each method for their "robustness" to type bacterial isolates. The "reproducibility" of a method is defined as the percentage of strains classified as the same subtype on repeated testing, preferably after a period of a few months. Some methods have a high reproducibility, such as PFGE and any sequence-based method, but others, such as REP-PCR, have a lower reproducibility and are more prone to variations. It is important to strictly adhere to the protocols when working with techniques with a lower reproducibility to avoid erroneous interpretations.

The "typability" of a method is defined as the proportion or percentage of strains that are typed by a given method. For instance, from a panel of 100 strains, DNA sequence methods will always yield 100% typability, while 2–4% of the strains may not be typable with PFGE. In general terms, methods that use restriction enzymes may not yield 100% typability.

8 Choice of Typing Methods

Many variables affect the choice of a typing method for a given project, especially if a test is to be incorporated in a laboratory for the first time. The choice will also depend on the specific question that we want to answer: Do we want a molecular technique to characterize the intraspecies variability of an organism in a large number of strains? Do we want a technique to compare relatedness among a few strains isolated from similar food samples?

In the first question a consideration to take into account is a large number of strains to compare, and therefore a method that is amenable to automation with high throughput may have some advantages. In addition, a combination of two complementary methods may be more appropriate to give in-depth answers. For the second question, a simple method whose results can be analyzed visually may be sufficient.

There are no "gold standard" methods, although the restriction and fragment migration in PFGE is a solid system for analysis. However, if we are looking for a method for population studies, MLST may be superior. The epidemiological concordance of a typing method is more important than its discriminatory power, and less labor-intensive methods are preferred when analyzing a large number of strains.

Typing methods are extremely useful to track bacterial pathogens to determine the source of contamination, to identify potential risk, to understand sources of transmission, and to study the dynamics of bacterial pathogens in geographically defined areas. Ideally, we would like to use these methods to determine pathovars, or strains with a high pathogenicity potential, versus nonpathogenic strains. However, we are still missing many different pieces of information around the circumstances that result in the appearance of foodborne diseases. Some of this information relates to the human host and its conditions; therefore, any further attempts to define "pathovar" or "nonpathogenic" strains will have to be carefully weighted in the light of new understandings of how foodborne pathogens produce disease.

9 Summary

DNA typing methods are invaluable tools for epidemiological studies and for the surveillance of foodborne pathogens. As our knowledge of the genomic makeup of bacterial and viral pathogens improves, new methods are being developed that vary from simple, inexpensive ones, such as PCR-RFLP, to complex, automated systems, such as the sequencing of the complete genome of the target organism. Typing methods can be categorized as amplification; amplification and restriction analysis; restriction analysis of the whole chromosome; and sequence-based methods. However, there is no single standard categorization for all known typing methods. Each method has its advantages and disadvantages, and in general, the simpler, less expensive methods provide simple results that can be visually analyzed. Many different algorithms have been applied to the analysis of the results, and the choice of analysis depends on whether the method yields bands, densitometric curves of DNA sequence data. PFGE, a method based on the restriction and band migration from the whole bacterial genome, is still a powerful method for epidemiological studies and one of the methods of choice for surveillance purposes. However, as sequence methods become less expensive and more available and computer systems more powerful for data analysis, sequence-based methods are bound to become the predominant techniques for molecular epidemiology in the near future.

References

Behringer, M., W.G. Miller, and O.A. Oyarzabal. 2011. Typing of *Campylobacter jejuni* and *Campylobacter coli* isolated from live broilers and retail broiler meat by *flaA*-RFLP, MLST, PFGE and REP-PCR. *Journal of Microbiological Methods* 84: 194–201.

Böhm, H., and H. Karch. 1992. DNA fingerprinting of *Escherichia coli* O157:H7 strains by pulsed-field gel electrophoresis. *Journal of Clinical Microbiology* 30: 2169–2172.

Botstein, D., R.L. White, M. Skolnick, and R.W. Davis. 1980. Construction of a genetic linkage map in man using restriction fragment length polymorphisms. *American Journal of Human Genetics* 32: 314–331.

Bouchet, V., H. Huot, and R. Goldstein. 2008. Molecular genetic basis of ribotyping. *Clinical Microbiology Reviews* 21: 262–273.

Carrico, J.A., C. Silva-Costa, J. Melo-Cristino, F.R. Pinto, H. de Lencastre, J.S. Almeida, and M. Ramirez. 2006. Illustration of a common framework for relating multiple typing methods by application to macrolide-resistant *Streptococcus pyogenes*. *Journal of Clinical Microbiology* 44: 2524–2532.

Danna, K., and D. Nathans. 1971. Specific cleavage of simian virus 40 DNA by restriction endonuclease of *Hemophilus influenzae*. *Proceedings of the National Academy of Sciences USA* 68: 2913–2917.

Foley, S.L., A.M. Lynne, and R. Nayak. 2009. Molecular typing methodologies for microbial source tracking and epidemiological investigations of Gram-negative bacterial foodborne pathogens. *Infection, Genetics and Evolution* 9: 430–440.

Griffin, H.G., and A.M. Griffin. 1993. DNA sequencing. In *Methods in molecular biology, DNA sequencing protocols*, ed. H.G. Griffin and A.M. Griffin, 1–8. Totowa: Humana Press.

Grimont, F., and P.A. Grimont. 1986. Ribosomal ribonucleic acid gene restriction patterns as potential taxonomic tools. *Annales de l'Institut Pasteur Microbiology* 137B: 165–175.

Harrington, C.S., F.M. Thomson-Carter, and P.E. Carter. 1997. Evidence for recombination in the flagellin locus of *Campylobacter jejuni*: Implications for the flagellin gene typing scheme. *Journal of Clinical Microbiology* 35: 2386–2392.

Harrington, C.S., L. Moran, A.M. Ridley, D.G. Newell, and R.H. Madden. 2003. Inter-laboratory evaluation of three flagellin PCR/RFLP methods for typing *Campylobacter jejuni* and *C. coli*: The CAMPYNET experience. *Journal of Applied Microbiology* 95: 1321–1333.

Hulton, C.S.J., C.F. Higgins, and P.M. Sharp. 1991. ERIC sequences, a novel family of repetitive elements in the genome of *Escherichia coli*, *Salmonella typhimurium* and other enterobacterial. *Molecular Microbiology* 5: 825–834.

Hunter, P.R. 1990. Reproducibility and indices of discriminatory power of microbial typing methods. *Journal of Clinical Microbiology* 28: 1903–1905.

Hunter, P.R., and M.A. Gaston. 1988. Numerical index of the discriminatory ability of typing systems: an application of Simpson's index of diversity. *Journal of Clinical Microbiology* 26: 2465–2466.

Jeffreys, A., and R. Flavell. 1977. A physical map of the DNA regions flanking the rabbit δ-globin gene. *Cell* 12: 429–439.

Johnson, J.R., and C. Clabots. 2000. Improved repetitive-element PCR fingerprinting of *Salmonella enterica* with the use of extremely elevated annealing temperatures. *Clinical and Diagnostic Laboratory Immunology* 7: 258–264.

Kennedy, S., and N. Oswald. 2011. *PCR troubleshooting and optimization: the essential guide*. Portland: Caister Academic.

Kostman, J.R., T.D. Edlind, J.J. LiPuma, and T.L. Stull. 1992. Molecular epidemiology of *Pseudomonas cepacia* determined by polymerase chain reaction ribotyping. *Journal of Clinical Microbiology* 30: 2084–2087.

Li, W.-H., and D. Graur. 1991. Evolutionary changes in nucleotide sequences. In *Fundamentals of molecular evolution*. Sunderland: Sinauer Associates.

Maiden, M.C., J.A. Bygraves, E. Feil, G. Morelli, J.E. Russell, R. Urwin, Q. Zhang, J. Zhou, K. Zurth, D.A. Caugant, I.M. Feavers, M. Achtman, and B.G. Spratt. 1998. Multilocus sequence typing: a portable approach to the identification of clones within populations of pathogenic microorganisms. *Proceedings of the National Academy of Sciences USA* 95(6): 3140–3145.

Mardis, E.R. 2008. An overview of new DNA sequencing technology. In *Genome sequencing technology and algorithms*, ed. S. Kim, H. Tang, and E.R. Mardis. Norwood: Artech House.

Maslow, J.N., M.E. Mulligan, and R.D. Arbeit. 1993a. Molecular epidemiology: application of contemporary techniques to the typing of microorganisms. *Clinical Infectious Diseases* 17: 153–162.

Maslow, J.N., A.M. Slutsky, and R.D. Arbeit. 1993b. The application of pulsed-field gel electrophoresis to molecular epidemiology. In *Diagnostic molecular microbiology*, ed. D.H. Persing, F.C. Tenover, T.F. Smith, and T.J. White, 563–572. Washington, DC: ASM Press.

Messing, J. 2001. The universal primers and the shotgun DNA sequencing method. In *Methods in molecular biology, Vol. 167: DNA sequencing protocols*, 2nd ed, ed. C.A. Graham and A.J.M. Hill, 13–31. Totowa: Humana Press.

Mullis, K.B., F. Faloona, S. Scharf, R. Saiki, G. Horn, and H. Erlich. 1986. Specific enzymatic amplification of DNA *in vitro*: the polymerase chain reaction. *Cold Spring Harbor Symposia on Quantitative Biology* 51: 263–273.

Mushegian, A.R. 2007. Phylogenetic inference and the era of complete genomes. In *Foundations of comparative genomics*, ed. A.R. Mushegian, 151–177. Burlington: Academic Press/Elsevier.

Nachamkin, I., K. Bohachick, and C.M. Patton. 1993. Flagellin gene typing of *Campylobacter jejuni* by restriction fragment length polymorphism analysis. *Journal of Clinical Microbiology* 31: 1531–1536.

Pinto, F.R., J.A. Carriço, M. Ramirez, and J.S. Almeida. 2007. Ranked adjusted rand: integrating distance and partition information in a measure of clustering agreement. *BMC Bioinformatics* 8: 44.

Pinto, F.R., J. Melo-Cristino, and M. Ramirez. 2008. A confidence interval for the Wallace coefficient of concordance and its application to microbial typing methods. *PloS One* 3(11): e3696. doi:10.1371/journal.pone.0003696.

Rasschaert, G., K. Houf, H. Imberechts, K. Grijspeerdt, L. De Zutter, and L. Heyndrickx. 2005. Comparison of five repetitive-sequence-based PCR typing methods for molecular discrimination of *Salmonella enterica* isolates. *Journal of Clinical Microbiology* 43: 3615–3623.

Riley, L.W. 2004a. Laboratory methods used for strain typing of pathogens: PCR-based strain typing methods. In *Molecular epidemiology of infectious diseases, principles and practices*, ed. L.W. Riley, 63–83. Washington, DC: ASM Press.

Riley, L.W. 2004b. Analysis of similarity and relatedness in molecular epidemiology. In *Molecular epidemiology of infectious diseases, principles and practices*, ed. L.W. Riley, 91–124. Washington, DC: ASM Press.

Schwartz, D.C., and C.R. Cantor. 1984. Separation of yeast chromosome-sized DNAs by pulsed field gradient gel electrophoresis. *Cell* 37: 67–75.

Southern, E.M. 1975. Detection of specific sequences among DNA fragments separated by gel electrophoresis. *Journal of Molecular Biology* 98: 503–517.

Southern, E.M., and J.K. Elder. 1995. Theories of gel electrophoresis of high molecular weight DNA. In *Pulsed field gel electrophoresis: a practical approach*, ed. A.P. Monaco, 1–19. New York: Oxford University Press.

Stern, M.J., G.F. Ames, N.H. Smith, E.C. Robinson, and C.F. Higgins. 1984. Repetitive extragenic palindromic sequences: a major component of the bacterial genome. *Cell* 37: 1015–1026.

Swaminathan, B., T.J. Barrett, S.B. Hunter, R.V. Tauxe, and The CDC PulseNet Task Force. 2001. PulseNet: the molecular subtyping network for foodborne bacterial disease surveillance, United States. *Emerging Infectious Diseases* 7: 382–389.

Tompkins, L.S., N. Troup, A. Labigne-Roussel, and M.L. Cohen. 1986. Cloned, random chromosomal sequences as probes to identify *Salmonella* species. *Journal of Infectious Diseases* 154: 156–162.

van Belkum, A., P.T. Tassios, L. Dijkshoorn, S. Haeggman, B. Cookson, N.K. Fry, V. Fussing, J. Green, E. Feil, P. Gerner-Smidt, S. Brisse, M. Struelens, and The European Society of Clinical Microbiology and Infectious Diseases (ESCMID) Study Group on Epidemiological Markers (ESGEM). 2007. Guidelines for the validation and application of typing methods for use in bacterial epidemiology. *Clinical Microbiology and Infection* 13: 1–46.

Vauterin, L., and P. Vauterin. 2006. Integrated databasing and analysis. In *Molecular identification, systematic and population structure of prokaryotes*, ed. E. Stackebrandt, 141–217. Berlin: Springer.

Versalovic, J., T. Koeuth, and J.R. Lupski. 1991. Distribution of repetitive DNA sequences in eubacteria and application to fingerprinting of bacterial genomes. *Nucleic Acids Research* 19: 6823–6831.

Vos, P., R. Hogers, M. Bleeker, M. Reijans, T. van de Lee, M. Hornes, A. Frijters, J. Pot, J. Peleman, M. Kuiper, and M. Zabeau. 1995. AFLP: a new technique for DNA fingerprinting. *Nucleic Acids Research* 23: 4407–4414.

Weissensteiner, T., H.G. Griffin, and A.M. Griffin. 2004. *PCR technology: current innovations*. Boca Raton: CRC Press.

Wise, M.G., G.R. Siragusa, J. Plumblee, M. Healy, P.J. Cray, and B.S. Seal. 2009. Predicting *Salmonella enterica* serotypes by repetitive sequence-based PCR. *Journal of Microbiological Methods* 76: 19–24.

Foodborne Viruses*

Daniel C. Payne, Umid Sharapov, Aron J. Hall, and Dale J. Hu

1 Introduction

The study and control of foodborne diseases have traditionally focused on bacterial and parasitic agents instead of viruses. Despite the fact that there are relatively few major viral agents responsible for foodborne illness, those that exist are responsible for approximately half of all foodborne disease outbreaks in the United States and can cause serious – even life-threatening – illnesses. The ease of viral transmission and the high level of infectiousness are mainly due to the small number of *viral particles* needed to cause the disease, the high volume of viral particles shed by infected persons, and the environmental stability of most viruses on contaminated objects. Acceptance is growing that the commonly used bacteria-based approaches to prevent foodborne illness are often insufficient to eliminate the risk of viral contamination, and therefore additional control measures are necessary. An understanding of foodborne viruses, separate from that of bacterial or parasitic agents, is needed to prevent viral transmission via food and food handlers.

Many reasons for these foodborne viruses to be overlooked arise from the inherent differences in bacteria and viruses. For example, techniques for detecting viruses are recent additions to public health laboratories and have been developed only in recent decades, as opposed to tools to detect bacterial and parasitic organisms that were first described more than a century ago. Also, viral foodborne illnesses are underappreciated because they have no specific medical treatment – *antibiotics* are useless against viral illness – and there is a prevalent mistaken impression that viruses cause only mild, transient diarrhea or vomiting (gastroenteritis). Though many foodborne illnesses of viral causation produce gastroenteritis (such as norovirus), others (such as hepatitis viruses) cause very different symptoms that may not involve diarrhea or vomiting. Instead, these illnesses may cause weakness and *jaundice* and may sometimes lead to long-lasting organ damage.

Recently, advances in the field of viral detection and epidemiology have changed the way we detect, understand, and prevent foodborne viruses. *Molecular biology* now holds the tools to diagnos

---

*The findings and conclusions of this chapter are those of the authors and do not necessarily represent the views of the Centers for Disease Control and Prevention.

D.C. Payne (✉) • A.J. Hall
National Center for Immunization and Respiratory Diseases, Division of Viral Diseases,
Epidemiology Branch, U.S. Centers for Disease Control and Prevention, Atlanta, GA, USA
e-mail: dvp6@cdc.gov

U. Sharapov • D.J. Hu
National Center for Immunization and Respiratory Diseases, Division of Viral Hepatitis,
Epidemiology and Surveillance Branch, U.S. Centers for Disease Control and Prevention, Atlanta, GA, USA

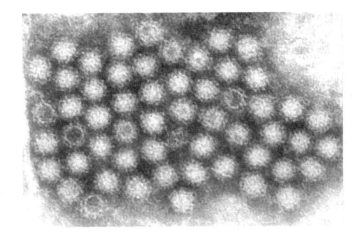

Fig. 1 Scanning electron micrograph of noroviruses (Reproduced from the Public Health Image Library (PHIL, http://phil.cdc.gov/Phil/home.asp) of the Centers for Disease Control and Prevention, Atlanta)

and type viruses in the laboratory. As a result, viral gastroenteritis is now recognized as a substantial health burden, and preventive public health measures that better target viral infections are now coming into widespread acceptance. The objective of this chapter is to review the viruses most commonly implicated in foodborne illnesses, their epidemiology, and their prevention and control.

2 Norovirus: Half of the Problem

Human noroviruses (formerly referred to as Norwalk-like viruses) are members of the *Caliciviridae* family of viruses (Fig. 1). Noroviruses were discovered in 1972 as the first viruses definitively associated with acute gastroenteritis and are now known to be the most common cause of infectious gastroenteritis among persons of all ages (Estes et al. 2006; Patel et al. 2008). Noroviruses are estimated to cause nearly 21 million cases of gastroenteritis in the United States annually (Scallan et al. 2011) and are responsible for approximately half of all foodborne outbreaks of gastroenteritis (Widdowson et al. 2005a).

2.1 Clinical Presentation

Within 12–48 h of exposure to the virus, individuals develop a spectrum of gastroenteritis symptoms, including diarrhea, vomiting, fever, and nausea, which last for a period of 1–3 days. During this period, diarrhea and/or vomiting can occur with extraordinary frequency and severity. The infected person often has trouble replacing fluids lost to this purging, sometimes resulting in dehydration. The vomitus and feces of those infected are rich in norovirus particles, making these fluids extremely infectious. Though there is no evidence that an infected person can become a long-term carrier of the virus, norovirus can be present in excreted materials from the day of the person's first symptom until several weeks after all symptoms have subsided.

Studies have demonstrated that as many as 30% of norovirus infections may be asymptomatic (Graham et al. 1994), and approximately 10% of cases may require medical care, and even brief hospitalization, to reverse fluid loss (Widdowson et al. 2005b). Recent outbreaks of norovirus have suggested that deaths may also be associated with infections, particularly among the elderly (CDC 2007a). Nonetheless, the role of norovirus in causing severe disease or death has been poorly investigated due to the perception that norovirus generally causes mild gastroenteritis and the lack of a rapid, specific, and sensitive antigen-detection assay for norovirus for use in clinical settings.

Until the early 1990s, when sensitive molecular assays such as *reverse transcription-polymerase chain reaction* (RT-PCR) and *nucleotide hybridization probes* emerged, the clinical diagnosis of norovirus infection was not reliable, and the pathogenicity of this viral infection was poorly understood (Hutson et al. 2004; Jiang et al. 1992; Moe et al. 1994). The development of these new sensitive and specific laboratory capabilities has enhanced the understanding of norovirus epidemiology and has particularly improved the detection and management of norovirus outbreaks. However, to date, all attempts to propagate human norovirus in cell culture have been unsuccessful (Duizer et al. 2004), and most information on the persistence of infectious virus after disinfection, stability in the environment, and basic virology questions of replication and immune protection has come from human challenge studies and studies using model or surrogate viruses.

2.2 Epidemiology

Although norovirus often occurs sporadically in the absence of an outbreak, noroviruses are the most frequent cause of gastroenteritis outbreaks worldwide and can be transmitted by contaminated foods, water, *fomites*, or person-to-person contact. In the United States each year there are estimated to be nearly 21 million cases of illness due to norovirus (Scallan et al. 2011), and approximately 50% of all foodborne gastroenteritis outbreaks are attributed to norovirus (Widdowson et al. 2005b). From 2006 to 2007, a total of 822 foodborne disease outbreaks attributed to norovirus were reported to the U.S. Centers for Disease Control and Prevention (CDC) (CDC 2009a, 2010). Among 660 norovirus outbreaks confirmed by the CDC during 1994 to 2006, the most common setting was long-term care facilities such as nursing homes (35%), followed by restaurants, parties, and events (31%); cruise ships and vacations (20%); and schools, child-care, and communities (13%) (Zheng et al. 2010). These findings are consistent with other studies that have found noroviruses to be responsible for a large proportion of foodborne-related gastroenteritis outbreaks and significant illnesses in nursing homes in the United States. Noroviruses contribute significantly to hospital illness in Europe (Green et al. 2002; Lopman et al. 2004; Widdowson et al. 2005a).

In 2006 and 2007, many U.S. state health departments reported increases in norovirus-like outbreaks and emergency department visits, which did not seem to be simply a result of improved reporting and testing. Instead, it is widely believed that these increases may have been due to the recent emergence of two new circulating strains of norovirus, which may be immunologically novel to the population, have increased pathogenicity, or have increased transmissibility compared to previous strains (CDC 2007a).

Norovirus transmission occurs via foodborne, waterborne, and person-to-person routes, as well as through contact with contaminated environmental surfaces. Rare for other pathogens, norovirus transmission is also believed to occur via droplets of vomitus (Marks et al. 2000, 2003). The act of vomiting produces aerosolized droplets that settle on surfaces and may be ingested involuntarily by people nearby. The low *infectious dose* of norovirus (less than 100 viral particles) required for transmission, in addition to the virus's environmental persistence and prolonged shedding after recovery, contribute to the high level of norovirus-related foodborne illness. Most foodborne norovirus outbreaks involve *ready-to-eat foods* such as salads, sandwiches, and fresh produce (Turcios et al. 2006). These items are usually contaminated at the point of service by an infected food handler, even in the absence of illness reported by the workers. However, several outbreaks involving raspberries have shown that contamination of fresh produce can also occur at their originating source (Gaulin et al. 1999). Additionally, a recent investigation found that ready-to-eat meats could be contaminated with norovirus before vacuum packaging (Barzilay et al. 2006).

Table 1 Pathogens indentified in outbreaks and cases involving U.S. food workers[a]

Category	Agents	No. (%) Cases	Outbreaks
Viruses	Norovirus	27,081 (33)	274 (34)
	Hepatitis A	5,046 (6)	84 (10)
	Probably norovirus	2,085 (3)	64 (8)
	Unknown	2,148 (3)	57 (7)
	Rotavirus	1,418 (2)	12 (1)
Bacteria	*Salmonella* (nontyphoidal)	9,136 (11.2)	130 (15.9)
	Salmonella Typhi	757 (0.9)	21 (2.6)
	Staphylococcus aureus	6,423 (8)	53 (6.5)
	Shigella spp.	15,276 (18.9)	33 (4.0)
	Streptococcus A and G	3,670 (4.5)	17 (2.1)
	Vibrio cholera	2,399 (3)	11 (1.3)
	Yersinia enterocolitica	532 (0.7)	7 (0.9)
	Campylobacter jejuni	238 (0.3)	5 (0.6)
	Escherichia coli O157:H7, O6:H16 and enterotoxigenic	105 (0.1)	3 (0.4)
Parasites	*Cyclospora cayetanensis*	3,393 (4.2)	11 (1.3)
	Giardia lamblia/intestinalis	302 (0.4)	9 (1.1)
	Cryptosporidium spp.	157 (0.2)	22 (2.7)
Unknown		516 (0.6)	22 (2.7)

[a]Modified from Greig et al. (2007). Reprinted with permission from *The Journal of Food Protection*. Copyright held by the International Association for Food Protection, Des Moines, IA

In environmental evaluations using a murine norovirus surrogate (human norovirus cannot yet be cultured for study), norovirus was able to survive at refrigeration temperature (either wet or dry) and at room temperature (wet) for at least 1 week (Cannon et al. 2006). Anecdotal evidence from outbreaks suggests that noroviruses can remain on surfaces and cause infections for even longer periods of time (Falkenhorst et al. 2005; Malek et al. 2009). Thus, contamination of foods with norovirus may occur at any point along the "farm-to-table" continuum: during growing, harvesting, processing, handling, and even after food has been refrigerated and cooked.

Until recently, only foodborne or waterborne outbreaks of gastrointestinal illness have required reporting to the CDC, while outbreaks determined to be transmitted from person to person are not (such as outbreaks in nursing homes). Many of these latter outbreaks are attributable to norovirus infection. Therefore, the burden of disease due to norovirus may be significantly underestimated by focusing exclusively on foodborne disease. An expansion of the CDC outbreak reporting system to include person-to-person outbreaks was implemented in 2009, so that more reliable estimates of the burden of norovirus disease outbreaks and the proportion attributable to each mode of transmission may soon be available.

The International Association for Food Protection's Committee on the Control of Foodborne Illness evaluated 816 outbreaks where food workers were implicated in the spread of foodborne illness in the literature through 2006 (Todd et al. 2007). This report found that norovirus (or probable norovirus) accounted for over 40% of foodborne outbreaks evaluated and involved approximately one third (36%) of the persons who fell ill (Greig et al. 2007) (Table 1). Over time, the proportion of foodborne outbreaks of norovirus origin appeared to be increasing. Several examples of norovirus outbreaks were recorded, including one on a cruise ship where fresh-cut fruit sickened 41% of the passengers before landing at the first port of call. Although two cooks who prepared the implicated fruit believed they were not ill, one did become ill on the second day of the outbreak and was likely infected but *asymptomatic* at the time he contaminated the food (i.e., he was infected and excreting the virus but did not realize he was sick at the time). Another example involved a cruise ship's ice machines that had not been cleaned and ice not discarded after the previous trip, during which norovirus illnesses were reported. The result was that the norovirus outbreak continued on to the subsequent cruise, with many passengers falling ill within 24 h of departure.

A person can be infected with norovirus more than once in his or her lifetime because there are many different norovirus genotypes, and infection with one type does not prevent infection with another type later. However, at least short-term immunity appears to be conferred upon repeated infections with the same type of norovirus, as demonstrated in human challenge studies (Johnson et al. 1990). In recent years, host genetic factors have been recognized as important determinants of infection and disease, and are suggested by an apparent association between blood group and norovirus susceptibility. In limited studies, persons having blood group B have been observed to be more resistant to infection and disease than those of blood group O (Hennessy et al. 2003; Hutson et al. 2002).

The favored method to detect norovirus remains through RT-PCR, although commercially available enzyme immunoassay (EIA) kits may have a role to play in the rapid detection of norovirus in outbreak settings where many stools are available for testing. However, EIA tests have exhibited inadequate sensitivity (≤50%) and lack sufficient coverage of the two dozen norovirus genotypes (Zheng et al. 2006) to be useful in the clinical diagnosis of individual cases. Two different types of RT-PCR can be used for the detection of norovirus, both available only at public health laboratories. Quantitative *real-time PCR* (qrt-PCR) is a rapid, quantitative test for conserved regions of the norovirus genome, which can classify the detected virus into genogroup I or II with the use of differential probes. Though less sensitive than qrt-PCR, conventional PCR is used to amplify specific regions of the genome that are subsequently sequenced, allowing for the identification of specific genetic clusters and strains. This molecular fingerprinting can be used to link case of illness to one another and to specific food or environmental specimens implicated in an outbreak.

3 Hepatitis Viruses

There are two distinct hepatitis viruses that have been characterized and are known to be transmitted through the fecal–oral route via contaminated water or food (Fig. 2). Hepatitis A virus (HAV) is an RNA virus of the *Picornaviridae* family that has been extensively studied, while hepatitis E virus (HEV) is an unrelated RNA virus of the *Hepeviridae* family that was characterized in the early 1980s (Purcell and Emerson 2008). Since HEV has been implicated primarily in large waterborne outbreaks in developing countries and has not been reported in any large foodborne outbreaks, most of the information presented in this section will refer to HAV. Nevertheless, relevant information on HEV is also presented here, given the recent publications documenting HEV human infections from eating raw or uncooked meat of wild and domesticated animals.

3.1 Clinical Presentation

Both hepatitis A and E viruses can cause an acute and usually self-limited infection. Illness from either infection is associated with fever, *malaise*, nausea, anorexia, abdominal discomfort, dark urine, and jaundice (Koff 1992; Okamoto et al. 2003). Most children with HAV will have no symptoms, and less than 10% of children under 6 years of age have jaundice (Gingrich et al. 1983). In contrast, 76–97% of young adults with HAV infection have symptoms, and 40–70% are jaundiced (Lednar et al. 1985). Symptomatic HEV infection is most common in young adults aged 15–40 years. Although HEV infection is frequent in children, it is mostly asymptomatic or causes a very mild illness without jaundice (anicteric) that goes undiagnosed.

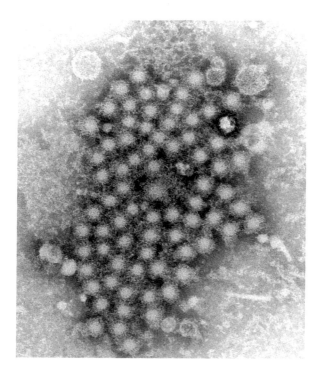

Fig. 2 Scanning electron micrograph of hepatitis viruses (Reproduced from the Public Health Image Library (PHIL, http://phil.cdc.gov/Phil/home.asp) of the Centers for Disease Control and Prevention, Atlanta)

The average time from exposure to onset of symptoms for HAV is approximately 28 days (range: 15–50 days) (Krugman and Giles 1970). The incubation period for HEV is slightly longer, on average, lasting approximately 40 days (range: 15–64 days) (Mushahwar 2008). HAV replicates in the liver, is excreted in bile, and is found in very high concentrations in the feces from 2 weeks before to 1 week after the onset of clinical illness. The severity of illness from either HAV or HEV infection is dose-dependent. While a single infectious dose of HAV is adequate to initiate a hepatitis A infection, the clinical response to HEV is thought to be dose-dependent, with low amounts of infectious virus resulting in unapparent infection (Balayan et al. 1983).

Neither HAV nor HEV results in chronic infection or chronic liver disease, except in rare cases of immunologically suppressed organ-transplant recipients infected with HEV (Kamar et al. 2008; Takahashi et al. 2007; Tamura et al. 2007). Fifteen to 20% of persons infected with HAV may have prolonged or relapsing disease lasting up to 6 months (Glikson et al. 1992; Sjogren et al. 1987). Overall, *fulminant hepatitis* is a rare complication of HAV infection, and the case fatality rate is 0.5% (with 1.8% among persons aged $\geq$50 years). HEV disease is often more severe and is associated with a higher mortality rate (1–4%), especially among pregnant women (up to 20%) (Dalton et al. 2008). Immunity after infection for HAV persists for life, while the long-term duration of immunity for HEV is less clear.

3.2 Epidemiology

HAV infection is primarily transmitted by the fecal–oral route, by either person-to-person contact or consumption of contaminated food or water. HAV-infected persons without symptoms or jaundice, especially children, are an important source of HAV transmission (Staes et al. 2000).

The epidemiology of HAV infection has been well described, and the prevalence of HAV infection is generally associated with socioeconomic development and standards of hygiene in different populations (Nainan et al. 2006). In developing countries, where infection is *endemic*, most persons are infected during childhood and many are asymptomatic; Therefore, HAV transmission is often not recognized. In developed countries, where the prevalence of HAV is lower, infection typically occurs at older ages when clinical symptoms become more apparent (Green et al. 1989).

In the United States, nearly half of all reported hepatitis A cases have no specific risk factor identified. Household contact or other close personal contact with an HAV-infected person is the most important reported risk factor. In contrast, relatively few reported cases (2–3% cases per year) are identified as part of common source outbreaks of diseases transmitted by food or water. Due to the relatively long incubation period of HAV and HEV, documentation of sporadic cases of infection is difficult unless transmission occurs over an extended period of time and many people are infected with clinical symptoms.

Produce might be contaminated by the hands of HAV-infected workers or children in the field, by contact with HAV-contaminated water during irrigation or rinsing after picking, or during the processing steps leading to packaging. For example, the removal of stems by infected workers in the field during picking is one potential mechanism for strawberry contamination.

Outbreaks usually result from one of two sources of contamination: an infected food handler or an infected food source. A local health department will investigate the cause of the outbreak, and health officials will try to determine the source of the contamination and the best ways to minimize health threats to the public. Foods can become contaminated at any point along the complex farm-to-table continuum.

In contrast to HAV, the epidemiology of HEV is not as well understood. HEV has been primarily implicated in large waterborne epidemics in Asia, Africa, and Central America where thousands of people developed acute hepatitis. In contrast, in developed countries, there have been sporadic cases from individuals who either traveled to an HEV endemic area or were associated with the consumption of raw or undercooked meat of animals infected with HEV. The consumption of raw or undercooked meat from wild boar and deer has been reported as a source of human HEV infection in Japan (Matsuda et al. 2003; Takahashi et al. 2004). In Japan, between 2001 and 2002, nine of ten patients having acute fulminant hepatitis E reported eating grilled or undercooked pig liver 2–8 weeks before the onset of their illness (Mushahwar 2008). HEV was also detected in samples of pig livers obtained from groceries in the United States (Feagins et al. 2007) and Japan (Yazaki et al. 2003).

As there are other common types of viral hepatitis (B and C) that are clinically indistinguishable, only laboratory tests can distinguish HAV and HEV from each other or from other forms of viral hepatitis in the absence of patient risk history or epidemiological findings. Acute HAV infection can be confirmed by a serologic test detecting immunoglobulin antibody (IgM) to capsid proteins of HAV (also known as IgM anti-HAV). IgM anti-HAV is typically detectable 5–10 days before the onset of symptoms. IgG anti-HAV appears early in the course of infection and remains detectable for the person's lifetime, providing lifelong protection against the disease. Nucleic acid amplification methods are also used to identify HAV RNA in the blood and stool during the acute phase of illness. Nucleic acid sequencing is used to help to determine the relatedness of HAV isolates during the epidemiologic investigations. However, these tests are performed by few research laboratories (Fiore et al. 2006). Similarly, hepatitis E infection is diagnosed by detection of IgM anti-HEV in the serum by ELISA, or by detection of viral genomic RNA in the serum or feces by nested or RT-PCR (Purcell and Emerson 2008).

4 Other Foodborne Viruses: A Mixed Bag

Many viruses have the potential to be spread via the foodborne route, including lesser-known agents, but these are often related to particular food preparation practices in specific geographic locations. For example, Ebola virus (a frequently fatal hemorrhagic viral disease) has been transmitted to humans during the preparation and consumption of African primates as bushmeat (Chastel and Charmot 2004), and Nipah virus, a frequently fatal acute respiratory viral disease often accompanied by altered mental status, has been associated with drinking raw date palm sap in South Asia, presumably contaminated by infected bats who also drink from the sap pots during the night (Luby et al. 2006).

Foodborne outbreaks of rotavirus (more commonly a cause of viral acute gastroenteritis disease) have been documented, including an outbreak at a U.S. university setting (CDC 2000) and at a Japanese restaurant (Anonymous 2000). On the other hand, sapovirus – a member of the same *Calicivirus* family as norovirus – has rarely been implicated in foodborne outbreaks (Blanton et al. 2006; Gaulin et al. 1999). The most likely explanation for the apparent lack of foodborne transmission of most viruses causing acute gastroenteritis is that they predominantly cause infections among children. Since children rarely prepare food for others and are less likely to consume products potentially contaminated in the environment, opportunities for foodborne transmission are far fewer. Noroviruses are the exception, however, as they are antigenically diverse and often elicit only short-term immunity; hence, noroviruses are common pathogens in both child and adult illness and therefore have increased opportunity for transmission.

5 Prevention and Control of Viral Foodborne Pathogens

5.1 Vaccines

No vaccine is currently licensed to prevent norovirus infection, although candidate vaccines are under development and testing (Table 2). It is technically challenging to develop a vaccine for any agent, such as norovirus, for which natural infection does not provide lasting immunity. Active research is focused on clarifying the apparently complex immunity to norovirus infection. The current emphases on improving norovirus reporting systems and collecting better data on norovirus disease burden are partially driven by the potential for vaccine development.

An effective and safe vaccine against hepatitis A became available in the United States during 1995–1996. It is currently recommended for use among persons between the ages of 1 and 40 years by the U.S. Advisory Committee on Immunization Practices (ACIP) recommendations (CDC 2007b; Fiore et al. 2006). In the United States, the annual number of symptomatic cases reported has dropped steadily since its last peak in 1995 until 2006, from 31,582 to 3,579, or from a national incidence of 12.0 to 1.2 per 100,000 population (Wasley et al. 2008), in part because of increasing hepatitis A immunizations (CDC 1999).

A candidate vaccine against HEV has been tested for safety and immunogenicity among U.S. and Nepalese Army volunteers, and later in an efficacy trial in Nepal (Shrestha et al. 2007). In addition, another candidate vaccine has been developed in China and is currently undergoing phase III trials (Zhang et al. 2009). However, no HEV vaccine has yet been licensed (Table 3).

Table 2 Applicability of specific preventive and control measures against norovirus and hepatitis A and E foodborne transmission

Preventive/control measures	Norovirus	Hepatitis A virus	Hepatitis E virus
Vaccine licensed and approved for use	X	✓	X
Postexposure prophylaxis (PEP) following known exposure	X	✓	X
Food-handler hygiene (processing, preparing, serving)	✓	✓	✓
Hygienic food growing/harvesting	✓	✓	✓
Proper cleanup of infected materials	✓	✓	✓

Table 3 Recommended measures for the prevention and control of norovirus infection[a]

1. Promote good hand hygiene
    - Wash hands frequently with soap and running water for a minimum of 20 s
    - If available, alcohol-based hand sanitizers (≥70% ethanol) can be used as an adjunct immediately following proper hand washing with soap and water
2. After initial cleaning to remove soiling, disinfect potentially contaminated environmental surfaces
    - Use a chlorine bleach solution with a concentration of 1,000–5,000 ppm [1:50–1:10 dilution of household bleach (5.25%)] or other Environmental Protection Agency (EPA)–approved disinfectant[b]
    - In healthcare settings, cleaning products and disinfectants used should be EPA-registered and have label claims for use in health care
    - Personnel performing environmental services should adhere to the manufacturer's instructions for dilution, application, and contact time
3. Exclude ill staff in sensitive positions (e.g., food, child-care, and patient-care workers) until 48–72 h after symptom resolution
    - In closed or institutional settings (e.g., long-term care facilities, hospitals, cruise ships), isolate ill residents/patients/passengers, until 24–48 h after symptom resolution
    - In licensed food establishments, approval from the local regulatory authority may be necessary before reinstating a food employee following a required exclusion
4. Reinforce effective preventive controls and employee practices such as eliminating bare hand contact with ready-to-eat foods and the proper cleaning and sanitizing of equipment and surfaces
5. Report all outbreaks of acute gastroenteritis to state and local health departments in accordance with local regulations, and the health department should then report to the CDC via the National Outbreak Reporting System (NORS)

[a] Agents registered as effective against norovirus by EPA are available at http://www.epa.gov/oppad001/list_g_norovirus.pdf. Evidence for efficacy against norovirus usually is based on studies using feline calicivirus (FCV) as a surrogate. However, FCV and norovirus exhibit different physiochemical properties, and whether inactivation of FCV reflects efficacy against norovirus is unclear
[b] Modified from CDC (2011)

5.2 Postexposure Prophylaxis (PEP)

Postexposure prophylaxis (PEP) refers to a health professional trying to prevent or provide early treatment against a disease after someone has already been exposed to the pathogen. For many pathogens (such as norovirus), PEP is not an option. Administering vaccinations and immunoglobulin are examples of these weapons against HAV transmission. If the risk of HAV transmission is present, the patrons may be notified via mass media channels and may be offered PEP. Large PEP interventions may include assessment of HAV risk transmission by restaurant inspection and food-handling practices, risk communication, media releases, interviews, setting up immunization clinics, and surveillance of secondary cases. Most people do not get sick when someone at a restaurant has hepatitis A. However, if an infected food handler is infectious and has poor hygiene, the risk increases for patrons of that restaurant. In such cases, health officials might try to identify patrons and provide hepatitis A vaccine or immune globulin if they can find them within 2 weeks of exposure. Persons who recently have been exposed to HAV and who were not previously vaccinated against hepatitis A should be administered a single dose of single-antigen vaccine or IG[54] as soon as possible. For healthy individuals aged 12 months–40 years, ACIP recommends single-antigen hepatitis A vaccine at the age-appropriate dose, which is preferred to IG because of the vaccine's advantages (including long-term protection, ease of administration, and higher availability) (CDC 2007b). Information about vaccine performance in persons older than 40 years is not available. Therefore, IG is preferred in this age group, but if IG is not available, vaccine can be used. IG also should be used for children younger than 12 months, immunocompromised persons, persons with chronic liver disease, and person for whom vaccine is contraindicated. The second dose of vaccine should be administered according to the vaccine manufacturer's licensed schedule.

5.3 Food-Handling Hygiene (Processing, Preparing, Serving)

People working with food who are sick with norovirus gastroenteritis are a particular risk to others, because they handle the food and drinks many other people will consume. Because so few viral particles are necessary to cause infection and the virus is so environmentally stable, a sick food handler can easily contaminate the food he or she is handling. Many of those eating the contaminated food may become ill, causing an outbreak.

The U.S. *Food Code* requires that food handlers and preparers with acute gastroenteritis be excluded from food processing, preparation, and serving activities for at least 1 day after the symptoms have resolved. Many local and state health departments require that food handlers and preparers with gastroenteritis not handle food for other people for 2 or 3 days after the symptoms have resolved. In addition, because the virus continues to be present in the stool for as long as 2–3 weeks after the person feels better, strict hand washing after using the bathroom and before handling food items is important in preventing the spread of this virus. Food handlers who were recently sick can be given different duties in the restaurant so that they do not have to handle food (for example, taking reservations and hosting).

Although both norovirus and hepatitis A infections and outbreaks have been clearly associated with the consumption of oysters and other shellfish (David et al. 2007; Hewitt and Greening 2004; Huppatz et al. 2008), such outbreaks are likely to occur with prepared salads, sandwiches, bakery products, and other foods that are handled and contaminated just before serving by infected food handlers (Greig et al. 2007). It is likely that outbreaks are more frequent than currently reported because norovirus and hepatitis A infections are relatively prevalent, and bare-handed contact with meat and produce during processing, preparing, and even serving is common. Once a food product is contaminated by an infected food worker, the viruses do not replicate further, unlike bacteria, and simply rest on the food. Nonetheless, millions of viral particles are shed and widely disseminated during a symptomatic infection, and relatively few are necessary to cause infection.

Asymptomatic and nonjaundiced HAV-infected persons, especially children, are an important source of HAV transmission. Asymptomatic food handlers who are IgM anti-HAV positive are sometimes identified during investigations and may potentially transmit the infection. Infected persons are highly infectious during the 2-week period before the onset of jaundice or elevation of liver enzymes, when the concentration of virus in stool is highest, while the virus concentration in stool declines after jaundice appears.

Thorough cooking will destroy norovirus, although some evidence points to resistance to temperatures up to 140 °F for 30 min (Dolin et al. 1972; McDonnell et al. 1997). Although direct study of the physical inactivation profile of norovirus is limited by the inability to culture the virus in the laboratory, such studies have been performed using surrogate viruses demonstrating inactivation at

145 °F after less than 30 s (Cannon et al. 2006). Noroviruses can survive refrigeration and freezing and can persist on hard food preparation surfaces for at least a week at room temperature (D'Souza et al. 2006). Similar to norovirus, HAV can live in the environment for months, depending on the conditions. HAV is killed by heating to 185°F (85°C) for 1 min (CDC 2009b); however, the virus can still be spread from cooked food if it is contaminated after cooking. HAV can survive on the surface of produce (lettuce, carrots, and fennel) for 4 to 9 days. Produce washing significantly reduces the level but does not fully eliminate detectable HAV (Croci et al. 2002).

Studies of oysters have shown that norovirus binds specifically to carbohydrate receptors related to human blood group A antigens in the oyster intestine (Le Guyader et al. 2006; Tian et al. 2006). This implies that purifying practices, such as placing oysters in clean water for some time to allow them to attempt to filter out impurities and pathogens and external power washing, are unlikely to decontaminate oysters. Thus, raw oysters can pose a risk of foodborne illness even if obtained from reputable sources, although appropriate documentation on the origin of oysters should be kept by food suppliers to facilitate rapid public health investigation in the event of an outbreak. All fresh fruits and raw vegetables should be washed in water thoroughly to remove soil and other contaminants (some chemicals may be used as specified under CFR 173.315, pp. 140–141).

The appropriate disposal of sewage and soiled diapers also helps to reduce the spread of norovirus and prevent illness. In small home-based catering businesses or family-operated restaurants, sick children should be excluded from food preparation areas. Hand gels formulated with alcohol may also be helpful, although washing with soap and water for at least 20 s remains the preferred method of hand hygiene (USDA 1999).

Most reported outbreaks of foodborne hepatitis A were linked to HAV-infected food handlers as a source of infection, or to a person who prepared food for social events. A food handler with hepatitis A can infect several hundred patrons. The Committee on Control of Foodborne Illnesses noted, in its report on 816 outbreaks where food workers were implicated, that hepatitis A was the second most common virus (after norovirus), causing 84 outbreaks involving 5,046 cases during 1927–2006 (Greig et al. 2007). Between 1992 and 2000, among 38,881 adults with hepatitis A, 8% were identified as food handlers, including 13% of the 3,292 persons aged 16–19 (Fiore 2004).This can be explained by the large number of persons of this age group employed by the food and beverage serving, and related, service industry (7.4 million in 2006) (Anonymous 2002). Although food handlers are not at higher risk of hepatitis A because of their occupation, they may represent a vulnerable population with a relatively lower income and young age who have a demonstrated higher incidence of hepatitis A (CDC 1999).

5.4 Hygienic Food Growing and Harvesting

The investigation of foodborne disease outbreaks has led to the development, implementation, and ongoing refinement of prevention and control measures in the food industry. Public health officials and regulatory agencies identify critical control points in the complex path from farm-to-table that can be monitored to reduce contamination by foodborne pathogens. Improvements at all levels of the food continuum (e.g., farm, processing plant, distribution center, restaurant, home) are expected to reduce the overall contamination in the food supply. Unlike bacterial and parasitic agents of foodborne disease, most foodborne viruses are not harbored by animals, and animal products are thereby not directly infected. However, any situation in which human fecal material comes in contact with food items can present an opportunity for contamination with foodborne viruses. At the farm level, this can occur through the irrigation of fresh produce with sewage contaminated water, handling by infected produce pickers during harvesting, or surface contamination of sorting and processing equipment.

Waterborne outbreaks of norovirus are believed to be rare and occur only during breakdown in the chlorination of drinking water (Kukkula et al. 1999) or recreational water (Podewils et al. 2006). Similarly, waterborne outbreaks of HAV are less common than foodborne outbreaks in the United States (Fiore 2004). In contrast, waterborne outbreaks play a major role in the transmission of HEV in developing countries, as mentioned previously. Nevertheless, unregulated and untreated water supplies are often used in agricultural settings and may pose a risk for contamination by all of the foodborne viruses discussed, particularly when used for produce that does not undergo disinfection after harvesting.

Per capita, total fruit and vegetable availability reached 687 lb in 2005, up 110 lb, or 19%, since 1970 (Wells and Buzby 2009). Fresh fruit and vegetable consumption has increased, and despite the nutritional health benefits of these food items, they are particularly vulnerable to contamination with foodborne viruses. Furthermore, people increasingly depend upon food prepared outside the home and seek qualities such as taste, variety, convenience, and enjoyment with regard to food consumption (Frazão et al. 2008). To meet these changes in food preferences and demands for convenience, food distribution has became more centralized and globalized. Produce (i.e., fruits and vegetable) can be collected from several farms in disparate geographic regions, commingled, repackaged, and later shipped nationwide from one distribution firm. Several multistate outbreaks have occurred because tainted food was distributed to several states nationwide, as exemplified by the largest foodborne outbreak in the United States reported to date, which involved tomatoes and peppers contaminated with *Salmonella* (CDC 2008). The contemporary food industry network is complex and requires collaboration among regulatory agencies, local and federal public health agencies, and the food industry to both prevent and detect viral contamination in food products.

The types of foods implicated in the transmission of HAV include shellfish, salads, fresh produce, sandwiches, vegetables, fruits, reconstituted orange juice, ice cream, cheese, rice pudding, iced cake, custard, milk, bread, cookies, and other raw or undercooked foods (Cliver 1997). Two large hepatitis A outbreaks are described below. In 1997 frozen strawberries were implicated as the source of a multistate outbreak involving 262 persons. Another large outbreak of hepatitis A occurred in 2003 in four states, with 601 persons sickened, three deaths, and at least 124 people hospitalized due to the consumption of green onions contaminated with HAV imported to the United States (Wheeler et al. 2005).

5.5 *Cleanup of Infected Materials*

People who are sick with norovirus illness can produce infectious vomit or stool; therefore, particular care must be taken to clean and disinfect surfaces touched by, or near, these substances using bleach solution and then rinse them thoroughly (USDA 1999) (Boxes 1, 2, and 3). Furthermore, food items that may have become contaminated with norovirus should be thrown out. Linens (clothing, napkins, tablecloths, etc.) contaminated to any extent with vomit or stool should be promptly washed at a high temperature.

Environments and fomites contaminated with norovirus remain persistent and continuing sources of norovirus infections. Common sources are often never confirmed in the course of an investigation but can range from kitchen tools or implements, to salt/pepper shakers, to door handles or handrails, etc. Environmental persistence has also been demonstrated in an outbreak of schoolchildren who simply sat in the same theater seats where ill children from another school had been seated on the previous day: The same norovirus strain was identified in these cases (Evans et al. 2002). The U.S. Environmental Protection Agency has approved over two dozen disinfectants for use against norovirus, based primarily on studies using feline *Calicivirus* as a surrogate (Anonymous 2009). More recently, murine norovirus has been suggested as a potentially more appropriate surrogate for human norovirus (Cannon et al. 2006), and an evaluation of disinfectants using this virus is currently ongoing. Quaternary ammonium and ethyl alcohol are generally not considered to be effective against norovirus. Due to the environmental resiliency of norovirus, the following procedure developed for healthcare settings has been shown to be most effective (Barker et al. 2004):

- The contaminated area must first be wiped clean with detergent and water and then disinfected with exposure to 5,000 parts per million (ppm) hypochlorite solution for at least 5 min. This solution is equivalent to about 1½ cups of household chlorine bleach in 1 gal of water. This concentration is much higher than recommended for sanitizing food contact surfaces in the U.S. Food Code, and may damage many materials, so this disinfection procedure requires care.
- If the area is a food contact area, this disinfection procedure must also include the following step: The disinfection solution must be rinsed off with clear water and wiped down with a sanitizing bleach solution, consisting of 200 ppm chlorine bleach.
- Environmental disinfection should include all heavy hand contact surfaces such as food preparation surfaces, self-service utensil handles, faucets, tables, chairs, counters, door handles, push plates, railings, elevator buttons, telephones, keyboards, vending machine keyboards, pens, pencils, casino chips, cards, slot machines, sports equipment, etc., as applicable.
- Public restroom surfaces, including faucet handles, soap dispensers, stall doors and latches, toilet seats/handles, and towel dispensers, are also important to disinfect as applicable.
- When norovirus contamination is suspected, cleaning procedures that increase the aerosolization of norovirus should not be utilized, such as vacuuming carpets or buffing hard surface floors.
- HAV infection can be prevented through careful hand washing with soap and proper drying, particularly after using the toilet and before handling food.

These practices should be strictly adhered to by all staff to reduce the risk of infection, including the use of disinfectant hand sanitizers and proper use of gloves. Written guidelines on cleaning and disinfection should be developed and implemented for use by the staff.

Depending on various conditions, HAV can persist outside the host and in the environment for months (McCaustland et al. 1982). For example, HAV can survive in water and sewage for months, while survival on *fomites* is more than 7 days at low room humidity at 41°F (5°C). The virus is killed by heating to 185°F (85°C) for 1 min. Disinfection of surfaces with a 1:100 dilution of household bleach in water or cleaning solutions containing quaternary ammonium and/or HCl is effective in inactivating HAV. HAV is also susceptible to 2% glutaraldehyde, formaldehyde (Favero and Bond 1998; Mbithi et al. 1990; Sattar et al. 2000).

Box 1 Outbreak Investigation #1: Come Together

In October 2006, an outbreak of acute gastroenteritis was reported among attendees at a family reunion. The information was promptly relayed to state and local public health authorities, which initiated an investigation. The objectives were to identify the cause and source of the outbreak, to identify risk factors for illness, and to recommend appropriate control measures. The preliminary investigation consisted of open-ended interviews with 11 attendees using a hypothesis-generating questionnaire, from which a comprehensive list of attendees and foods served was developed. A total of 53 attendees were identified, including residents of six states, requiring collaboration with multiple public health agencies. There were reports that a child was present at the reunion with symptoms of vomiting and diarrhea and other immediate family members of that child had been previously ill. The pot-luck meal that was served included 31 food items, and preliminary responses from attendees implicated a store-bought rotisserie chicken.

(continued)

Box 1 (continued)

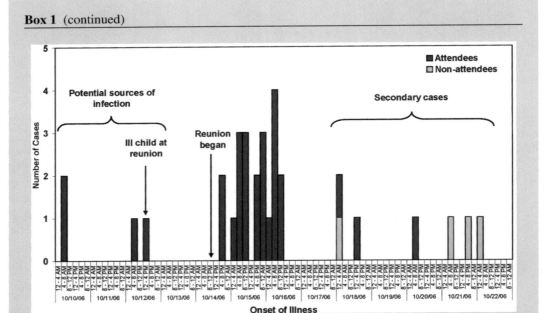

Fig. 3 Epidemic curve of acute gastroenteritis outbreak among attendees at a family reunion ($n=28$), including secondary cases who were nonattendees ($n=4$)

A cohort study was performed using a brief, standardized questionnaire administered to all reunion attendees. Questions included illness onset and symptoms, consumption of specific foods, contact with ill persons, and identification of secondary cases not present at the reunion. A concurrent laboratory investigation was performed to test stool samples from ill attendees for common *enteric* bacterial and viral pathogens. Of the 53 attendees, 48 (91%) were interviewed, of whom 28 reported illness, for an attack rate of 58%. Six of the ill attendees sought medical care, although none were hospitalized. The median incubation period was 36 h, with a median duration of illness of 54 h. Symptoms were characterized predominantly by diarrhea, vomiting, and abdominal cramps. Most cases occurred within 48 h of the reunion, although four attendees were ill prior to the reunion and three attendees had the illness onset occur more than 72 h after the reunion (Fig. 3). Two food items were implicated as significant risk factors for illness: store-bought rotisserie chicken and homemade scalloped potatoes. Direct contact with an ill person at the reunion was also identified as a significant risk factor. Laboratory analyses identified norovirus (Genogroup II) by RT-PCR in 12 (92%) of the 13 stool specimens submitted. Genetic sequencing of the PCR products identified two different strains among the ill attendees, suggesting multiple concurrent sources of infection.

This investigation concluded that a combination of person-to-person and foodborne transmission of two strains of norovirus, likely introduced by persons from two different states and subsequently at least two food items, was the probable cause of these illnesses. The rotisserie chicken was brought by the immediate family of the child that was symptomatic at the reunion, which also included two other attendees with illness prior to the reunion who may still have

(continued)

Box 1 (continued)

been shedding virus. The chicken and scalloped potatoes were likely contaminated by these ill persons during transport, preparation, and self-service at the reunion. Illness was also propagated through direct person-to person spread, affecting household contacts who did not attend the reunion. The convergence of two norovirus strains in this outbreak occurred during a season of increased norovirus activity nationwide, attributable to the emergence of new viral strains. The conclusions of this investigation highlight many of the challenges of investigating and controlling norovirus outbreaks. Particularly during periods of peak norovirus activity such as described in this investigation, appropriate hand washing to prevent direct transmission and contamination of food, along with the exclusion of ill persons from social gatherings, are critical for prevention.

Box 2 Outbreak Investigation #2: Strawberry Fields

In 1997, a total of 242 cases of hepatitis A associated with the consumption of frozen strawberries were reported from 36 schools in Michigan and Maine. The genetic sequences of viruses from 126 patients in Michigan and Maine were identical to one another and to those from five patients in Wisconsin and seven patients in Arizona. All of these patients attended schools where frozen strawberries from the same processor had been served. Additionally, HAV with identical sequences was isolated from two more patients from Louisiana who had consumed commercially prepared products containing frozen strawberries from the same processor. Viral genetic analyses enabled the investigators to find apparently sporadic cases in other states. The strawberries implicated in this outbreak were grown in Mexico, processed and frozen in a California plant, and distributed through the Department of Agriculture for school-lunch programs and through distributors for commercial use. The processor shipped frozen strawberries to the Michigan and Maine school-lunch programs. In each state, containers of frozen strawberries were opened in the school kitchens, thawed, and incorporated into shortcakes or served in cups. Officials of the California Department of Health Services, the Food and Drug Administration (FDA), and the Department of Agriculture surveyed the plant where the strawberries had been processed, packed, and frozen. The investigation did not identify any record of illness consistent with hepatitis A among the employees during the time the strawberries were processed. Strawberries were washed in a chlorine solution of 12 ppm, mechanically sliced, combined with a sucrose solution, packed, and frozen. Hand contact with the berries was limited to the rejection of unacceptable berries as they passed on the conveyor belt. FDA officials visited three of the four growing fields in Mexico and found that water for drip irrigation was piped from a river and filtered in sand tanks. Few slit latrines were available for use by pickers. The only hand-washing facilities were on trucks that circulated through the fields. The pickers did not wear gloves and removed the strawberry stems with their fingernails. No records were kept of illnesses among the pickers.

Box 3 Outbreak Investigation #3: Green Onion (Wheeler et al. 2005)

An unusually large foodborne outbreak of hepatitis A in 2003 involved 601 patients, of whom three persons died and at least 124 were hospitalized. An intensive outbreak investigation revealed that of 425 patients who recalled a single dining date at the restaurant, 356 (84%) had dined at Restaurant A in Pennsylvania between October 3 and October 6. Among 240 patients in the case–control study, 218 had eaten mild salsa (91%), as compared with 45 of 130 controls (35%) (odds ratio, 19.6; 95% confidence interval, 11.0–34.9), which means that those who became sick were almost 20 times more likely have eaten salsa than those who did not became sick. All restaurant workers were tested for recent infection. Infected persons were not the source of the outbreak. The genetic sequences of hepatitis A virus from all 170 patients who were tested were identical.

Mild salsa contained green onions grown in Mexico. It was prepared in large batches at the restaurant and provided to all patrons. Tainted green onions were included in large batches and served to all customers, which contributed to the size of the outbreak.

Green onions were purchased by Restaurant A from a distributor, which purchased them from a produce supplier. Shipping boxes contained 4 kg (8.5 lb) of green onions packed on ice in bundles of six to eight onions each. Restaurant A workers placed bundles into $30 \times 15 \times 23$-cm ($12 \times 6 \times 9$-in.) metal pans, which were stored in the refrigerator for up to 5 days. Green onions were chopped as needed. Bundles were rinsed with tap water, the roots were cut, the rubber band around the bundle was removed, and the onions were chopped with the use of an electric dicer. Chopped onions were refrigerated in plastic containers for up to 2 days. The dicer was used solely for dicing green onions and was cleaned each day. The dicer was discarded before the investigation commenced. In this large foodborne outbreak of hepatitis A, illness was strongly associated with the consumption of menu items containing green onions at a single restaurant in Beaver County, Pennsylvania. Foodservice workers at the restaurant were tested for recent infection with hepatitis A virus, and infected workers were shown not to be the source of the outbreak. The implicated green onions were apparently contaminated with hepatitis A virus before or during packing into shipping boxes on farms in northern Mexico during September 2003.

When green onions were implicated as a source, the FDA issued both an import ban on green onions from four farms (including the two that supplied green onions to Restaurant A) in Mexico and a consumer alert.

The unprecedented size of the Pennsylvania outbreak and the high attack rate observed may have been a result of several contributing factors. First, nearly 2,000 people were estimated to have dined at Restaurant A during the 4 days of the peak exposure period. All were offered mild salsa, the food item most strongly associated with illness. Uncooked green onions were also used in more than 50 other menu items. Second, rinsing green onions while they are still bundled could have contributed to the size of the outbreak. Uncontaminated and contaminated green onions stored in a common container were intermingled and might have further increased the chance of cross-contamination. In addition, liquid on the surface of onions from rinsing or from melting ice might have facilitated the diffusion of the hepatitis A virus. Contaminated onions were then dispersed throughout the mild salsa. Although none of the reported food-handling practices of Restaurant A employees was specifically linked to transmission, many of the preparation practices could have resulted in cross-contamination.

Contamination of green onions could occur by contact with workers (or workers' children) who are infected with hepatitis A virus during harvesting and preparation or by contact with contaminated water during irrigation, rinsing, processing, cooling, and icing.

Summary

Foodborne viruses can cause serious illnesses and are among the most common sources of foodborne disease outbreaks. Controlling norovirus remains challenging since the viruses have a low infectious dose, contaminate widely, persist in the environment, and are more resistant to disinfectants than bacteria. While hepatitis A control measures are supplemented by the existence of an approved and effective vaccine, the long incubation period makes the identification of disease outbreaks and infected food handlers more difficult to document. Nonetheless, standard hygienic precautions such as regular hand washing and care in the disposal of contaminated materials are instrumental in preventing foodborne viral infections.

Unlike bacterial and parasitic foodborne illnesses, the sources of viral contamination are predominantly from human handling and food preparation, and viral contamination does not usually arise from within the food or animal product itself. Recent illustrations exist of the modern complexities in investigating and controlling foodborne illnesses in our increasingly far-reaching and intertwined networks of food production and transport, and these factors complicate the identification of contamination sources. Therefore, the most effective prevention strategy is for everyone involved in the farm-to-table continuum to proactively avoid viral contamination of food and to ensure that food handlers who are actively infected by these viruses do not handle or prepare food.

References

Anonymous. 2000. Japan ministry of health and welfare, national institute of infectious diseases. 2000. An outbreak of group A rotavirus infection among adults from eating meals prepared at a restaurant, April 2000 – Shimane. *Infectious Agents Surveillance Report* 21: 145.

Anonymous. 2002. Bureau of Labor Statistics. Occupational outlook handbook, 2002–03 edition, food and beverage serving and related workers. Available at: http://www.bls.gov/oco/ocos162.htm. Accessed 6 Feb 2009.

Anonymous. 2009. Selected EPA-registered disinfectants. http://www.epa.gov/oppad001/chemregindex.htm. Accessed 15 Dec 2010.

Balayan, M.S., et al. 1983. Evidence for a virus in non-A, non-B hepatitis transmitted via the fecal-oral route. *Intervirology* 20: 23–31.

Barker, J., I.G. Vipond, and S.F. Bloomfield. 2004. Effects of cleaning and disinfection in reducing the spread of norovirus contamination via environmental surfaces. *The Journal of Hospital Infection* 58: 42–49.

Barzilay, E.J, Malek, M.A., Kramer, A., et al. 2006. An outbreak of gastroenteritis among rafters on the Colorado River caused by norovirus contamination of commercially packaged deli meat. Paper presented at the 5th International Conference on Emerging Infectious Diseases, March 19–22, Atlanta.

Blanton, L.H., S.M. Adams, R.S. Beard, G. Wei, S.N. Bulens, M.A. Widdowson, R.I. Glass, and S.S. Monroe. 2006. Molecular and epidemiologic trends of Caliciviruses associated with outbreaks of acute gastroenteritis in the US, 2000–2004. *The Journal of Infectious Diseases* 193: 413–421.

Cannon, J.L., E. Papafragkou, G.W. Park, J. Osborne, L.A. Jaykus, and J. Vinje. 2006. Surrogates for the study of norovirus stability and inactivation in the environment: A comparison of murine norovirus and feline Calicivirus. *Journal of Food Protection* 69: 2761–2765.

CDC. 2009a. Surveillance for foodborne disease outbreaks, US, 2006. *Morbidity and Mortality Weekly Report* 58: 609–615.

CDC. 2009b. Centers for disease control and prevention. Available at: http://www.cdc.gov/hepatitis/A/aFAQ.htm#prevention. Accessed 10 Jan 2009.

CDC. 2010. Surveillance for foodborne disease outbreaks – US, 2007. *Morbidity and Mortality Weekly Report* 59: 973–979.

CDC. 2011. Norovirus outbreak management and disease prevention – Updated CDC guidelines. *Morbidity and Mortality Weekly Report* 60 (RR-3): 1–18.

CDC (Centers for Disease Control and Prevention). 2007a. Norovirus activity – US, 2006–2007. *Morbidity and Mortality Weekly Report* 56: 842–846.

CDC (Centers for Disease Control and Prevention). 2007b. Update: Prevention of hepatitis A after exposure to hepatitis A virus and in international travelers. Updated recommendations of the Advisory Committee on Immunization Practices (ACIP). *Morbidity and Mortality Weekly Report* 56: 1080–1084.

CDC (Centers for Disease Control and Prevention). 1999. Prevention of hepatitis A through active or passive immunization: Recommendations of the advisory committee on immunization practices (ACIP). *Morbidity and Mortality Weekly Report* 48(RR-12): 1–37.

CDC (Centers for Disease Control and Prevention). 2008. Outbreak of *Salmonella* serotype Saintpaul infections associated with multiple raw produce items – US, 2008. *Morbidity and Mortality Weekly Report* 57: 929–934.

CDC (Centers for Disease Control and Prevention). 2000. Foodborne outbreak of Group A rotavirus gastroenteritis among college students – District of Columbia, March–April 2000. *Morbidity and Mortality Weekly Report* 49: 1131–1133.

Chastel, C., and G. Charmot. 2004. Bacterial and viral epidemics of zoonotic origin: the role of hunting and cutting up wild animals. *Bulletin de la Société de Pathologie Exotique (1990)* 97: 207–212.

Cliver, D.O. 1997. Virus transmission via food. *World Health Statistics Quarterly* 50: 90–101.

Croci, L., D. De Medici, C. Scalfaro, A. Fiore, and L. Toti. 2002. The survival of hepatitis A virus in fresh produce. *International Journal of Food Microbiology* 73: 29–34.

D'Souza, D.H., A. Sair, K. Williams, E. Papafragkou, J. Jean, C. Moore, and L. Jaykus. 2006. Persistence of Calicivirus on environmental surfaces and their transfer to food. *International Journal of Food Microbiology* 108: 84–91.

Dalton, H.R., R. Bendall, S. Ijaz, and M. Banks. 2008. Hepatitis E: an emerging infection in developed countries. *The Lancet Infectious Diseases* 8: 698–709.

David, S.T., L. McIntyre, L. MacDougall, D. Kelly, S. Liem, K. Schallie, et al. 2007. An outbreak of norovirus caused by consumption of oysters from geographically dispersed harvest sites, British Columbia, Canada, 2004. *Foodborne Pathogens and Disease* 4: 349–358.

Dolin, R., N.R. Blacklow, H. DuPont, R.F. Buscho, R.G. Wyatt, J.A. Kasel, R. Hornick, and R.M. Chanock. 1972. Biological properties of Norwalk agent of acute infectious nonbacterial gastroenteritis. *Proceedings of the Society for Experimental Biology and Medicine* 140: 578–583.

Duizer, E., K.J. Schwab, F.H. Neill, R.L. Atmar, R.P. Koopmans, and M.K. Estes. 2004. Laboratory efforts to cultivate noroviruses. *The Journal of General Virology* 85: 79–87.

Estes, M.K., B.V. Prasad, and R.L. Atmar. 2006. Noroviruses everywhere: has something changed? *Current Opinion in Infectious Diseases* 19: 467–474.

Evans, M.R., R. Meldrum, W. Lane, D. Gardner, C.D. Ribeiro, C.L. Gallimore, and D. Westmoreland. 2002. An outbreak of viral gastroenteritis following environmental contamination at a concert hall. *Epidemiology and Infection* 129: 355–360.

Falkenhorst, G., L. Krusell, M. Lisby, S.B. Madsen, B. Bottiger, and K. Molbak. 2005. Imported frozen raspberries cause a series of norovirus outbreaks in Denmark, 2005. *Euro Surveillance* 10: E050922 2.

Favero, M.S., and W.W. Bond. 1998. Disinfection and sterilization. In *Viral hepatitis*, ed. A.J. Zuckerman and H.C. Thomas, 627–635. London: Churchill Livingstone.

Feagins, A.R., T. Opriessnig, D.K. Guenette, P.G. Halbur, and X.J. Meng. 2007. Detection and characterization of infectious hepatitis E virus from commercial pig livers sold in local grocery stores in the USA. *The Journal of General Virology* 88: 912–917.

Fiore, A.E. 2004. Hepatitis A transmitted by food. *Clinical Infectious Diseases* 38: 705–715.

Fiore, A.E., A. Wasley, and B.P. Bell. 2006. Prevention of hepatitis A through active or passive immunization: recommendations of the Advisory Committee on Immunization Practices (ACIP). *Morbidity and Mortality Weekly Report* 55(RR-7): 1–23.

Frazão, E., Meade, B., and Regmi, A. 2008. Converging patterns in global food consumption and food delivery systems. http://www.ers.usda.gov/AmberWaves/February08/Features/CovergingPatterns.htm. Accessed 15 Dec 2010.

Gaulin, C.D., D. Ramsay, P. Cardinal, and M.-A. D'Halevyn. 1999. Epidemic of gastroenteritis of viral origin associated with eating imported raspberries. *Canadian Journal of Public Health* 90: 37–40.

Gingrich, G.A., S.C. Hadler, H.A. Elder, et al. 1983. Serologic investigation of an outbreak of hepatitis A in a rural day-care center. *American Journal of Public Health* 73: 1190–1193.

Glikson, M., E. Galun, R. Oren, R. Tur-Kaspa, and D. Shouval. 1992. Relapsing hepatitis A. Review of 14 cases and literature survey. *Medicine (Baltimore)* 71: 14–23.

Graham, D.Y., X. Jiang, T. Tanaka, A.R. Opekun, H.P. Madore, and M.K. Estes. 1994. Norwalk virus infection of volunteers: new insights based on improved assays. *The Journal of Infectious Diseases* 170: 34–43.

Green, M.S., C. Block, and P.E. Slater. 1989. Rise in the incidence of viral hepatitis in Israel despite improved socioeconomic conditions. *Reviews of Infectious Diseases* 11: 464–469.

Green, K., G. Belliot, J. Taylor, J. Valdesuso, J. Lew, A. Kapikian, et al. 2002. A predominant role for Norwalk-like viruses as agents of epidemic gastroenteritis in Maryland nursing homes for the elderly. *The Journal of Infectious Diseases* 185: 133–146.

Greig, J.D., E.C. Todd, C.A. Bartleson, and B.S. Michaels. 2007. Outbreaks where food workers have been implicated in the spread of foodborne disease. Part 1. Description of the problem, methods and agents involved. *Journal of Food Protection* 70: 1752–1761.

Hennessy, E.P., A.D. Green, M.P. Connor, R. Darby, and P. MacDonald. 2003. Norwalk virus infection and disease is associated with ABO histo-blood group type. *The Journal of Infectious Diseases* 188: 176–177.

Hewitt, J., and G.E. Greening. 2004. Survival and persistence of norovirus, hepatitis A virus, and feline Calicivirus in marinated mussels. *Journal of Food Protection* 67: 1743–1750.

Huppatz, C., S.A. Munnoch, T. Worgan, T.D. Merritt, C. Dalton, P.M. Kelly, et al. 2008. A norovirus outbreak associated with consumption of NSW oysters: implications for quality assurance systems. *Communicable Diseases Intelligence* 32: 88–91.

Hutson, A.M., R.L. Atmar, D.Y. Graham, and M.K. Estes. 2002. Norwalk virus infection and disease is associated with ABO histo-blood group type. *The Journal of Infectious Diseases* 185: 1335–1337.

Hutson, A.M., R.L. Atmar, and M.K. Estes. 2004. Norovirus disease: changing epidemiology and host susceptibility factors. *Trends in Microbiology* 12: 279–287.

Jiang, X., J. Wang, D.Y. Graham, and M.K. Estes. 1992. Detection of Norwalk virus in stool by polymerase chain reaction. *Journal of Clinical Microbiology* 30: 2529–2534.

Johnson, P.C., Mathewson, J.J., DuPont, H.L., and Greenberg, H.B. 1990. Multiple challenge study of host susceptibility to Norwalk gastroenteritis in US adults. *Journal of Infectious Diseases* 161: 18–21.

Kamar, N., J. Selves, et al. 2008. Hepatitis E virus and chronic hepatitis in organ-transplant recipients. *The New England Journal of Medicine* 358: 811–817.

Koff, R.S. 1992. Clinical manifestations and diagnosis of hepatitis A virus infection. *Vaccine* 10(Suppl 1): S15–S17.

Krugman, S., and J.P. Giles. 1970. Viral hepatitis: new light on an old disease. *Journal of the American Medical Association* 212: 1019–1029.

Kukkula, M., L. Maunula, E. Silvennoinen, and C.H. von Bonsdorff. 1999. Outbreak of viral gastroenteritis due to drinking water contaminated by Norwalk-like viruses. *The Journal of Infectious Diseases* 180: 1771–1776.

Le Guyader, L.F., R.L. Atmar, A.M. Hutson, M.K. Estes, N. Ruvoen-Clouet, M. Pommepuy, and J. Le Pendu. 2006. Norwalk virus-specific binding to oyster digestive tissues. *Emerging Infectious Diseases* 12: 931–936.

Lednar, W.M., S.M. Lemon, J.W. Kirkpatrick, et al. 1985. Frequency of illness associated with epidemic hepatitis A virus infections in adults. *American Journal of Epidemiology* 122: 226–233.

Lopman, B.A., M.H. Reacher, I.B. Vipond, J. Sarangi, and D.W. Brown. 2004. Clinical manifestation of norovirus gastroenteritis in health care settings. *Clinical Infectious Diseases* 39: 318–324.

Luby, S.P., M. Rahman, M.J. Hossain, L.S. Blum, M.M. Husain, E. Gurley, et al. 2006. Foodborne transmission of Nipah virus, Bangladesh. *Emerging Infectious Diseases* 12: 1888–1894.

Malek, M., E. Barzilay, A. Kramer, B. Camp, L.A. Jaykus, B. Escudero-Abarca, et al. 2009. Outbreak of norovirus infection among river rafters associated with packaged delicatessen meat, Grand Canyon, 2005. *Clinical Infectious Diseases* 48: 31–37.

Marks, P.J., I.B. Vipond, D. Carlisle, D. Deakin, R.E. Fey, and E.O. Caul. 2000. Evidence for airborne transmission of Norwalk-like virus (NLV) in a hotel restaurant. *Epidemiology and Infection* 124: 481–487.

Marks, P.J., I.B. Vipond, F.M. Regan, K. Wedgwood, R.E. Fey, and E.O. Caul. 2003. A school outbreak of Norwalk-like virus: evidence for airborne transmission. *Epidemiology and Infection* 131: 727–736.

Matsuda, H., K. Okada, K. Takahashi, and S. Mishiro. 2003. Severe hepatitis E virus infection after ingestion of uncooked liver from a wild boar. *The Journal of Infectious Diseases* 188: 944.

Mbithi, J.N., V.S. Springthorpe, and S.A. Sattar. 1990. Chemical disinfection of hepatitis A virus on environmental surfaces. *Applied and Environmental Microbiology* 56: 3601–3604.

McCaustland, K.A., W.W. Bond, D.W. Bradley, et al. 1982. Survival of hepatitis A virus in feces after drying and storage for 1 month. *Journal of Clinical Microbiology* 16: 957–958.

McDonnell, S., K. Kirkland, W.B. Hlady, C. Aristeguieta, R.S. Hopkins, S.S. Monroe, and R.I. Glass. 1997. Failure of cooking to prevent shellfish-associated viral gastroenteritis. *Archives of Internal Medicine* 157: 111–116.

Moe, C.L., J. Gentsch, T. Ando, et al. 1994. Application of PCR to detect Norwalk virus in fecal specimens from outbreaks of gastroenteritis. *Journal of Clinical Microbiology* 32: 642–648.

Mushahwar, I.K. 2008. Hepatitis E virus: molecular virology, clinical features, diagnosis, transmission, epidemiology, and prevention. *Journal of Medical Virology* 80: 646–658.

Nainan, O.V., G. Xia, G. Vaughan, and H.S. Margolis. 2006. Diagnosis of hepatitis A virus infection: a molecular approach. *Clinical Microbiology Reviews* 19: 63–79.

Okamoto, H., M. Takahashi, and T. Nishizawa. 2003. Features of hepatitis E virus infection in Japan. *Internal Medicine* 42: 1065–1071.

Patel, M.M., M.A. Widdowson, R.I. Glass, K. Akazawa, J. Vinje, and U.D. Parashar. 2008. Systematic literature review of role of noroviruses in sporadic gastroenteritis. *Emerging Infectious Diseases* 14: 1224–1231.

Podewils, L.J., L. Zanardi Blevins, M. Hagenbuch, D. Itani, A. Burns, C. Otto, L. Blanton, S. Adams, S.S. Monroe, M.J. Beach, and M.A. Widdowson. 2006. Outbreak of norovirus illness associated with a swimming pool. *Epidemiology and Infection* 135: 827–833.

Purcell, R.H., and S.U. Emerson. 2008. Hepatitis E: an emerging awareness of an old disease. *Journal of Hepatology* 48: 494–503.

Sattar, S.A., T. Jason, S. Bidawid, and J. Farber. 2000. Foodborne spread of hepatitis A: recent studies on virus survival, transfer and inactivation. *Canadian Journal of Infectious Diseases* 11: 159–163.

Scallan, E., R.M. Hoekstra, F.J. Angulo, et al. 2011. Foodborne illness acquired in the US – major pathogens. *Emerging Infectious Diseases* 17: 7–15.

Shrestha, M.P., R.M. Scott, D.M. Joshi, M.P. Mammen Jr., G.B. Thapa, N. Thapa, K.S. Myint, M. Fourneau, R.A. Kuschner, S.K. Shrestha, M.P. David, J. Seriwatana, D.W. Vaughn, A. Safary, T.P. Endy, and B.L. Innis. 2007. Safety and efficacy of a recombinant hepatitis E vaccine. *The New England Journal of Medicine* 356: 895–903.

Sjogren, M.H., H. Tanno, O. Fay, S. Sileoni, B.D. Cohen, D.S. Burke, et al. 1987. Hepatitis A virus in stool during clinical relapse. *Annals of Internal Medicine* 106: 221–226.

Staes, C.J., T.L. Schlenker, I. Risk, et al. 2000. Sources of infection among persons with acute hepatitis A and no identified risk factors during a sustained community-wide outbreak. *Pediatrics* 106: E54.

Takahashi, K., N. Kitajima, N. Abe, and S. Mishiro. 2004. Complete or near complete sequences of hepatitis E virus genome recovered from a wild boar, a deer and four patients who ate the deer. *Virology* 330: 501–505.

Takahashi, M., T. Tanaka, M. Azuma, E. Kusano, T. Aikawa, T. Shibayama, et al. 2007. Prolonged fecal shedding of hepatitis E virus (HEV) during sporadic acute hepatitis E: evaluation of infectivity of HEV in fecal specimens in a cell culture system. *Journal of Clinical Microbiology* 45: 3671–3679.

Tamura, A., Y.K. Shimizu, T. Tanaka, K. Kuroda, Y. Arakawa, K. Takahashi, et al. 2007. Persistent infection of hepatitis E virus transmitted by blood transfusion in a patient with T-cell lymphoma. *Hepatology Research* 37: 113–120.

Tian, P., A.H. Bates, H.M. Jensen, and R.E. Mandrell. 2006. Norovirus binds to blood group A-like antigens in oyster gastrointestinal cells. *Letters in Applied Microbiology* 43: 645–651.

Todd, E.C., J.D. Greig, C.A. Bartleson, and B.S. Michaels. 2007. Outbreaks where food workers have been implicated in the spread of foodborne disease. Part 2. Description of outbreaks by size, severity and settings. *Journal of Food Protection* 70: 1975–1993.

Turcios, R.M., M.A. Widdowson, A.C. Sulka, P.S. Mead, and R.I. Glass. 2006. Reevaluation of epidemiological criteria for identifying outbreaks of acute gastroenteritis due to norovirus: US, 1998–2000. *Clinical Infectious Diseases* 42: 964–969.

USDA. 1999. U.S. Food Safety and Inspection Service, U.S. Department of Agriculture. Sanitation performance standards compliance guide. Available at: http://www.fsis.usda.gov/OPPDE/RDAD/FRPubs/San_Guide.pdf. Accessed 15 Dec 2010.

Wasley, A., S. Grytdal, and K. Gallagher. 2008. Surveillance for acute viral hepatitis – US, 2006. *Morbidity and Mortality Weekly Report* 57(SS-2): 1–24.

Wells, F.H., and Buzby, J.C. 2009. Dietary Assessment of Major Trends in U.S. Food Consumption. 1970–2005. EIB-33 Economic Information Bulletin Number 33. Available at: http://www.ers.usda.gov/Publications/EIB33/EIB33.pdf. Accessed 6 Feb 2009.

Wheeler, C., T.M. Vogt, G.L. Armstrong, G. Vaughan, A. Weltman, O.V. Nainan, V. Dato, G. Xia, K. Waller, J. Amon, T.M. Lee, A. Highbaugh-Battle, C. Hembree, S. Evenson, M.A. Ruta, I.T. Williams, A.E. Fiore, and B.P. Bell. 2005. An outbreak of hepatitis A associated with green onions. *The New England Journal of Medicine* 353: 890–897.

Widdowson, M.A., S.S. Monroe, and R.I. Glass. 2005a. Are noroviruses emerging? *Emerging Infectious Diseases* 11: 735–737.

Widdowson, M.A., A. Sulka, S.N. Bulens, R.S. Beard, S.S. Chaves, R. Hammond, E.D. Salehi, E. Swanson, J. Totaro, R. Woron, P.S. Mead, J.S. Bresee, S.S. Monroe, and R.I. Glass. 2005b. Norovirus and foodborne disease, US, 1991–2000. *Emerging Infectious Diseases* 11: 95–102.

Yazaki, Y., H. Mizuo, M. Takahashi, T. Nishizawa, N. Sasaki, Y. Gotanda, and H. Okamoto. 2003. Sporadic acute or fulminant hepatitis E in Hokkaido, Japan, maybe foodborne, as suggested by the presence of HEV in pig liver as food. *The Journal of General Virology* 84: 2351–2357.

Zhang, J., et al. 2009. Randomized-controlled phase II clinical trial of a bacterially expressed recombinant hepatitis E vaccine. *Vaccine* 27: 1869–1874.

Zheng, D.P., T. Ando, R.L. Fankhauser, R.S. Beard, R.I. Glass, and S.S. Monroe. 2006. Norovirus classification and proposed strain nomenclature. *Virology* 346: 312–323.

Zheng, D.P., M.A. Widdowson, R.I. Glass, and J. Vinje. 2010. Molecular epidemiology of genogroup II-genotype 4 noroviruses in the US between 1994 and 2006. *Journal of Clinical Microbiology* 48: 168–177.

Part II
Safety of Major Food Products

Safety of Produce

Maha N. Hajmeer and Beth Ann Crozier-Dodson*

1 Introduction

The term "produce," used as a noun, refers to all fresh fruits and vegetable grown for the market. In the last two decades, the consumption of produce has increased in the United States, and the geographic sources and distribution of fresh produce have expanded greatly. Alongside this increase in consumption, there has been an increase in the number of reported produce-associated foodborne illness outbreaks, as documented by public health officials (Tauxe et al. 1997). This chapter will focus on presenting basic terms and information pertaining to the safety of produce. Brief discussions are provided about produce as vehicles of foodborne *pathogens* with an emphasis on bacterial foodborne *pathogens*.

2 Produce Safety

Produce encompasses a wide range of fruits and vegetables. Table 1 shows important examples of common produce and their categories. Depending on the nature of the fruits and vegetables, they may be available to consumers in varying forms. Fruits and vegetables may enter the market whole or in a value-added form, such as fresh-cut, cored, peeled, chopped, sliced, shredded, mixed, washed, or triple-washed. Some fresh or processed products may require preparation in part by the consumer, whereas others are considered ready-to-eat (RTE) foods.

Although there has been an increase in reported foodborne illness outbreaks associated with produce, this does not necessarily mean that these healthy and nutritious foods are less safe than other products. Any food, including those that undergo a microbial inactivation or killing step (e.g., by cooking), has the potential to become contaminated if it is improperly prepared, processed, or handled. For produce to become contaminated, the microbial agent must be pathogenic to humans and must survive and/or proliferate throughout the food chain and the complex distribution system. In addition,

---

*The opinions appearing in this chapter are the responsibility of the authors and not the entities with whom the authors are affiliated.

M.N. Hajmeer (✉)
Food and Drug Branch, California Department of Public Health, Sacramento, CA, USA
e-mail: Maha.Hajmeer@cdph.ca.gov

B.A. Crozier-Dodson
Food Safety Consulting, LLC, Manhattan, KS, USA

Table 1 Common produce and their categories

Produce	Examples
Fruits	Melons, watermelon, cantaloupe, honeydew
Vegetables	Tomatoes, carrots, cabbage, broccoli, cauliflower, mushrooms, green onions
Leafy green vegetables	Arugula, cabbage, chard, endive, escarole, kale, lettuce (including butter, iceberg, red lead, baby leaf, green leaf, and romaine lettuce), spinach

the pathogen, or its toxin(s), has to be present at a sufficient level to cause an illness in susceptible individuals.

Raw, fresh, or fresh-cut fruits and vegetables are high-risk foods. Fresh produce does not lend itself as well to rigorous inactivation treatments or interventions that would minimize, control, or eliminate pathogens before consumption. Furthermore, the intrinsic composition of produce has little or no killing effect on foodborne pathogens. For example, organic acids in fruits have a detrimental effect on pathogens, but the acidity is not always sufficient, as evidenced by outbreaks of *Escherichia coli* O157:H7 infections caused by unpasteurized apple juice (Tauxe 1997).

This association between produce and foodborne pathogens has led to an increased interest in preventive measures and antimicrobial interventions or treatments before and after harvest to help reduce and control foodborne illnesses. The increased consumption of fruits and vegetables may also be contributing to the noticeable increase in produce-related outbreaks. Reports indicate that the consumption of fruits and vegetables per capita increased by almost 50% between 1976 and 1996 and from 91.6 kg in 1982 to 121.1 kg in 1997 in the United States (Beuchat and Ryu 1997; USDA-ARS 1999). There are also improved sampling procedures, new and improved testing methods, enhanced surveillance programs and epidemiological work, and enhanced investigative approaches for the identification of potential food vehicles and/or reservoirs associated with the incident(s).

3 Potential Hazards Associated with Produce

The possible hazards associated with produce are not much different from those associated with foods in general. Physical hazards include foreign matter (e.g., pieces of metal, glass, wood) and inedible parts of some foods (e.g., fruit stones). Chemical hazards include residues from agricultural chemicals (e.g., pesticides), naturally occurring toxicants (e.g., mycotoxins), and environmental and/or industrial contaminants (e.g., mercury, lead, dioxin, or polychlorinated biphenyls). Biological hazards mainly refer to pathogens that can be bacterial or nonbacterial in nature (e.g., viruses, parasites, protozoa) and that would pose a public health threat.

Within the biological hazards, the microbial pathogens can include agents known as predominately zoonotic, foodborne, and environmental. *Campylobacter* and *Salmonella* spp. are classic foodborne pathogens and can be zoonotic (i.e., transmitted from animals to humans). *Listeria monocytogenes* has an animal origin and therefore can also be considered zoonotic. Yet *L. monocytogenes* is known to survive and multiply in the food processing environments and therefore can be considered to be an environmental contaminant to foods. Some of these microbial pathogens are well recognized, whereas others are emerging. The ever-changing environment of food from farm to table has led some pathogens to emerge in different foods, such as the hepatitis A virus and *E. coli* O157:H7 in produce. For survival, other pathogens have developed resistance mechanisms to certain environments. Over the years, microbiological hazards have gained more public attention compared with physical or chemical hazards (Box 1). We now understand the complexity of the food chain and the potential contamination sources that can be introduced during the production and processing of produce.

> **Box 1** Nonbacterial Foodborne Illnesses
>
> An example of a prominent microbiological nonbacterial food incident is the massive 1993 waterborne cryptosporidiosis outbreak (Corso et al. 2003). About 403,000 illnesses were reported and 100 AIDS patients died in that occurrence. The estimated total cost of illnesses associated with this outbreak up to $96.3 million, $31.7 million in medical costs and $64.6 million in productivity losses, with an average total costs for persons with mild, moderate, and severe illness at $116, $475, and $7,808, respectively. This outbreak was caused by a protozoan that was spread by water (Corso et al. 2003).

4 Foodborne Illness Associated with Produce

Current statistics indicate that the number of outbreaks associated with produce have continued to rise (Lynch et al. 2009). Between 1995 and 2006, a total of 22 produce-associated outbreaks were documented in the United States, with nearly half ($n=9$) traced back to lettuce or spinach grown in California (Cooley et al. 2007). The reported outbreaks are mostly associated with fresh or raw fruits and vegetables, which are sometimes labeled as RTE (e.g., bagged salad greens). Data indicate that the most frequently implicated produce includes salad, lettuce, melon, sprouts, and berries. Also, among 103 produce-associated outbreaks with a known pathogen, about 60% of the common outbreaks were caused by bacterial agents (Box 2).

A study of the 1999–2002 outbreak data from FoodNet attempting to identify contributing factors in the scenarios revealed that from a total of 890 outbreaks, 469 (53%) had an identified etiology, while 457 (51%) had only a food implicated and 365 (41%) had contributing factors identified. Produce were implicated in 97 (21%) of the 457 outbreaks, and 63 outbreaks had a confirmed or suspected etiology. The conclusion of this study was that raw product or ingredient contamination from an animal or environmental source was the most often cited contributing factor (Shakir et al. 2006). It has also been reported that leafy greens are the most likely produce category to be associated with an outbreak (Calvin 2007). Leafy greens, due to microbial contamination, have accounted for 34% of all produce-related outbreaks reported since 1996. Additionally, *E. coli* O157:H7 has been associated with 20 of the 24 leafy greens outbreaks in the United States since 1996 (Box 3).

> **Box 2** Produce-associated Outbreaks Reported from 1973 to 1997 and from 1998 to 2002
>
> Analysis of the data from the CDC's Foodborne Outbreak Surveillance System revealed that from 190 produce-associated outbreaks reported for the 1973–1997 period, there were an associated 16,058 illnesses, 598 hospitalizations, and eight deaths (Sivapalasingam et al. 2004). Of 62 outbreaks caused by bacterial agents, 30 (48%) were associated with *Salmonella*. For the period 1998–2002, there were 6,647 foodborne outbreaks reported, with an average of 1,200–1,300 outbreaks per year and 128,370 illnesses. Again, in these reports bacterial pathogens caused the largest percentage of outbreaks (55%). Among bacterial pathogens, *Salmonella* serotype Enteritidis accounted for the largest number of outbreaks, *L. monocytogenes* accounted for the majority of deaths, and *E. coli* O157:H7 was associated with multistate outbreaks (Lynch et al. 2006).

> **Box 3** Outbreak Alert!
>
> According to data from the Center for Science in the Public Interest (CSPI), from 1990 to 2003, fresh produce was the second most common vehicle for foodborne illness, with 428 produce-associated outbreaks and 23,857 illnesses reported. In a later report by the CSPI, produce outbreaks accounted for 13% (713/5,416) of incidents and 21% (34,049/161,089) of associated illnesses from 1990 to 2005. This information is based on a comprehensive survey that the CSPI conducted of outbreaks with an identified food source. The study indicated that 50% of the produce outbreaks were caused by food from restaurants and other food establishments, while private homes accounted for 13% of the outbreaks – the workplace, catered events, and schools were other reported locations. Furthermore, in terms of microbial agents, norovirus and *Salmonella* were the two major associated pathogens, with norovirus accounting for 40% and *Salmonella* accounting for 18% of all of these outbreaks. The CSPI survey also discussed *E. coli*, a bacterium that received much attention over the last few years because of its association with a number of outbreaks, several of which included produce and, more specifically, leafy greens. According to the survey, *E. coli* caused 8% of produce-related outbreaks. When assessing data in terms of the most common vehicles associated with produce-related foodborne illness outbreaks, greens-based salads, lettuce, potatoes, unspecified fruits, and sprouts were identified. The survey further indicated that produce items causing the most illnesses linked to the outbreaks during the 1990–2005 period were greens-based salads, berries, tomatoes, lettuce, and sprouts (CSPI 2004; DeWaal and Bhuiya 2007).

4.1 Bacterial Pathogens Associated to Produce

As outlined, pathogenic agents linked to outbreaks involving produce have included bacteria, viruses, and protozoa. Tauxe et al. (1997) indicated that in the periods of 1973–1987 and 1988–1991, the etiologic agent was unknown in more than half of the produce-associated foodborne outbreaks reported to the Centers for Disease Control and Prevention (CDC), but most of the outbreaks with an identified etiology were from bacterial origins. Under the bacterial category, *Salmonella* spp. was the pathogen most commonly reported. In addition to *Salmonella*, produce-associated outbreaks have been linked to *E. coli* O157:H7, *L. monocytogenes,* and *Shigella*, although the latter two pathogens are not the most frequent (Harris 2002). In 1981, an outbreak of *L. monocytogenes* was tied to coleslaw (chopped cabbage and carrots sold premixed at retail) in the Maritime Provinces of Canada. This outbreak resulted in 41 illnesses and 18 deaths. Investigations revealed that the suspect vehicle was the cabbage used to make the coleslaw. The cabbage was potentially contaminated with the pathogen from uncomposted sheep manure collected from sheep suspected to have had *Listeria meningitis*. It was further indicated that the cabbage was stored for extended periods of time prior to shredding, and no antilisterial processes were applied (Harris 2002).

Shigella has been associated with incidents linked to raw produce, including chopped, uncooked curly parsley and iceberg lettuce (CDC 1999; Davis et al. 1988; Kapperud et al. 1995; Frost et al. 1995). Field investigations revealed that municipal water supplied to the packing shed was unchlorinated and vulnerable to contamination. This water was used for chilling the parsley immediately after harvest and for making ice with which the parsley was packaged for transport. Because the parsley was chilled in a hydrocooler where the water was recirculated, there were opportunities for *Shigella* to survive in the unchlorinated water supply or on the parsley. Other findings indicated that workers had limited hygiene education and limited sanitary facilities available on the farm at the time of the outbreak.

Safety of Produce

In the next section, we will describe selected bacterial pathogens linked to outbreaks in fruits and vegetables, primarily *E. coli* O157:H7, *L. monocytogenes*, *Salmonella* spp., and *Shigella* spp. Information on the characteristics of each of the bacteria presented can be reviewed in chapters "Clinical Presentations and Pathogenicity of Foodborne Diseases by Bacteria" and "Methods for Identification of Bacterial Foodborne Pathogens." We also provide a brief description of the history of the occurrence and contemporary problems, pathogenesis, ability to survive and grow in the environment, reservoirs, and the produce that the pathogen has most contaminated.

4.1.1 *Escherichia coli* O157:H7

E. coli O157:H7 has become well known for its association with foodborne illness outbreaks related to produce. Unpasteurized apple juice, apple cider, cantaloupe, sprouts, lettuce, and spinach are some of the suspected or implicated vehicles in outbreaks of *E. coli* O157:H7 (CDC 1996, 1997a, b, 2006a, b; Besser et al. 1993; Tu et al. 2004; Griffin 1995).

The 1991 outbreak of *E. coli* O157:H7 in fresh-pressed apple cider identified transmission by a seemingly unlikely vehicle (Besser et al. 1993). In this outbreak, it was suspected that manure may have contaminated the dropped apples used to make the juice or cider (CDC 1996). Besser et al. (1993) indicated that the use of dropped apples to make apple cider is a common practice. Studies have indicated that the addition of preservatives to apple cider does not consistently kill some pathogens, including *E. coli* O157:H7 (Zhao et al. 1993; Miller and Kaspar 1994). Furthermore, although apple cider and juice are usually acidic (pH of 3–4), the strains of *E. coli* O157:H7 associated with these outbreaks were acid-tolerant. Reports indicated that *E. coli* O157:H7 can survive in apple cider for up to 4 weeks (Zhao et al. 1993; Millard et al. 1994).

According to CDC surveillance data, leaf lettuce was implicated in two separate outbreaks in 1995 (Armstrong et al. 1996). In those situations, the mode of lettuce contamination remained unknown, yet it was learned that *E. coli* O157:H7 can grow on lettuce at temperatures as low as 12°C. In 2006, there were two multistate outbreaks of *E. coli* O15:H7 linked to iceberg lettuce. One of the outbreaks occurred in the northeastern U.S. and resulted in 71 illnesses and 53 hospitalizations, where eight patients developed hemolytic-uremic syndrome (HUS) – a type of kidney failure (CDC 2006a). The second outbreak was linked to restaurants in Minnesota and Iowa (CDPH 2008), where approximately 80 people were sickened, and three patients developed HUS.

For some time, pathogenic strains of *E. coli* were thought to be human-specific, but this opinion changed with the discovery of *E. coli* O157:H7. Originally, *E. coli* O157:H7 was identified in an investigation by the CDC of two outbreaks of severe bloody diarrhea associated with the same fast-food restaurant chain (Riley et al. 1983; Wells et al. 1983; Armstrong et al. 1996). *E. coli* O157:H7 gained wide attention following a 1993 large multistate outbreak in the western United States associated with eating undercooked hamburgers at another fast-food restaurant chain. This outbreak resulted in more than 700 illnesses, 195 hospitalizatoins (55 HUS or thrombotic thrombocytopenic purpura, TTP), and four deaths. The Council of State and Territorial Epidemiologists (CSTE) in 1993 recommended that clinical laboratories begin culturing all bloody stools for *E. coli* O157:H7 since many clinical laboratories (at the time) did not routinely test stool samples for the organism (CSTE 1993). A resolution was passed in 2000 by the CSTE in which all Shiga toxin-producing *E. coli* were made nationally notifiable under the name EHEC, and in 2001 national surveillance for EHEC began (CDC 2005).

Among the EHEC, *E. coli* O157:H7 is the serotype responsible for the greatest proportion of infections (Fratamico and Smith 2006). Estimates indicate that *E. coli* O157 causes about 70,000 infections each year in the United States (Mead et al. 1999; CDC 2008b). It is estimated that a similar number of persons have diarrhea caused by non-O157 STEC each year in the United States (CDC 2008b).

According to the CDC, the incidence of Shiga-toxin producing *E. coli* (STEC) O157 infections increased in 2005 and 2006 after substantial declines in 2003 and 2004 (CDC 2007). The decline in incidence was temporally associated with measures taken by the Food Safety and Inspection Services of the U.S. Department of Agriculture (USDA-FSIS) and the beef processing industry to reduce the contamination of ground beef (CDC 2007). These declines were accompanied by a reduction in the frequency of isolation of STEC O157 from ground beef in 2003 and 2004 (Naugle et al. 2006). In 2005 and 2006, however, there was an increase in human STEC O157 infections (CDC 2007). Although the reasons for the increase are not known, STEC O157 outbreaks caused by contaminated spinach and lettuce in 2006 highlight the need to more effectively prevent the contamination of produce that is consumed raw.

Enteric pathogens can have a broad range of hosts. In the case of *E. coli* O157:H7, cattle have been identified as reservoirs for the agent. This pathogen will colonize the intestinal tract of cattle and show no clinical signs of the disease. However, *E. coli* O157:H7 can cause serious illness in humans, at times leading to death, especially in susceptible populations such as the very young and the elderly. Deer and sheep can also be asymptomatic carriers of *E. coli* O157:H7.

In general, infections with *E. coli* O157:H7 cause a gastrointestinal illness, including the severe hemolytic colitis, and severity ranges from mild to life-threatening. The infectious dose has been reported to be less than 1,000 *E. coli* O157:H7 and as few as ten microorganisms (Griffin et al. 1994; FDA 2001c). The incubation period is about 3–4 days, with a range of 1–10 days, and the illness lasts from 5 to 7 days. It is estimated that about 5–10% of individuals diagnosed with STEC infection develop HUS, which develops about 7 days after the first symptoms (CDC 2008b). HUS can leave permanent damage and have lifelong complications. It can also result in death. Symptoms are characterized by cramps or abdominal pains (which can be severe), diarrhea that initially can be watery and advances to bloody in severe cases, and vomiting (CDC 2008b; FDA 2001c).

4.1.2 *Listeria monocytogenes*

The infective dose of *L. monocytogenes* is not completely known, but the Food and Drug Administration (FDA) in 1992 reported that fewer than 1,000 colony-forming units (CFU) may be sufficient to cause the disease in susceptible persons (based on data from cases that contracted the diseases through raw or supposedly pasteurized milk). The incubation period is usually 4–21 days, but can be 3–70 days (IAFP 2007). Other documents indicate the incubation period to be from 24 h to 91 days (Harris 2002). The duration of foodborne illness caused by this pathogen can be 1–2 days.

Listeria can be found in various environments at pre- and postharvest, including soil, vegetation, water, fecal matter, processing equipment, and drains. This pathogen was detected in a number of foods in the raw and processed states and in particular with RTE products. Outbreaks of listeriosis have been associated with a number of commodities, including coleslaw, hotdogs and deli meats or cold cuts, pasteurized milk, and Mexican-style soft cheese. It has been indicated that *L. monocytogenes* is widely distributed on raw fruits and vegetables and plant material, and that plant and plant parts used as salad vegetables play a key role in the dissemination of the pathogen from natural habitats to the human supply (Harris 2002).

4.1.3 *Salmonella* spp.

In 2008, *Salmonella*, and specifically *Salmonella* serotype Saintpaul, received massive media attention as it was determined to be the cause of a large international outbreak that resulted in over 1,400 illnesses in the United States and Canada. Certain types of raw tomatoes were initially suspected, but later the CDC reported that jalapeño and Serrano peppers grown, harvested, or packed in Mexico were the vehicles for the outbreak and the cause of some clusters. The FDA also announced that samples of jalapeño peppers, Serrano peppers, and reservoir water used for irrigation were found positive and isolates were genetic matches with the *Salmonella* Saintpaul strain causing the outbreak (FDA 2008b).

The Center for Science in the Public Interest reported that *Salmonella* outbreaks have been traced to a number of produce commodities, including lettuce, salads, melons, sprouts, and tomatoes (CSPI 2005). Further, in 2004 three separate outbreaks involving 561 *Salmonella* infections were linked to contaminated Roma tomatoes. Also, from 2000 to 2002, *Salmonella*-contaminated cantaloupe imported from Mexico sickened 155 and killed two individuals (CSPI 2005). Other literature reveals that between 1989 and 2001, there were several *Salmonella* outbreaks in the United States associated with cantaloupes and most of these incidents were linked to product traced back to Mexican farms (CDC 1991, 2002; Mohle-Boetani et al. 1999; Alvarado-Casillas et al. 2007). *Salmonella* outbreaks have also been associated with watermelons (Gayler et al. 1955; CDC 1979; Blostein 1991).

Foods associated with outbreaks caused by *Salmonella* include fruit juices and fruits and vegetables, such as tomatoes, jalapenos, almonds, and sprouts. The infective dose of *Salmonella* may be as few as 15–20 CFU, and the incubation period of the microorganism is 6–48 h (FDA 1992). Signs and symptoms usually develop 12–72 h after infection, but they may appear as soon as 6 h and can last up to a week. The duration of the illness is estimated at 1–2 days, but it may last longer and resolve in 5–7 days. Colonized and infected individuals may shed *Salmonella* for 1–3 months, and a few (1–3%) may shed for a year.

4.1.4 *Shigella* spp.

Shigella is a human-specific pathogen that is widespread and potentially deadly. This pathogen can cause serious fatal infections called **Shigellosis** and bacillary dysentery that account for less than 10% of the reported outbreaks of foodborne illness in the United States (FDA 2001c). The Council for Agricultural Science and Technology estimates that about 90,000–163,000 cases occur per year, with an average cost of $390 per case (CAST 1994). Deaths are estimated at 180 or less per year. Reports from the CDC indicate that about 14,000 cases of shigellosis are reported in the United States every year; however, the actual number of infections may be 20 times greater (CDC 2008b). FoodNet data revealed that in 2007, of the total 17,883 identified laboratory-confirmed cases of infection in FoodNet surveillance areas, the number of cases and the incidence per 100,000 population for *Shigella* were 2,848 and 6.26, respectively (CDC 2008a) – ranking this microorganism third following *Salmonella* and *Campylobacter*.

Shigella is not as prevalent as *Salmonella*. The mortality rate of shigellosis, however, is higher than that of salmonellosis. *Shigella* is very important in waterborne diseases, especially in tropical and subtropical countries where sanitation conditions are poor and especially after natural disasters. The organism is transmitted by water, food, humans, and animals (Hajmeer and Fung 2006). Water polluted with human feces tends to be a reservoir for this pathogen. Foods associated with *Shigella* problems usually involve RTE commodities, where the implicated food has been contaminated by an infected person. Some of the food vehicles implicated with outbreaks include produce and potato salad.

As described by Hajmeer and Fung (2006), infections caused by *Shigella* occur when foods like produce are handled with pathogen-contaminated water and then ingested. The infective dose can be as low as a ten bacterial cells depending on the age and health condition of the target individual. If the pathogen is able to successfully attach and penetrate the epithelial cells of the intestinal tract, about 1–4 days after ingestion of the organisms there will be an inflammation of the walls of the large intestines and ileum. The pathogen multiplies intracellularly, spreads to other adjacent epithelial cells, and, in the process, destroys the tissue. Bloody stool will occur due to superficial ulceration.

The lysed cell wall of *Shigella* will release endotoxin. *S. dysenteriae* can also produce an exotoxin and Shiga-toxin like STEC. People inflicted with shigellosis may experience abdominal pain, diarrhea (with possible blood, pus, or mucus in stools), cramps, vomiting, fever, and tenesmus. The fatality rate may be as high as 10–15%. Some people may experience complications such as Reiter's disease, HUS, and reactive arthritis in the aftermath of shigellosis. The incubation period can be from 0.5 to 7 days (typically 1–2 days), with an onset of illness from 12 to 50 h.

5 Routes and Sources of Produce Contamination

The increase in foodborne illness outbreaks associated with produce may be due to several factors that would influence the contamination of such commodities at the pre- and postharvest levels. The potential routes for produce contamination are numerous, and product contamination can occur at any point in the food chain. This applies to domestic commodities as well as imported products. Contamination of produce can occur in the field, and sources may include soil, water, air, and dust, all of which can potentially carry pathogens. Wildlife and farm and/or domestic animal activity in proximity to produce-growing fields; use of green, contaminated, or improperly composted manure; and improper health and hygienic practices of workers pre- and postharvest can also contribute to the contamination of produce.

At preharvest, soil may contribute to producing contamination, as it is a habitat for some bacterial pathogens, including *L. monocytogenes* and *Clostridium botulinum*. Soil can also be contaminated with fecal matter, contaminated water, or improperly treated amendments and fertilizers.

Water used for irrigation as well as for pesticides or foliar treatment applications may become contaminated with bacterial pathogens and eventually impact produce safety. The irrigation method (e.g., sprinkle versus furrow) also plays a role. With sprinkle irrigation, for example, contaminated water containing pathogens may come in contact with the edible portion of the fruits or vegetables. Watchel et al. (2002) reported that contaminated irrigation water led to the contamination of commercial produce with *E. coli* O157:H7. In that situation, a sewage spill released unchlorinated tertiary-treated effluent into a creek used to irrigate commercial young cabbage transplants. *E. coli* O157:H7 was isolated from the roots of the plants removed from plots of the suspected field for testing.

Field practices during harvest, cooling, and transport may also lead to product contamination and microbial proliferation. McEvoy et al. (2009) reported that iceberg lettuce may be cored in the field and the outer leaves of the lettuce heads removed at harvest to reduce shipping waste and maximize production yield. These practices may increase the potential for contamination during field procedure, and the problem may be further augmented at postharvest with potential improper practices such as temperature abuse in transport and/or at processing and/or extended storage time under inappropriate conditions (McEvoy et al. 2009).

The implementation of proper Good Agricultural Practices is important at preharvest. According to Delazari et al. (2006), these practices include the use of properly composted manure; the use of clean water for irrigation, pesticide, and other applications where water comes in close contact with the fruits and vegetables, especially RTE products; the control of domestic animals and wildlife; the control of insects, which can serve as vectors for foodborne pathogens; and the provision, use, and maintenance of toilets and facilities for hand washing or hand disinfection for field workers.

At postharvest, factors that contribute to food safety problems with produce include unsanitary conditions of vehicles used for transport or temperature abuse of products in transport, processing problems, contaminated processing equipment or food contact surfaces, the presence of animal activities in food processing or preparation areas, improper sanitary and hygienic practices of workers or food handlers, contaminated water or ice, inadequate air quality, cross-contamination, problems with packaging materials, and packaging methods that extend shelf life but do not necessarily improve product safety. Francis and O'Beirne (2006) indicated that fresh-cut vegetables packaged under a modified atmosphere can support the growth of numerous species of *Listeria*, including *L. monocytogenes*. At postharvest, the *Hazard Analysis and Critical Control Points* (HACCP) system seems a feasible approach that would assist in the control, minimization, and elimination of pathogens. Interventions such as irradiation and further research and assessments may be needed to minimize the risk.

6 Measures to Reduce the Risk of Foodborne Illness

With increasing numbers of outbreaks associated with produce, a number of measures have been proposed and/or implemented. In 1997, the Produce Safety Initiative was introduced after outbreaks associated with both domestic and imported fresh fruits and vegetables occurred (FDA 1999a). The import of products potentially produced under suboptimal safety conditions at pre- and/or postharvest has raised concerns and led to more scrutiny and testing of imported goods. After the Produce Safety Initiative was introduced, the FDA issued a guidance document for produce safety (FDA 1998a). In 2007, the FDA published a draft guidance advising processors on approaches to minimize microbial food safety hazards common to the processing of most fresh-cut fruits and vegetables. A more recent version of this document is now available online (FDA 2008a).

With the dramatic increase in the number of outbreaks due to eating sprouts and elevated concerns regarding the safety of sprouted seeds, the FDA issued an interim advisory about eating sprouts (FDA 1998b). The FDA also recommended that contaminated seeds be disinfected with solutions of calcium hypochlorite and microbiological testing be conducted on the used irrigation water from each batch or production lot (Anonymous 1999).

Outbreaks associated with juices resulted in the FDA's mandating HACCP for juices and in 2001 publishing the final rule for "Juice HACCP" to increase the safety of juices (FDA 2001a). This came after a 1996 *E. coli* O157:H7 outbreak linked to apple juice and two citrus juice outbreaks attributed to *Salmonella* spp. in 1999 and 2000 (FDA 2001b) (Box 4).

Leafy greens have also been problematic. The FDA introduced the Lettuce Safety Initiative as a response to the recurring outbreaks of *E. coli* O157:H7 associated with fresh and fresh-cut lettuce and in an attempt to reduce public health risks by focusing on the product, agents, and areas of greatest concern (FDA 2006). In 2006, members of the leafy greens industry introduced commodity-specific guidelines such as the "Commodity Specific Food Safety Guidelines for Lettuce and Leafy Greens Supply Chain" published by the International Fresh-Cut Produce Association, Produce Marketing Association, United Fresh Fruit and Vegetable Association, and the Western Growers Association (WGA). In 2007, the Leafy Greens Marketing Agreement (LGMA) metrics program was initiated in California. The overriding objective of the LGMA is to protect the public health by reducing potential sources of contamination in California-grown leafy greens (LGMA 2011). According to the WGA, an agricultural trade organization, the LGMA is a response to several years of decreasing confidence in the leafy green market on the part of buyers, consumers, legislators, and federal and state regulatory agencies (WGA 2006). The LGMA references "Best Practices," which is a set of guidelines that provide specific criteria and target values for controls and monitoring as compared to the commodity-specific document(s). Similarly, in 2005, because of repeated problems linked to melons, a "Commodity Specific Food Safety Guidelines for the Melon Supply Chain" was introduced.

> **Box 4** HACCP and Produce
>
> With the notable increase in foodborne illness outbreaks associated with produce over the last few years, interest has grown in using HACCP at preharvest to reduce the risk of foodborne illness from produce. However, the implementation of a true HACCP system at preharvest may not be achievable as it is on the postharvest side. This is primarily because of the nature of an open environment at the field/farm level, where the control of hazards is more challenging compared to more restricted settings. Implementation of all HACCP principles may not be achievable at preharvest either.

Much has been accomplished over the last few years to improve produce safety. However, efforts and measures to further improve the safety of these commodities and reduce the risk of foodborne illness continue to be introduced. Doyle and Erickson (2008) reported that there are several chemical and biological interventions that can be used to reduce surface contamination on produce and minimize cross-contamination. However, they added that these interventions are ineffective on internalized pathogens, and internalization is a significant route of contamination in the field. Currently, the most commonly used commercial antimicrobial intervention for fresh produce during processing is wash water containing 50–200 parts per million (ppm) of chlorine (Stopforth et al. 2008).

The issues of the health and hygiene of workers handling produce, the potential for produce contamination in food service or at retail, and the health and hygienic practices of consumers are additional factors that impact the safety of produce. Throughout the food chain, these individuals should exercise care when handling produce to avoid and minimize the risk of foodborne illness. Measures that consumers can exercise include proper washing of fruits and vegetables before consumption; proper hygienic practices, such as washing hands before handling food; proper cleaning and sanitizing of food contact surfaces and utensils; separation of all raw meat, poultry, and fish products that are not RTE from produce; following product handling and washing instructions; following label directions on usage and storage; selecting fruits and vegetables without excessive soil on the edible portion and produce that is whole and intact since bruising or cutting exposes the edible portion of the food and may allow for contamination; and selecting fruits and vegetables that look wholesome without any signs of decay, off odors, or discoloration.

7 Summary

Concerns about the safety of produce have increased over the last few years due to the notable numbers of reported produce-associated foodborne illness outbreaks. Produce encompasses a wide range of fruits and vegetables, which may be available to consumers as fresh or processed. The potential hazards associated with produce are not any different from those associated with foods in general and can be physical, chemical, or microbial in nature. Between 1995 and 2006, 22 produce-associated outbreaks were documented in the United States, with bacterial pathogens, mainly *Salmonella* spp. and *E. coli* O157:H7, making up most of the known etiological agents associated with foodborne disease in produce. Several approaches have been used to control or reduce bacterial pathogens in produce, but water chlorination remains the most widely used approach. As health authorities continue to enhance the quality of their reporting of outbreaks and offer better education on safe eating, disease prevention and control, epidemiologic investigations, and laboratory testing, it is expected that we will improve our understanding of the transmission of bacterial foodborne diseases by produce.

References

Alvarado-Casillas, S., S. Ibarra-Sanchez, O. Rodriguez-Garcia, N. Martinez-Gonzales, and A. Castillo. 2007. Comparison of rinsing and sanitizing procedures for reducing bacterial pathogens on fresh cantaloupes and bell peppers. *Journal of Food Protection* 70(3): 655–660.

Anonymous. 1999. Microbiological safety evaluations and recommendations on sprouted seeds. *International Journal of Food Microbiology* 52: 123–153.

Armstrong, G.L., J. Hollingsworth, and J.G. Morris Jr. 1996. Emerging foodborne pathogens: *Escherichia coli* O157:H7 as a model of entry of a new pathogen into the food supply of the developed world. *Epidemiologic Reviews* 18: 29–51.

Besser, R.E., S.M. Lett, J.T. Weber, M.P. Dolye, J.G. Wells, and P.M. Griffin. 1993. An outbreak of diarrhea and hemolytic uremic syndrome from *Escherichia coli* O157:H7 in fresh pressed apple cider. *The Journal of the American Medical Association* 269: 2217–2220.

Beuchat, L.R., and J.-H. Ryu. 1997. Produce handling and processing practices. *Emerging Infectious Diseases* 3: 459–465.

Blostein, J. 1991. An outbreak of *Salmonella javiana* associated with consumption of watermelon. *J. Environ. Health* 56: 29–31.

Calvin, L. 2007. Outbreak linked to spinach forces reassessment of food safety practices. *USDA-ERS Amber Waves* 5: 25–31.

CAST. 1994. Foodborne Pathogens: Risks and Consequences. Task Force Report no. 122. CAST, Ames, IA.

CDC. 1979. *Salmonella oranienburg* gastroenteritis associated with consumption of precut watermelons. *Morbidity and Mortality Weekly Report* 28: 522–523.

CDC. 1991. Multi-state outbreak of *Salmonella* Poona infections – United States and Canada. *Morbidity and Mortality Weekly Report* 40: 549–552.

CDC. 1996. Outbreak of *Escherichia coli* O157:H7 infections associated with drinking unpasteurized commercial apple juice – British Columbia, California, Colorado, and Washington, October 1996. *Morbidity and Mortality Weekly Report* 45: 975.

CDC. 1997a. Outbreaks of *Escherichia coli* O157:H7 infection and cryptosporidiosis associated with drinking unpasteurized apple cider – Connecticut and New York, October 1996. *Morbidity and Mortality Weekly Report* 46: 4–8.

CDC. 1997b. Outbreaks of *Escherichia coli* O157:H7 infection associated with eating alfalfa sprouts – Michigan and Virginia, June–July 1997. *Morbidity and Mortality Weekly Report* 46: 741–744.

CDC. 1999. Outbreaks of *Shigella sonnei* infection associated with eating fresh parsley – United States and Canada, July–August 1998. *Morbidity and Mortality Weekly Report* 48: 285–289.

CDC. 2002. Multistate outbreaks of *Salmonella* serotype Poona infections associated with eating cantaloupe from Mexico – United States and Canada, 2000–2002. *Morbidity and Mortality Weekly Report* 51: 1044–1047.

CDC. 2005. Summary of notifiable diseases – United States, 2003. *Morbidity and Mortality Weekly Report* 52: 1–85.

CDC. 2006a. Multistate Outbreak of E. coli O157 infections, November–December 2006. http://www.cdc.gov/ecoli/2006/december/121406.htm. Accessed 21 March 2011.

CDC. 2006b. Updates from the multi-state outbreak of E. coli O157:H7 infections from fresh spinach, September–October 2006. http://www.cdc.gov/ecoli/2006/september/updates/. Accessed 21 March 2011.

CDC. 2007. Preliminary FoodNet data on the incidence of infection with pathogens transmitted commonly through food – 10 states, 2006. *Morbidity and Mortality Weekly Report* 56: 336–339.

CDC. 2008a. Preliminary FoodNet data on the incidence of infection with pathogens transmitted commonly through food – 10 states, 2007. *Morbidity and Mortality Weekly Report* 57: 366–370.

CDC. 2008b. DFDMB disease listing. http://www.cdc.gov/nczved/dfbmd/disease_listing.html. Accessed 21 March 2011.

CDPH. 2008. Investigation of the Taco John's *Escherichia coli* O157:H7 outbreak associated with iceberg lettuce. The California Food Emergency Response Team. http://www.cdph.ca.gov/pubsforms/Documents/fdb%20eru%20IceLet%20TacoJohn022008.pdf. Accessed 21 March 2011.

Cooley, M., D. Carychao, L. Crawford-Miksza, M.T. Jay, C. Myers, C. Rose, C. Keys, J. Farrar, and R.E. Mandrell. 2007. Incidence and tracking of *Escherichia coli* O157:H7 in a major produce production region in California. *PloS One* 2(11): e1159.

Corso, P.S., M.H. Kramer, K.A. Blair, D.G. Addiss, J.P. Davis, and A.C. Haddix. 2003. Cost of illness in the 1993 waterborne *Cryptosporidium* outbreak, Milwaukee, Wisconsin. *Emerging Infectious Diseases* 9: 426–431.

CSPI. 2004. Contaminated produce top food poisoning culprit. http://www.cspinet.org/new/200404011.html. Accessed 21 March 2011.

CSPI. 2005. *Salmonella* outbreaks linked to produce on the rise. http://www.cspinet.org/new/200511211.html. Accessed 21 March 2011.

CSTE. 1993. CSTE position statement #4: National surveillance of *Escherichia coli* O157:H7. Council of State and Territorial Epidemiologists, Atlanta.

Davis, H., J.P. Taylor, J.N. Perdue, G.N. Stelma Jr., J.M. Humphreys Jr., R. Rowntree III, and K.D. Greene. 1988. A shigellosis outbreak traced to commercially distributed shredded lettuce. *American Journal of Epidemiology* 128: 1312–1321.

Delazari, I., H.P. Riemann, and M.N. Hajmeer. 2006. Food safety. In *Foodborne infections and intoxications*, 3rd ed, ed. H.P. Riemann and D.O. Cliver, 833–877. New York: Elsevier.

DeWaal, C., and Bhuiya, F. 2007. Outbreaks by the numbers: Fruits and vegetables 1990–2005. International Association for Food Protection, Orlando, FL. Poster Presentation P3–03, July 8–11.

Doyle, M.P., and M.C. Erickson. 2008. Summer meeting 2007 – the problems with fresh produce: an overview. *Journal of Applied Microbiology* 105: 317–330.

FDA. 1992. Listeria monocytogenes. Bad bug book: Foodborne pathogenic microorganisms and natural toxins handbook. Food and Drug Administration. Center for Food Safety and Applied Nutrition, College Park, MD.

FDA. 1998a. Guide to minimize microbial food safety hazards for fresh fruits and vegetables. Food and Drug Administration. Center for Food Safety and Applied Nutrition, College Park, MD. www.fda.gov/downloads/Food/.../UCM169112.pdf. Accessed 21 March 2011.

FDA. 1998b. Interim Advisory on Alfalfa Sprouts. Food and Drug Administration. Washington, DC. August 31, 1998.

FDA. 1999a. The food safety initiative. Food and Drug Administration, Center for Food Safety and Applied Nutrition, College Park, MD.

FDA. 2001a. Juice HACCP. http://www.cfsan.fda.gov/~comm/haccpjui.html. Accessed 21 March 2011.

FDA. 2001b. FDA publishes final rule to increase safety of fruit and vegetable juices. News Release P01-03, January 18, 2001. Food and Drug Administration, Washington, DC.

FDA. 2001c. Bad Bug Book: Foodborne Pathogenic Microorganisms and Natural Toxins Handbook. Food and Drug Administration. Center for Food Safety and Applied Nutrition, College Park, MD.

FDA. 2006. Lettuce safety initiative. http://www.fda.gov/Food/FoodSafety/Product-SpecificInformation/FruitsVegetablesJuices/FDAProduceSafetyActivities/ucm115906.htm. Accessed 21 March 2011.

FDA. 2008a. Guide to minimize microbial food safety hazards of fresh-cut fruits and vegetables. http://www.fda.gov/food/guidancecomplianceregulatoryinformation/guidancedocuments/produceandplanproducts/ucm064458.htm. Accessed 21 March 2011.

FDA. 2008b. *Salmonella* Saintpaul outbreak. http://www.fda.gov/NewsEvents/PublicHealthFocus/ucm179116.htm. Accessed 21 March 2011.

Francis, G.A., and D. O'Beirne. 2006. The incidence of *Listeria monocytogenes* in modified atmosphere packaged fresh-cut vegetables collected in Ireland. *Journal of Food Protection* 69: 2524–2528.

Fratamico, P.M., and J.L. Smith. 2006. *Escherichia coli* infections. In *Foodborne infections and intoxications*, 3rd ed, ed. H.P. Riemann and D.O. Cliver, 205–258. New York: Elsevier.

Frost, J.A., M.B. McEvoy, C.A. Bentley, and Y. Andersson. 1995. An outbreak of *Shigella sonnei* infection associated with consumption of iceberg lettuce. *Emerging Infectious Diseases* 1: 26–29.

Gayler, G.E., R.A. MacCready, P. Reardon, and B.E. McKernan. 1955. An outbreak of salmonellosis traced to watermelon. *Public Health Reports* 70: 311–313.

Griffin, P.M. 1995. Symposium on new and emerging pathogenic *Escherichia coli* O157:H7 – Beyond the burger. In: Abstracts of the 35th Interscience Conference on Antimicrobial Agents and Chemotherapy, San Francisco, California. Sept. 17–20. American Society for Microbiology.

Griffin, P.M., B.P. Bell, P.R. Cieslak, J. Tuttle, T.J. Barrett, M.P. Doyle, A.M. McNamara, A.M. Shefer, and J.G. Well. 1994. Large outbreak of *Escherichia coli* O157:H7 infections in the western United States: the big picture. In *Recent advances in verocytotoxin-producing Escherichia coli infections*, ed. M.A. Karmali and A.G. Goglio, 7–12. Amsterdam: Elsevier Science B.V.

Hajmeer, M.N., and D.Y.C. Fung. 2006. Infections with other bacteria. In *Foodborne infections and intoxications*, 3rd ed, ed. H.P. Riemann and D.O. Clive, 341–365. London: Academic Press (Elsevier).

Harris, L. 2002. Listeria monocytogenes. In *Foodborne diseases*, 2nd ed, ed. D.O. Cliver and H.P. Riemann, 137–150. San Diego: Academic.

IAFP. 2007. Procedures to Investigate Foodborne Illness, 5th ed. International Association for Food Protection, Chicago.

Kapperud, G., L.M. Rorvik, V. Hasseltvedt, E.A. Høiby, B.G. Iversen, K. Staveland, G. Johnsen, J. Leitao, H. Heriksted, Y. Andersson, G. Langeland, B. Gondrosen, and J. Lassen. 1995. Outbreak of *Shigella sonnei* infection traced to imported iceberg lettuce. *Journal of Clinical Microbiology* 33: 609–614.

LGMA. 2011. California leafy green products handler marketing agreement. http://www.caleafygreens.ca.gov/. Accessed 21 March 2011.

Lynch, M.F., J. Painter, R. Woodruff, and C. Braden. 2006. Surveillance for foodborne-disease outbreaks – United States, 1998–2002. *CDC Surveillance Summaries* 55: 1–34.

Lynch, M.F., R.V. Tauxe, and C.W. Hedberg. 2009. The growing burden of foodborne outbreaks due to contaminated fresh produce: Risks and opportunities. *Epidemiology and Infection* 137: 307–315.

McEvoy, J.L., Y. Luo, W. Conway, B. Zhou, and H. Feng. 2009. Potential of *Escherichia coli* O157:H7 to grow on field-cored lettuce as impacted by postharvest storage time and temperature. *International Journal of Food Microbiology* 128(3): 506–509.

Mead, S., L. Slutsker, V. Dietz, L.F. McCaig, J.S. Bresee, C. Shapiro, P.M. Griffin, and R.V. Tauxe. 1999. Food-related illness and death in the United States. *Emerging Infectious Diseases* 5: 607–625.

Millard, P.S., K.F. Gensheimer, D.G. Addiss, D.M. Sosin, G.A. Beckett, A. Houck-Jankoski, and A. Hudson. 1994. An outbreak of cryptosporidiosis from fresh-pressed apple cider. *The Journal of the American Medical Association* 272: 1592–1596.

Miller, L.G., and C.W. Kaspar. 1994. *Escherichia coli* O157:H7 acid tolerance and survival in apple cider. *Journal of Food Protection* 57: 460–464.

Mohle-Boetani, J.C., R. Reporter, S.B. Werner, S. Abbott, J. Farrar, S.H. Waterman, and D.J. Vugia. 1999. An outbreak of *Salmonella* serogroup Saphra due to cantaloupes from Mexico. *The Journal of Infectious Diseases* 180: 1361–1364.

Naugle, A.L., K.G. Holt, P. Levine, and R. Eckel. 2006. Sustained decrease in the rate of *Escherichia coli* O157:H7-positive raw ground beef samples tested by the Food Safety and Inspection Service. *Journal of Food Protection* 69: 480–481.

Riley, L.W., R.S. Rennis, S.D. Helgerson, H.B. McGee, J.G. Wells, B.R. Davis, R.J. Hebert, E.S. Olcott, L.M. Johnson, N.T. Hargrett, P.A. Blake, and M.L. Cohen. 1983. Hemorrhagic colitis associated with a rate *Escherichia coli* serotype. *The New England Journal of Medicine* 308(12): 681–685.

Shakir, F. K., McCarthy, P. V., Guzewich, J. J., Braden, C. R., Klontz, K. C., Hedberg, C. W., Fullerton, K. E., Bogard, A., Dreyfuss, M., Larson, K., Vugia, D., Nicholas, D. C., Radke, V. J., and Jones, T. F. 2006. Contributing Factors (CFs) Identified in Produce-Associated Outbreaks from CDC's National Electronic Foodborne Outbreak Reporting System (eFORS), FoodNet Sites, 1999–2002. Poster. International Conference on Emerging Infectious Diseases. March 19–22. Atlanta.

Sivapalasingam, S., C.R. Friedman, L. Cohen, and R.V. Tauxe. 2004. Fresh produce: A growing cause of outbreaks of foodborne illness in the United States, 1973 through 1997. *Journal of Food Protection* 67(10): 2342–2353.

Stopforth, J.D., T. Mai, B. Kottapalli, and M. Samadpour. 2008. Effect of acidified sodium chlorite, chlorine, and acidic electrolyzed water onto leafy greens. *Journal of Food Protection* 71(3): 625–628.

Tauxe, R.V. 1997. Emerging foodborne diseases: an evolving public health challenge. *Dairy, Food, and Environmental Sanitation* 17(12): 788–795.

Tauxe, R., H. Kruse, C. Hedberg, M. Potter, J. Madden, and K. Wachsmuth. 1997. Microbial hazards and emerging issues associated with produce: a preliminary report to the National Advisory Committee on Microbiologic Criteria for Foods. *Journal of Food Protection* 60(11): 1400–1408.

Tu, S.-I., Uknalis, J., and Gehring, A. 2004. Optical methods for detecting *Escherichia coli* O157:H7 spiked on cantaloupes. In: Nondestructive Sensing for Food Safety, Quality, and Natural Resources (Y.-R. Chen and S.-I. Tu, Eds.). Proceedings SPIE, 5587:83–189.

USDA-ARS. 1999. Fruit and tree nut situation and outlook report. FTS–287. http://usda.mannlib.cornell.edu/MannUsda/viewDocumentInfo.do?documentID=1378. Accessed 21 March 2011.

Watchel, M.R., L.C. Whitehand, and R.R. Mandrell. 2002. Prevalence of *Escherichia coli* associated with a cabbage crop inadvertently irrigated with partially treated sewage wastewater. *Journal of Food Protection* 65(3): 471–475.

Wells, J.G., B.R. Davis, I.K. Wachsmuth, L.W. Riley, R.S. Remis, R. Sokolow, and G.K. Morris. 1983. Laboratory investigation of hemorrhagic colitis outbreaks associated with a rare *Escherichia coli* serotype. *Journal of Clinical Microbiology* 18(3): 512–520.

WGA. 2006. Effects of *E. coli* outbreak on consumers. http://www.wga.com. Accessed 21 March 2011.

Zhao, T., M.P. Doyle, and R.E. Besser. 1993. Fate of enterohemorrhagic *Escherichia coli* O157:H7 in apple cider with and without preservatives. *Applied and Environmental Microbiology* 59: 2526–2530.

Safety of Fruit, Nut, and Berry Products

Mickey Parish, Michelle Danyluk, and Jan A. Narciso*

1 Introduction

Fruits, nuts, berries, and their associated packed and processed products are known sources of foodborne illnesses. Documented outbreaks from *Salmonella*, pathogenic *Escherichia coli*, *Cryptosporidium*, *Cyclospora*, viruses, and some unknown agents have occurred with alarming frequency in recent decades, possibly due to an increased consumption of these products by a larger population, along with more scrutiny of outbreaks by public health officials (Tables 1 and 2). Other factors, such as changes in microbial virulence potential, changed human resistance to infections and intoxications, greater international trade, and altered processing and handling regimes, are also viable though somewhat speculative alternatives to explain the increased numbers of foodborne outbreaks documented since 1990. As an introduction to the safety of these foods, this chapter will provide basic information about the safety of three commodity groups: whole fruit and fruit products, berries and berry products, and nuts and nut products.

Figure 1 shows the number of documented illnesses attributed to fruit, berries, nuts, and corresponding products for the years 1998–2008 as reported to the Centers for Disease Control's (CDC) National Outbreak Reporting System by state and local health departments (Anonymous 2011). Documented illnesses range from a low of about 500 in 2002 to more than 2,000 in 2008. Over this 11-year period, the number of illnesses per year averaged more than 1,100, indicating that these products remain problematic in regard to foodborne outbreaks.

---

*The opinions and conclusions expressed in this manuscript are solely the views of the authors and do not necessarily reflect those of the U.S. Food and Drug Administration.

M. Parish (✉)
U.S. Food and Drug Administration, College Park, MD, USA
e-mail: mickey.parish@fda.hhs.gov

M. Danyluk
Citrus Research and Education Center, University of Florida, Lake Alfred, FL, USA

J.A. Narciso
USDA/ARS/CSPRU, US Horticultural Research Laboratory, Fort Pierce, FL, USA

Table 1 Analysis of CDC Foodborne Outbreak Online Database (FOOD) entries between 1998 and 2008 that identify events showing fruit, nut, or berry products as possible vehicles of illness[a]

Year	Number of outbreak-related events	Illnesses		Average per outbreak	Products implicated from epidemiological surveys[b]	Disease agents (number of outbreak-related events)
		Number	Range			
1998	22	868	2–270	39	Banana, fruit salad, guacamole, lemonade, mango, melon, pineapple, strawberries, unspecified fruit, unspecified nuts	*Campylobacter* (1), *E. coli* O157:H7 (1), hepatitis A (1), norovirus (3), *Salmonella* Oranienburg (1), *Shigella boydii* (1), unconfirmed (3), unknown (11)
1999	33	1420	2–398	43	Apple cider, blackberries, cantaloupe, fruit punch, fruit salad, grapes, lemonade, mamey, mango, melon, orange juice, peaches, raspberries, strawberries, unspecified fruit	*Cyclospora cayetanensis* (1), *E. coli* O157:H7 (1), mycotoxins (1), norovirus (8), *Salmonella* Anatum (1), *S.* Enteritidis (2), *S.* Javiana (1), *S.* Muenchen (1), *S.* Newport (1), *S.* Typhi (1), unconfirmed (9), unknown (6)
2000	35	1893	2–736	54	Blackberries, coconut milk, fruit salad, grapes, guacamole, lemonade, melon, orange, orange juice, pineapple, pineapple juice, raspberries, strawberries, unspecified fruit	*Cyclospora cayetanensis* (2), *E. coli* O157:H7 (2), hepatitis A (2), *Salmonella* Enteritidis (2), *S.* Heidelberg (1), *S.* Poona (1), *Shigella sonnei* (1), norovirus (9), unconfirmed (4), unknown (10), heavy metal (1)
2001	34	1161	2–250	34	Almonds, ambrosia, avocado, fruit punch, fruit salad, grapes, grapefruit, guacamole, kiwi, lemon, lemonade, mango, melon, pear, pineapple, plantains, strawberries, unspecified fruit	*Campylobacter jejuni* (1), *Escherichia coli* O157:H7 (1), norovirus (8), *Salmonella* Enteritidis (1), *S.* Poona (2), *S.* Saintpaul (1), *S.* Senftenberg (1), *Salmonella* spp. (1), unconfirmed (10), unknown (8)
2002	26	495	2–51	19	Apple cider, fruit punch, fruit salad, grape, guacamole, melon, peanut butter, pineapple, plantains, raspberries, unspecified fruit, unspecified nuts	*Campylobacter jejuni* (1), norovirus (3), *Salmonella* Berta (1), *S.* Heidelberg (1), *S.* Newport (1), *S.* Poona (1), *Shigella sonnei* (1), *Staphylococcus aureus* (1), unconfirmed (7), unknown (6)
2003	17	541	3–144	32	Almonds, apple cider, applesauce, banana, fruit punch, fruit salad, guacamole, mango, melon, pineapple, strawberries, unspecified fruit	*Cryptosporidium parvum* (1), norovirus (3), *Salmonella* spp. (1), *S.* Enteritidis (1), *S.* Muenchen (1), *S.* Newport (1), *S.* Saintpaul (1), *S.* Typhimurium (1), *Shigella sonnei* (1), unconfirmed (4), unknown (2)
2004	17	817	2–212	48	Apple cider, fruit salad, grapes, guacamole, melon, orange juice, plantains, strawberries, unspecified berries, unspecified fruit	*Escherichia coli* O111 (1), *Cryptosporidium parvum* (1), norovirus (5), refrigerant chemical (1), unconfirmed (7), unknown (2), unspecified natural toxin (1)

Year				Foods	Pathogens
2005	24	1314	2–162	55	Fruit salad, grapes, guacamole, melon, orange juice, peanuts, pineapple, strawberries, unspecified fruit, unspecified fruit juice	*Escherichia coli* O157:H7 (2), norovirus (9), *Salmonella* Enteritidis (1), *S.* Newport (1), *S.* Saintpaul (1), *S.* Typhimurium (2), unconfirmed (7), unknown (2)
2006	29	1608	2–715	55	Applesauce, blueberries, fruit salad, guacamole, lemonade, melon, peanuts, peanut butter, plum, strawberries, susumber berries, unspecified fruit	*Campylobacter jejuni* (1), *Cyclospora cayatenensis* (1), *E. coli* O26 (1), norovirus (9), *Salmonella* Newport (2), *S.* Oranienburg (1), *S.* Tennessee (1), *S.* Thompson (1), *S.* Typhimurium (1), unconfirmed (6), unknown (4), unspecified chemical (1)
2007	30	546	2–60	18	Açaí, apple cider, avocado, banana, fruit salad, grapes, guacamole, lemon, lemonade, melon, pineapple, plantain, strawberries, sugar cane juice, unknown fruit, unspecified berries, unspecified fruit, unspecified nuts	*Clostridium perfringens* (1), *Escherichia coli* O157:H7 (2), hepatitis A (1), norovirus (7), *Salmonella* Javiana (1), *S.* Litchfield (2), *S.* Newport (1), unconfirmed (6), unknown (9)
2008	35	2100	2–716	60	Apple cider, applesauce, fruit salad, fruit tea, grapes, guacamole, lemon, lime, melon, peanut butter/paste, pineapple, plantain, unspecified berries, unspecified fruit	*Cyclospora cayatenensis* (2), *E. coli* O157:H7 (3), norovirus (12), *Salmonella* 1,4,[5],12:i:- (1), *S.* Aberdeen (1), *S.* Javiana (2), *S.* Litchfield (3), *S.* Newport (1), *S.* Panama (1), *S.* Typhimurium (1), unconfirmed (4), unknown (4)

[a] Available at http://wwwn.cdc.gov/foodborneoutbreaks/Default.aspx (accessed March 30, 2011)
[b] Excludes tomatoes. "Melon" includes cantaloupes, watermelon, honeydew, and other melons. Foods were implicated by state and local epidemiological surveys and may or may not have been confirmed as an outbreak vehicle

Table 2 Outbreaks associated with the consumption of nuts

Variety	Product	Pathogen	Year	Outbreak location(s)	Source
Almond	Raw whole	*Salmonella* Enteritidis PT 30	2000–2001	Canada, USA	Isaacs et al. (2005)
	Raw whole	*Salmonella* Enteritidis PT 9c	2004	Canada, USA	Anonymous (2004)
	Raw whole	*Salmonella* Enteritidis	2005–2006	Sweden	Ledet Muller et al. (2007)
Coconut	Desiccated	*Salmonella typhi*, *Salmonella* Senftenberg and possibly others	1953	Australia	Wilson and Mackenzie (1955)
	Desiccated	*Salmonella* Java PT Dundee	1999	United Kingdom	O'Brien et al. (1999)
	Milk	*Vibrio cholerae*	1991	USA	Anonymous (1991); Taylor et al. (1993)
Hazelnut	Conserve (for yogurt)	*Clostridium botulinum*	1989	United Kingdom	O'Mahony et al. (1990)
Peanut	Canned	*Clostridium botulinum*	1986	Taiwan	Chou et al. (1988)
	Savory snack	*Salmonella* Agona PT 15	1994–1995	United Kingdom, Israel	Killalea et al. (1996); Shohat et al. (1996)
	Peanut butter	*Salmonella* Mbandaka	1996	Australia	Scheil et al. (1998)
	Flavored or roasted in-shell	*Salmonella* Stanley and *Salmonella* Newport	2001	Australia, Canada, United Kingdom	Kirk et al. (2004)
	Peanut butter	*Salmonella* Tennessee	2006–2007	USA	Anonymous (2007)
	Peanut butter, peanut butter–ontaining products	*Salmonella* Typhimurium	2008–2009	USA	Anonymous (2009)
Sesame seed	Halva	*Salmonella* Typhimurium DT 104	2001	Australia, Sweden, Norway, United Kingdom, Germany	O'Grady et al. (2001)
		Salmonella Montevideo	2002	Australia	Tauxe et al. (2008)

Source: Adapted from Danyluk et al. (2007a)

wider investigation in Canada, England, Wales, and Scotland (Kirk et al. 2004). *Salmonella* cells have been determined by the Most Probable Number (MPN) method and ranged from <0.03 to approximately 2 MPN/g, and testing of opened and unopened jars of peanut butter had populations of less than three *Salmonella* per g in three samples and four *Salmonella* organisms per g in one sample (Scheil et al. 1998). In recalled lots of almonds from an outbreak in 2000–2001, 84% of 100-g samples tested were positive for the outbreak strain. Populations of <0.3 MPN/g were reported upon initial testing, and 6–9 MPN/100 g were reported upon retesting using larger sample sizes (Danyluk et al. 2007b). Importantly, these low numbers of *Salmonella* detected during outbreak investigations may have been sufficient to cause illness, similar to reports seen for a number of other dried foods (Grocery Manufacturers Association 2009).

In early 2009, a *Salmonella* outbreak linked to peanut butter and peanut butter–containing products led to the recall of over 2,745 products, making it one of the largest recalls in U.S. history (Anonymous 2009b). *Salmonella enterica* serotype Typhimurium matching the outbreak strain was isolated from a number of these products (Anonymous 2009a). This ingredient-driven outbreak raises awareness of the possibility for nuts and nut products to cause widely distributed outbreaks (Anonymous 2009a).

In addition to *Salmonella*, *Vibrio cholerae* and *Clostridium botulinum* have produced outbreaks associated with nuts and nut products. Temperature abuse was involved in an outbreak of cholera linked to the consumption of coconut milk (Taylor et al. 1993). Besides *Vibrio cholera*, other *Vibrio* spp. and three serotypes of *Salmonella* were isolated from the same lot of coconut milk, indicating a lack of sanitation and insufficient heat treatment of the product. Inadequate processing is cited as a possible explanation for outbreaks of botulism implicating canned peanuts (Chou et al. 1988), and a combination of high pH (5.0–5.5), a formulation change from sugar to aspartame, and insufficient heating are cited as possible causes for a botulism outbreak from a hazelnut conserve used in yogurt (O'Mahony et al. 1990). *C. botulinum* is capable of growing and producing toxin in peanut spread stored aerobically when the a_w 0.96 (Clavero et al. 2000) because under aerobic conditions, molds growing on the surface of the spread are thought to provide microenvironments suitable for toxin production. On the other hand, anaerobic conditions allow the growth of lactic acid bacteria, which lower the pH even more and inhibit the growth of *C. botulinum*. While commercially manufactured peanut butter has an a_w 0.70, the potential for microbial growth should be considered when formulating nut-based products.

A number of studies have documented the presence of *Salmonella* on nuts; however, the incidence of other bacterial foodborne pathogens on nuts or in nut products is not well documented. A survey of raw almonds arriving at processors in California from 2001–2005 determined that *Salmonella* was present in 81 of 9,274 100-g samples tested (0.87%). *Salmonella* was not isolated upon retesting in 59 of 65 positive samples. The counts when *Salmonella* was detected were 1.2–2.9 MPN/100 g. Of the 81 total isolates, 35 different serotypes of *Salmonella* were represented (Danyluk et al. 2007b). In addition, *Salmonella* was isolated from 11 of 117 (9.4%) sesame seed samples and ready-to-eat sesame products and sesame seed paste collected from retail and delicatessen stores in Germany (Brockmann et al. 2004). Seventeen different *Salmonella* serotypes have been isolated in Australia by the National Enteric Pathogen Surveillance Scheme from sesame seeds and sesame seed products, including hummus, tahini, and halva tested on 30 occasions between 1985 and 2001 (O'Grady et al. 2001). A survey of 250-g preroasted peanut, almond, cashew, hazelnut, and Brazil nut kernels received into Australian nut processing facilities over a 3-year period (921 total samples) yielded only one *Salmonella*-positive sample of almond (Eglezos et al. 2008). *Salmonella* has also been isolated from macadamia nuts (St. Clair and Klenk 1990), walnuts (Riyaz-Ul-Hassan et al. 2003) cashews and Brazil kernels (Freire and Offord 2002), and dried coconut meats (Schaffner et al. 1967).

Available data indicate that the survival of enteric pathogens on nuts or nut products is extremely high. Generic *E. coli* or *Salmonella* inoculated onto nuts or nut products were detected for months to years (Beuchat and Heaton 1975; Burnett et al. 2000; King and Jones 2001; King et al. 1970; Kokal 1965; Kotzekidou 1998; Uesugi et al. 2006). A significantly better survival of *Salmonella* was seen on pecan halves, in-shell pecans (Beuchat and Heaton 1975), almonds (Uesugi et al. 2006), and peanut butter (Burnett et al. 2000; Park et al. 2008) stored at refrigeration rather than ambient or elevated temperatures. A temperature effect was not observed when *Salmonella* was inoculated into halva, and an average two-log reduction was observed in hummus that contains tahini at pH 4.0–4.3 stored at 4°C and 22°C over 28 days, respectively (Kotzekidou 1998). *Salmonella* cells did adapt to acidic conditions in the hummus, as evidenced by the induction and expression of three genes involved in the acid tolerance response (*ompR*, *cadC*, and *fur*) (Laycock et al. 2008). During the 2006–2007 U.S. peanut butter outbreak, *Salmonella* Tennessee matching the outbreak strains were identified from jars of peanut butter with production dates ranging from July 2006–January 2007 (Anonymous 2007). These data collectively demonstrate the potential for *Salmonella* survival through the typical 1- to 2-year shelf life of nuts and nut products.

Processing steps that involve thermal treatment are frequently used to alter the texture and appearance of the nut and result in reduced microbial contaminations. However, microorganisms in dry environments and dry processing conditions are significantly more heat-resistant than in liquids or high-a_w foods (Riemann 1968). Traditional heat treatments of nuts include roasting by hot air or immersion in hot oil and blanching in hot water or steam. Blanching provides wet processing conditions that allow for a more rapid reduction of *Salmonella* on almonds than either oil or dry roasting (Uesugi and Harris 2005; Du and Harris 2005; Kim and Harris 2006). Almonds tolerate higher temperatures than many other nuts, such as coconut, walnut, macadamia, and Brazil nuts (Schaffner et al. 1967), which darken more rapidly during roasting. A reduction of several logs of *Salmonella* on coconut pieces has been observed without deleterious effects on the sensory quality (Schaffner et al. 1967), and coconut meat is routinely pasteurized before or after shredding (Wareing et al. 2000). No published results are available so far for the microbial reductions achieved during normal roasting processes of nuts other than almonds and coconut. Thermal treatments of peanut butter are inadequate to consistently reduce high levels of *Salmonella*, and extending treatment times or elevating the temperature does not significantly improve the pasteurization process (Shachar and Yaron 2006). Alternative processing treatments to decrease microbial populations, specifically *Salmonella* on nuts, are documented, especially for almonds. These techniques, successful to varying degrees, include acidic sprays (Pao et al. 2006), high hydrostatic pressure (Goodridge et al. 2006), infrared heat (Brandl et al. 2008), nonthermal plasma (Deng et al. 2007), and propylene oxide (Beuchat 1973; Danyluk et al. 2005).

3.1 Berries and Berry Products

"Berry" is a generic term for a small, edible fruit. Most of the common fruits termed "berries" (e.g., blueberries, strawberries, and raspberries), however, are not true berries by botanical classification, as are, for example, gooseberries, huckleberries, and currants. Berries have become a significant addition to the increased consumption of fresh produce in the United States (Sivapalasingam et al. 2004; Notermans et al. 2004; Zhao 2007). Major berry crops grown in the United States (not including fresh grapes) are blueberries, raspberries, strawberries, gooseberries, blackberries, black currants, and cranberries (Zhao 2007; NASS 2009). Most berries are cultivated although some berries are also harvested from the wild (e.g., blueberries, blackberries, raspberries, cranberries). Other berries are only minor crops grown commercially in specific geographic areas (e.g., lingonberries, chokeberries, and elderberries) or are locally important crops (Zhao 2007).

Berries have gained popularity due to their high nutritional composition and because they contain many diverse nutrients that are beneficial to human health. In addition to a variety of vitamins and minerals, berries contain a high *antioxidant* capacity provided by their high flavonoid and phenolic contents (Parry et al. 2005; Puupponen-Pimiä et al. 2005a, b; Luther et al. 2007; Zhao 2007). Lignans, sterols, and stilbenes, all health-promoting *phytochemicals*, are also found in rich supply in berry fruits (Zhao 2007).

Berries are available to the consumer as a packed fresh, processed, or "U-harvest" product (Notermans et al. 2004; Zhao 2007). An increase in exports from emerging countries as well as larger markets established in fruit trade have made it possible to enjoy fresh berries in all seasons in most places. In 2007 in the United States, the highest production of the major cultivated berries was strawberries, followed by cranberries and blueberries. The total berry production includes both fresh market and processed fruits (i.e., dried, juiced, pureed, or frozen) (Table 3).

Although berries have enjoyed a relatively good record of food safety, berries have been found to be the source of several foodborne pathogenic outbreaks in recent years. Table 4 summarizes some of the food safety hazards associated with berries. It was thought that the acidity of the fruit (pH 3.0–4.5) would deter the survival of pathogenic organisms (Bagamboula et al. 2002; Kärenlampi and Hänninen 2004; Zhao 2005; Siro et al. 2006). However, cleaning harvested fruit with contaminated water; unsanitary field, harvesting, and storage conditions; and improper handling by workers can all introduce microorganisms that are capable of remaining in fresh or processed (pureed or juiced) berries for as long as 5 days in numbers above 5 log/g (Beuchat and Ryu 1997; De Roever 1998; Zhao 2005; Todd et al. 2007). Berries sent to fresh market are handpicked and field packed and shipped without cleaning (Notermans et al. 2004; Flessa et al. 2005). Berries for processing are transported (sometimes at ambient temperature) to a facility where they are washed with potable water, perhaps sanitized, and then sliced or frozen. Fruits sent for nonthermal processing are handled by employees, a practice that can introduce additional avenues of contamination (Beuchat and Ryu 1997; Greening 2006; Todd et al. 2007).

3.2 Role of Contaminating Protozoa

Surveys of foodborne diseases have revealed nine berry outbreaks between 1985 and 1997 (Sivapalasingam et al. 2004). There were 1,815 reported illnesses, and 24 people were hospitalized, but there were no deaths. Several outbreaks in 1996 were attributed to the consumption of fresh raspberries and blackberries contaminated with *Cyclospora cayetanensis,* a protozoan in the family *Eimeriida* (García-López et al. 1996; Osterholm 1997; Anonymous 1998a; Chalmers et al. 2000; Hammond et al. 2001; Zhao 2005). These outbreaks occurred in at least 15 states and in some provinces in Canada. Traceback investigations indicated that the berries involved were imported from Guatemala but did not reveal the mode of contamination (Beuchat and Ryu 1997; Anonymous 1998a; Hammond et al. 2001), possibly because of a lack of protocols for, and the difficulty of, detecting *Cyclospora* on plant tissues. It has been suggested that water or birds might be likely sources of contamination to berries (García-López et al. 1996; Osterholm 1997; Anonymous 1998a).

The occurrence of *cyclosporiasis* from berries raised concerns at the international level, and a call emerged to establish well-defined criteria for evaluating epidemiologic data, quick consumer alerts, rapid product information reviews, and scientific data generated by local, state, and federal agencies (Osterholm 1997; De Roever 1998; Chalmers et al. 2000). Some exporting countries, such as Guatemala, in cooperation with the U.S. Food and Drug Administration (FDA), began to voluntarily introduce control measures to prevent the recurrence of *Cyclospora* in berries (Anonymous 1998a).

Table 3 Total major fresh berry production/utilization in the United States in 2007

Crop	Total production (1,000 t fresh equivalent)	States with highest production	Processed fruit (1,000 lb)[a]
Blackberry	32.0	Oregon	56,200
Blueberry: cultivated	141.7	Michigan, New Jersey	137,450
Blueberry: wild	38.6	Maine	76,800
Raspberry	81.5	California, Washington	62,250
Strawberry	1249.7	California, Florida	471,900
Cranberry	327.7	Wisconsin, Massachusetts	619,400
Boysenberry	3.9	California, Oregon	3,800

Source: All data from NASS (2009)
[a] Includes canned, frozen, pureed

Table 4 Possible food safety and quality hazards associated with berries

Possible safety or quality hazard	Berry product	Source
Parasites		
Cyclospora cayetanensis	Raspberries	Chalmers et al. (2000)
Toxoplasma gondii	Raspberries, blueberries	Kniel et al. (2002)
Viruses		
Norovirus	Strawberries, frozen raspberries	Greening (2006); Notermans et al. (2004)
Calicivirus	Frozen raspberries	Pönkä et al. (1999)
Poliovirus	Frozen strawberries, raspberries	Kurdziel et al. (2001)
Hepatitis A	Strawberries, raspberries (fresh and frozen), blueberries	Hammond et al. (2001); Calder et al. (2003); Fiore (2004)
Fungi (decay)		
Botrytis cinerea, Rhizopus stolonifer, Colletotrichum spp., *Penicillium* spp., *Fusarium* spp., *Cladosporium* spp., *Alternaria* spp.	Most berries in storage	Farr et al. (1995)
Fungi (mycotoxin)		
Alternaria alternata	Blueberries, raspberries, strawberries, raspberry and cranberry drinks	Scott (2001); Tournas and Stack (2001)
Bacteria		
Salmonellae	Strawberries	Notermans et al. (2004)
Listeria monocytogenes	Strawberries	Fless et al. (2005)
Shigella spp.	Strawberries, fruit salad	Bagamboula et al. (2002)
Campylobacter jejuni	Strawberries	Kärenlampi and Hänninen (2004)
Pesticide residues	Various berries and products	Wang et al. (2005)

A study with blueberries and raspberries inoculated with *Toxoplasma gondii*, a coccidial parasite related to *Cyclospora*, showed that the *oocysts* can adhere to the berry epidermis and remain on the berry surface in excess of 8 weeks (Kniel et al. 2002). It was shown that *T. gondii* can serve as a model for *Cyclospora* in food safety studies, as it is easier to detect (Kniel et al. 2002). It is expected that production tests with *Toxoplasma* might allow researchers to develop a protocol for eliminating *Cyclospora* from imported berries.

3.3 Viruses in Berries

Berry fruits can also be contaminated with diverse enteric viruses. There are no standard methods for the guaranteed detection of viruses in foods, so that data are not always comprehensive when looking into virus contamination trends (Pönkä et al. 1999; Kurdziel et al. 2001; Calder et al. 2003; Fiore 2004; Greening 2006). In the United States, approximately 67% of foodborne diseases are attributed to viruses, and virus contamination in foods has become a major global concern (Le Guyader et al. 2004; Greening 2006). While there are many types of enteric viruses, noroviruses are most commonly associated with cases of foodborne virus infections: About 40% of norovirus infection in the United States per year is foodborne (Greening 2006). Enteric viruses cause acute gastrointestinal distress (Le Guyader et al. 2004; Notermans et al. 2004; Greening 2006) and are spread by the fecal–oral route via irrigation with sewage-contaminated water or improper handling of the fruit by workers (Beuchat and Ryu 1997; Gaulin et al. 1999; Calder et al. 2003; Fiore 2004; Le Guyader et al. 2004; Greening 2006; Todd et al. 2007). Enteric viruses are resistant to most environmental stresses (including stomach acid) and most chemical degradation, and are not altered by freezing (Gaulin et al. 1999; Pönkä et al. 1999; Greening 2006; Buttot et al. 2008).

Berry-related virus epidemics in Finland between 1998 and 2001 resulted in a ban on unheated berries in catering and large-scale kitchens (Pönkä et al. 1999; Le Guyader et al. 2004). In the United States, strawberries were implicated in a 1999 norovirus outbreak, and in 2001, frozen raspberries were found to be the cause of norovirus infection from pastries in Sweden (Le Guyader et al. 2004; Notermans et al. 2004). Although noroviruses are the predominant enteric virus in foodborne epidemics, other viruses cause berries to be at risk as well. In Canada and Finland, outbreaks of calicivirus were associated with frozen raspberries from imported fruit (Gaulin et al. 1999; Pönkä et al. 1999). Freezing allows the extended survival of viruses in berries, and imported berries can be frozen and stored for long periods of time before consumption.

Poliovirus was the first virus to be linked with foodborne transmission (Greening 2006). Kurdziel et al. (2001) found that polioviruses on fresh raspberries and frozen strawberries stored at household refrigeration temperatures could last for more than 1 week. Fresh raspberries had no decline in virus, but the frozen strawberries had a 90% decline in approximately 8 days. Ward and Trying (1987) found that poliovirus survived for 76 days on celery and 55 days on spinach in storage at 4°C. It was found that for all studies, high humidity on the produce surface helped determine the survival of the poliovirus (Kurdziel et al. 2001).

Another example is hepatitis A, an enteric virus that survives desiccation and other environmental stress better than poliovirus (Kurdziel et al. 2001). Foodborne pathogenic events with hepatitis A–contaminated berries have occurred often. There have been several reported outbreaks in frozen strawberries (1990, 1997, 1998), fresh strawberries (1997, 1998, 1999), raspberries (1987), and frozen raspberries and blueberries (Calder et al. 2003; Fiore 2004; Notermans et al. 2004; Greening 2006). In one case involving frozen strawberries, the mode of contamination was wastewater irrigation and in another, involving fruits imported from Mexico, the picker hygiene (Calder et al. 2003; Fiore 2004; Notermans et al. 2004). In New Zealand in 2002, an outbreak of hepatitis A on raw blueberries was attributed to probable contamination by pickers (Calder et al. 2003). Foodborne virus illness from berries prompted action for the formation of a *Hazard Analysis Critical Control Point (HACCP)* plan for fresh berries to ensure that even berries from imported areas are free of contamination (Beuchat and Ryu 1997; De Roever 1998; Pönkä et al. 1999; Greening 2006).

3.4 Important Bacterial Contaminants in Berries

Bacterial pathogens are a threat for contamination because most berries for the fresh market are handpicked into containers in the field and normally do not go through any sanitation protocol (Kärenlampi and Hänninen 2004; Notermans et al. 2004; Flessa et al. 2005; Zhao 2007). In random testing of imported strawberries, the FDA found *Salmonella* spp. in 1 of 143 samples (Zhao 2005). From Norwegian retail markets, *Listeria monocytogenes* was isolated in 1 of 173 samples tested of fresh strawberries (Flessa et al. 2005).

To investigate the potential threat of bacterial foodborne pathogens on berries, *Shigella flexneri, S. sonnei, Listeria monocytogenes*, and *Campylobacter jejuni* were inoculated on strawberries in three separate studies (Bagamboula et al. 2002; Kärenlampi and Hänninen 2004; Flessa et al. 2005). *Campylobacter* showed a decreased survival on strawberries, possibly due to the low pH of the fruit, but survived sufficiently long enough (72 h at 7°C) to be a risk to consumers (Kärenlampi and Hänninen 2004). Pathogens survived better on product surfaces when the storage temperature was 10°C or below (Flessa et al. 2005). *Listeria* inoculated on whole and cut strawberries survived for up to 7 days at 4°C. *Listeria* on frozen strawberries survived also, but with a higher death rate in fruit pieces that were not mixed with 20% sucrose when frozen after 28 days in storage (Flessa et al. 2005). Experiments with *Shigella* using various *inocula* on strawberries and fruit salad kept refrigerated showed that acid fruits support the survival of this microorganism (Bagamboula et al. 2002). Although *S. sonnei* was recovered from strawberries with all inoculum levels at 4 and 48 h at 4°C, *S. flexneri* was not. However, *S. flexneri* was recovered from inoculated fruit salad samples after 4 and 48 h. Results from these bacterial studies showed that there is a large variability in survival capabilities by different strains of the same organism on an individual commodity depending on the physiological state of the fruit, inoculum level, physiological properties of the organism, and conditions in the microenvironment, such as moisture availability, pH, and temperature (De Roever 1998; Bagamboula et al. 2002; Kärenlampi and Hänninen 2004; Flessa et al. 2005; Zhao 2007).

3.5 Role of Fungi in Berry Contamination

Fungi are problematic on berry production because they cause a general decay of fruit tissues, and not because of their pathogenicity to humans. The fungi most commonly found on a variety of berries are *Botrytis cinerea* (gray mold), *Rhizpous stolonifer* (soft rot), *Colletotrichum* spp. (anthracnose), *Penicillium* spp, *Fusarium* spp. *Cladosporium* spp., and *Alternaria* spp. (Ellis et al. 1991; Caruso and Rumsdell 1995; Farr et al. 1995). These organisms attach to the surface of the fruit, and when fruit are not properly sanitized, they become incorporated in salads and minimally processed fruits, such as presliced and fresh juiced (Tournas et al. 2006). However, some molds may pose a health hazard, as in the case of *mycotoxin* production from organisms from the genus *Alternaria* and other genera. *Alternaria* produce a number of toxins that have been found to be mutagenic and carcinogenic (Scott 2001). It has been shown that various berries support the production of *Alternaria* toxins; they have been found in blueberries, raspberries, strawberries, and in fruit drinks containing raspberry and cranberry (Scott 2001; Tournas and Stack 2001).

3.6 Other Contaminants

The contamination of foods is not always microbial. Pesticide residues have also been found in berry fruit mixes (pureed) (Wang et al. 2005). Pesticides are utilized at various stages in crop cultivation and can be carried to the final food product, including baby foods and other processed products

(Wang et al. 2005; Zhao 2007). In the future, it will be important to establish good agricultural practices in the field to ensure that pesticide residues are not present at harvest and to develop reliable confirmatory methods to monitor the limits of these chemicals in foods. But it is also interesting to note that berries contain antimicrobial substances (e.g., phenolic polymers) that are selectively active against some human pathogens (Puupponen-Pimiä et al. 2005a, b). The modes of action by berry compounds to inhibit bacterial cells vary, including a direct action on microbial metabolism and the inhibition of extracellular microbial membranes (Puupponen-Pimiä et al. 2005a). Some antimicrobial activities may also determine the adherence to the epithelial cells of the fruit, a prerequisite for colonization and infection.

The increase in foodborne pathogen outbreaks involving berries is related to many factors, including better reporting, increased consumption of fresh fruit and vegetables, and a lack of comprehensive sanitary measures for fresh produce (Beuchat and Ryu 1997; Hazen et al. 2001; Notermans et al. 2004; Zhao 2005, 2007). Berries that are imported are more of a challenge as there are no fixed international sanitary standards, and different countries and trading regions have different systems to monitor the safety of foods (Anonymous 1998a; Lupien 2005; Zhao 2005). Processing and storage methods, such as freezing, do not always destroy or inactivate pathogens on the berries such as enteric viruses (Gaulin et al. 1999; Pönkä et al. 1999; Le Guyader et al. 2004; Greening 2006; Buttot et al. 2008). However, freezing does significantly reduce *E. coli* and *Staphylococcus* spp. *colony-forming units (CFU)* on blueberries (Hazen et al. 2001). Modified atmosphere packaging does not reduce *E. coli, Listeria*, or *Salmonella* on raspberries and strawberries in storage from 5 to 7 days (Siro et al. 2006). For some mycotoxin producing fungi, such as *Aspergillus carbonarius*, cold storage with sulfur dioxide reduces fungal CFUs and inhibits their ability to sporulate (Guzev et al. 2008). These studies exemplify the differences in results of postharvest pathogen control methods. Each pathogen and its relationship with the fruit surface pose a challenge to growers and processors.

No uncooked product is risk-free from contamination. The only "comprehensive" method to increase the level of safety of fresh fruits and vegetables is the incorporation of *good agricultural practices (GAP)* during production (preharvest) and to follow with voluntary good management and record keeping during harvesting, processing, transport, storage, and marketing (Beuchat and Ryu 1997; Anonymous 1998b; De Roever 1998; Zhao 2005, 2007). With responsible actions on the parts of the growers, processors, marketers, and consumers, berries will continue to be a healthy, safe, and high-quality addition to a balanced diet.

4 Summary

Most food safety problems associated with fruits and berries arise from fresh-cut fruits and raw juices produced to meet consumer demand for minimally processed products. Outbreaks from *Salmonella*, pathogenic *Escherichia coli, Cryptosporidium, Cyclospora,* and viruses have been linked to the consumption of fruits, nuts, berries, or associated products. Nuts become contaminated with pathogenic bacteria in the field or by cross-contamination during processing and handling. Nuts may also become contaminated with mycotoxins due to fungal growth during storage. Avenues of contamination vary for fresh commodities throughout processing and distribution, including consumer mishandling, and problems have been linked to both domestically grown and imported fruits, berries, nuts, and associated products. There is no comprehensive protocol to guarantee that raw products will be free of pathogens, but careful management, record keeping, and continuous consumer education can reduce the chances of illness from contaminated fresh commodities.

References

Anonymous. 1991. Cholera associated with imported frozen coconut milk – Maryland, 1991. Morbid. Mortal. Wkly. Rep. Dispatch Dec 13. http://www.cdc.gov/mmwr/preview/mmwrhtml/00015726.htm. Accessed 3 Dec 2010.

Anonymous. 1998a. Outbreak of cyclosporiasis – Ontario, Canada, May 1998. Morbid. Mortal. Wkly. Rep. 47:806–809.

Anonymous. 1998b. U.S. Food and drug administration, guide to minimize microbial food safety hazards for fresh fruits and vegetables. Washington, DC.

Anonymous. 2004. Outbreak of *Salmonella* serotype Enteritidis infections associated with raw almonds – United States and Canada, 2003–2004. Morbid. Mortal. Wkly. Rep. Dispatch June 4. http://www.cdc.gov/mmwr/preview/mmwrhtml/mm53d604a1.htm. Accessed 3 Dec 2010.

Anonymous. 2007. Multistate outbreak of *Salmonella* serotype Tennessee infections associated with peanut butter – United States, 2006–2007. Morbid. Mortal. Wkly. Rep. Dispatch June 1. http://www.cdc.gov/MMWR/preview/mmwrhtml/mm5621a1.htm. Accessed 3 Dec 2010.

Anonymous. 2009a. Multistate outbreak of *Salmonella* infections associated with peanut butter and peanut butter-containing products – United States, 2008–2009. Morbid. Mortal. Wkly. Rep. Dispatch January 20. http://www.cdc.gov/MMWR/preview/mmwrhtml/mm58e0129a1.htm. Accessed 3 Dec 2010.

Anonymous. 2009b. Peanut product recall: *Salmonella* Typhimurium. http://www.fda.gov/oc/opacom/hottopics/salmonellatyph.html. Accessed 3 Dec 2010.

Anonymous. 2011. National Outbreak Reporting System surveillance data, United States. Centers for disease control and prevention. http://www.cdc.gov/foodborneoutbreaks/outbreak_data.htm. Accessed 3 Dec 2010.

Bagamboula, C., M. Uyttendaele, and J. Debevere. 2002. Acid tolerance of *Shigella sonnei* and *Shigella flexneri*. *Journal of Applied Microbiology* 93: 479–486.

Bartz, J.A., and R.K. Schowalter. 1981. Infiltration of tomatoes by bacteria in aqueous suspension. *Phytophathology* 71: 515–518.

Beuchat, L.R. 1973. *Escherichia coli* on pecans: Survival under various storage conditions and disinfection with propylene oxide. *Journal of Food Science* 38: 1063–1066.

Beuchat, L.R., and E.K. Heaton. 1975. *Salmonella* survival on pecans as influenced by processing and storage conditions. *Applied Microbiology* 29: 795–801.

Beuchat, L., and J. Ryu. 1997. Produce handling and processing practices. *Emerging Infectious Diseases* 3: 459–465.

Brandl, M.T., Z. Pan, S. Huynh, Y. Zhu, and T.H. McHugh. 2008. Reduction of *Salmonella* Enteritidis population sizes on almond kernels with infrared heat. *Journal of Food Protection* 71: 897–902.

Brockmann, S.O., I. Piechotowski, and P. Kimmig. 2004. *Salmonella* in sesame seed products. *Journal of Food Protection* 67: 178–180.

Buchanan, R., S. Edelson, R. Miller, and G. Sapers. 1999. Contamination of intact apples after immersion in an aqueous environment containing *Escherichia coli* O157:H7. *Journal of Food Protection* 62: 444–450.

Burnett, S.L., E.R. Gehm, W.R. Weissinger, and L.R. Beuchat. 2000. Survival of *Salmonella* in peanut butter and peanut butter spread. *Journal of Applied Microbiology* 89: 472–477.

Buttot, S., T. Putallaz, and G. Sánchez. 2008. Effects of sanitation, freezing and frozen storage on enteric viruses in berries and herbs. *International Journal of Food Microbiology* 126: 30–35.

Calder, L., G. Simmons, C. Thornley, P. Taylor, K. Pritchard, G. Greening, and J. Bishop. 2003. An outbreak of hepatitis A associated with consumption of raw blueberries. *Epidemiology and Infection* 131: 745–751.

Caruso, F., and D. Rumsdell. 1995. *Compendium of blueberry and cranberry diseases*. St. Paul: American Phytopathological Society Press.

Chalmers, R., G. Nichols, and R. Rooney. 2000. Foodborne outbreaks of cyclosporiasis have arisen in North America. Is the United Kingdom at risk? *Communicable Disease Public Health* 3: 50–55.

Chou, J.H., P.H. Hwang, and M.D. Malison. 1988. An outbreak of type A foodborne botulism in Taiwan due to commercially preserved peanuts. *International Journal of Epidemiology* 17: 899–902.

Clavero, M.R.S., R.E. Brackett, L.R. Beuchat, and M.P. Doyle. 2000. Influence of water activity and storage conditions on survival and growth of proteolytic *Clostridium botulinum* in peanut spread. *Food Microbiology* 17: 53–61.

Danyluk, M.D., A.R. Uesugi, and L.J. Harris. 2005. Survival of *Salmonella* Enteritidis PT 30 on inoculated almonds after commercial fumigation with propylene oxide. *Journal of Food Protection* 68: 1613–1622.

Danyluk, M.D., L.J. Harris, and W.H. Sperber. 2007a. Nuts and cereals. In *Food microbiology: fundamentals and frontiers*, ed. M.P. Doyle and L.R. Beuchat, 171–183. Washington, DC: ASM Press.

Danyluk, M.D., T.M. Jones, S.J. Abd, F. Schlitt-Dittrich, M. Jacobs, and L.J. Harris. 2007b. Prevalence and amounts of *Salmonella* found on raw California almonds. *Journal of Food Protection* 70: 820–827.

De Roever, C. 1998. Microbiological safety evaluations and recommendations on fresh produce. *Food Control* 9: 321–347.

Deng, S., R. Ruan, C.K. Mok, G. Huang, X. Lin, and P. Chen. 2007. Inactivation of *Escherichia coli* on almonds using nonthermal plasma. *Journal of Food Science* 72: M62–M66.

Du, W.-X., and Harris, L. J. 2005. Survival of *Salmonella* Enteritidis PT 30 on almonds after exposure to hot oil. Program and abstract book. Proceedings of the 92nd annual meeting of the international association of food protection, Baltimore.

Eglezos, S., B. Huang, and E. Stuttard. 2008. A survey of the bacteriological quality of preroasted peanut, almond, cashew, hazelnut, and Brazil nut kernels received into three Australian nut-processing facilities over a period of 3 years. *Journal of Food Protection* 71: 402–404.

Eisenstein, A., R. Aach, W. Jacobson, and A. Goldman. 1963. An epidemic of infectious hepatitis in a general hospital. *Journal of the American Medical Association* 185: 171.

Ellis, M., K. Converse, R. Williams, and B. Williamson. 1991. *Compendium of raspberry and blackberry diseases and insects*. St. Paul: American Phytopathological Society Press.

Farr, D., G. Bills, G. Chamuris, and A. Rossman. 1995. *Fungi on plants and plant products in the United States*. St. Paul: American Phytopathological Publications.

Fiore, A. 2004. Hepatitis A transmitted by food. *Clinical Infectious Diseases* 38: 705–715.

Flessa, S., D. Lusk, and L. Harris. 2005. Survival of *Listeria monocytogenes* on fresh and frozen strawberries. *International Journal of Food Microbiology* 101: 255–262.

Frank, C., J. Walter, M. Muehlen, A. Jansen, U. van Treeck, A. Hauri, I. Zoellner, M. Rakha, M. Hoehne, O. Hamouda, E. Schreier, and K. Stark. 2007. Major outbreak of hepatitis A associated with orange juice among tourists, Egypt, 2004. *Emerging Infectious Diseases* 13(1): 156–158.

Freire, F.C.O., and L. Offord. 2002. Bacterial and yeast counts in Brazilian commodities and spices. *Brazilian Journal of Microbiology* 33: 145–148.

García-López, H., L. Rodriguez-Tovar, and C. Medina-De La Garza. 1996. Widespread foodborne cyclosporiasis outbreaks present major challenges. *Emerging Infectious Diseases* 2: 354–357.

Gaulin, C., D. Ramsay, P. Cardinal, and M. D'Halevyn. 1999. Épidémie de gastro-entérite d'origine virale associée à la consummation de framboises importées [in French]. *Canadian Journal of Public Health* 90: 37–40.

Goodridge, L.D., J. Willford, and N. Kalchayanand. 2006. Destruction of *Salmonella* Enteritidis inoculated onto raw almonds by high hydrostatic pressure. *Food Research International* 39: 408–412.

Greening, G.E. 2006. Enteric viruses – a new hazard in our food? *Food Australia* 58: 589–592.

Grocery Manufacturers Association. 2009. Control of Salmonella in low-moisture foods. Washington, DC

Guzev, L., A. Danshin, T. Zahavi, A. Ovadia, and A. Lichter. 2008. The effects of cold storage of table grapes, sulphur dioxide and ethanol on species of black *Aspergillus* producing ochratoxin A. *International Journal of Food Science & Technology* 43: 1187–1194.

Hammond, R., D. Bodager, K. Ward, and A. Rowan. 2001. Case studies in foodborne illness in Florida from fresh produce. *Hortscience* 36: 22–25.

Hazen, R., A. Bushway, and K. Davis-Dentici. 2001. Evaluation of the microbiological quality of IQF processed Maine wild blueberries: an "in plant study". *Small Fruits Review* 1: 47–59.

Isaacs, S., J. Aramini, B. Ceibin, J.A. Farrar, R. Ahmed, D. Middleton, A.U. Chandran, L.J. Harris, M. Howes, E. Chan, A.S. Pichette, K. Campbell, A. Gupta, L.J. Lior, M. Pearce, C. Clark, F. Rodgers, F. Jamieson, I. Brophy, and A. Ellis. 2005. An international outbreak of salmonellosis associated with raw almonds contaminated with a rare phage type of *Salmonella* Enteritidis. *Journal of Food Protection* 68: 191–198.

Kärenlampi, R., and M. Hänninen. 2004. Survival of *Campylobacter jejuni* on various fresh produce. *International Journal of Food Microbiology* 97: 187–195.

Killalea, D., L.R. Ward, D. Roberts, J. de Louvois, F. Sufi, J.M. Stuart, P.G. Wall, M. Susman, M. Schwieger, P.J. Sanderson, I.S.T. Fisher, P.S. Mead, O.N. Gill, C.L.R. Bartlett, and B. Rowe. 1996. International epidemiological and microbiological study of outbreak of *Salmonella* Agona infection from a ready to eat savoury snack – I: England and Wales and the United States. *British Medical Journal* 313: 1105–1107.

Kim, B.U., and L.J. Harris. 2006. The effect of pre-treatmens on the reduction of *Salmonella* Enteritidis PT 30 on almonds during dry roasting. Proceedings of the 93rd annual meeting of the international association of food protection. Calgary, AB, Canada.

King Jr., A.D., and T. Jones. 2001. Nut meats. In *Compendium of methods for the microbiological examination of foods*, ed. F.P. Downes and K. Ito, 561–563. Washington, DC: American Public Health Association.

King Jr., A.D., M.J. Miller, and L.C. Eldridge. 1970. Almond harvesting, processing, and microbial flora. *Applied Microbiology* 20: 208–214.

Kirk, M.D., C.L. Little, M. Lem, M. Fyfe, D. Genobile, A. Tan, J. Threlfall, A. Paccagenella, D. Lightfoot, H. Lyi, L. McIntyre, L. Ward, D.J. Brown, S. Surnam, and I.S.T. Fisher. 2004. An outbreak due to peanuts in their shell caused by *Salmonella enterica* serotypes Stanley and Newport – sharing molecular information to solve international outbreaks. *Epidemiology and Infection* 132: 571–577.

Kniel, K., D. Lindsay, S. Sumner, C. Hackney, M. Pierson, and J. Dubey. 2002. Examination of attachment and survival of *Toxoplasma gondii* oocysts on raspberries and blueberries. *Journal of Parasitology* 88: 790–793.

Kokal, D. 1965. Viability of *Escherichia coli* on English walnut meats (*Juglans regia*). *Journal of Food Science* 30: 325–332.

Kotzekidou, P. 1998. Microbial stability and fate of *Salmonella* Enteritidis in halva, a low-moisture confection. *Journal of Food Protection* 61: 181–185.

Kurdziel, A., N. Wilkinson, S. Langton, and N. Cook. 2001. Survival of poliovirus on soft fruit and salad vegetables. *Journal of Food Protection* 64: 706–709.

Laycock, A.J., T.L. Jaico, M. Sharma, and K.E. Kniel. 2008. Survival of *Salmonella* Heidelberg in hummus. Presented at Institute of Food Technologists Annual Meeting, New Orleans.

Le Guyader, F., C. Mittelholzer, L. Haugarreau, K. Hedlund, R. Alsterlund, M. Pommepuy, and L. Svensson. 2004. Detection of noroviruses in raspberries associated with gastroenteritis outbreak. *International Journal of Food Microbiology* 97: 179–186.

Ledet Muller, L., M. Hjertquvist, L. Payne, H. Pettersson, A. Olsson, L. Plym Forshell, and Y. Andersson. 2007. Cluster of *Salmonella* Enteritidis in Sweden 2005–2006 – Suspected source: Almonds. Eurosurveillance Vol. 12. http://www.eurosurveillance.org/em/v12n06/1206-225.asp. Accessed 3 Dec 2010.

Lupien, J. 2005. Food quality and safety: traceability and labeling. *Critical Reviews in Food Science and Nutrition* 45: 119–123.

Luther, M., J. Parry, J. Moore, J. Meng, Y. Zhang, Z. Cheng, and L. Yu. 2007. Inhibitory effect of Chardonnay and black raspberry seed extracts on lipid oxidation in fish oil and their radical scavenging and antimicrobial properties. *Food Chemistry* 104: 1065–1073.

Lynch, M., R. Tauxe, and C. Hedberg. 2009. The growing burden of foodborne outbreaks due to contaminated fresh produce: risks and opportunities. *Epidemiology and Infection* 137: 307–315.

NASS. 2009. National Agricultural Statistical Service. USDA, Washington, DC. http://usda.mannlib.cornell.edu/MannUsda/viewDocumentInfo.do?documentID=1113. Accessed 3 Dec 2010.

Notermans, S., J. van Zandvoort-Roelofsen, A. Barendsz, and J. Beczner. 2004. Risk profile for strawberries. *Food Protection Trends* 24: 730–739.

O'Brien, S., S. Brustin, G. Duckworth, and L. Ward. 1999. *Salmonella* Java phage type Dundee – Rise in cases in England: Update. http://www.eurosurveillance.org/ew/1999/990318.asp. Accessed 3 Dec 2010.

O'Grady, K. A., J. Powling, A. Tan, M. Valcanis, D. Lightfoot, J. Gregory, K. Lalor, R. Guy, B. Ingle, R. Andrews, S. Crerar, and R. Stafford. 2001. *Salmonella* Typhimurium DT104 – Australia, Europe. Archive no. 20010822.1980. http://www.promedmail.org. Accessed 3 Dec 2010.

O'Mahony, M., E. Mitchell, R.J. Gilbert, N.D. Hutchinson, N.T. Begg, J.C. Rodhouse, and J.E. Morris. 1990. An outbreak of foodborne botulism associated with contaminated hazelnut yoghurt. *Epidemiology and Infection* 104: 385–395.

Osterholm, M. 1997. Cyclosporiasis and raspberries – lessons for the future. *The New England Journal of Medicine* 336: 1597–1598.

Pao, S., A. Kalantri, and G. Huang. 2006. Utilizing acidic sprays for eliminating *Salmonella enterica* on raw almonds. *Journal of Food Science* 71: M14–M19.

Parish, M. 1997. Public health and nonpasteurized fruit juices. *Critical Reviews in Microbiology* 23: 109–119.

Park, E.-J., S.-W. Oh, and D.-H. Kang. 2008. Fate of *Salmonella* Tennessee in peanut butter at 4 and 22°C. *Journal of Food Science* 73: M82–M86.

Parry, J., L. Su, M. Luther, K. Zhou, M.P. Yurawecz, P. Whittaker, and L. Yu. 2005. Fatty acid composition and antioxidant properties of cold-pressed marionberry, boysenberry, red raspberry, and blueberry seed oils. *Journal of Agricultural and Food Chemistry* 53: 566–573.

Penteado, A., B. Eblen, and A. Miller. 2004. Evidence of *Salmonella* internalization into fresh mangos during simulated postharvest insect disinfestation procedures. *Journal of Food Protection* 67: 181–184.

Pereira, K., F. Schmidt, A. Guaraldo, R. Franco, V. Dias, and L. Passos. 2009. Chagas' disease as a foodborne illness. *Journal of Food Protection* 72: 441–446.

Pitt, J.I., A.D. Hocking, K. Bhudhasamia, B.F. Miscamble, K.A. Wheeler, and P. Tanboon-Ek. 1993. The normal microflora of commodities from Thailand. 1. Nuts and oilseeds. *International Journal of Food Microbiology* 20: 211–216.

Pönkä, A., L. Maunula, C.-H. von Bonsdorff, and O. Lyytikäinen. 1999. An outbreak of calicivirus associated with the consumption of frozen raspberries. *Epidemiology and Infection* 123: 469–474.

Puupponen-Pimiä, R., L. Nohynek, H. Alakomi, and K. Oksman-Caldentey. 2005a. Bioactive berry compounds – novel tools against human pathogens. *Applied Microbiology and Biotechnology* 67: 8–18.

Puupponen-Pimiä, R., L. Nohynek, S. Hartmann-Schmidlin, M. Kähkönen, M. Heinonen, K. Määttä-Riihinen, and K. Oksman-Caldentey. 2005b. Berry phenolics selectively inhibit the growth of intestinal pathogens. *Journal of Applied Microbiology* 98: 991–1000.

Riemann, H. 1968. Effect of water activity on the heat resistance of *Salmonella* in "dry" materials. *Applied Microbiology* 16: 1621–1622.

Riyaz-Ul-Hassan, S., V. Verma, A. Malik, and G.N. Qazi. 2003. Microbiological quality of walnut kernels and apple juice concentrate. *World Journal of Microbiology and Biotechnology* 19: 845–850.

Rodriguez-Palacios, A., H. Staempfli, T. Duffield, and J.S. Weese. 2007. *Clostridium difficile* in retail ground meat. *Canadian Emerging Infectious Diseases* 13: 485–487.

Rupnik, M. 2007. Is *Clostridium difficile*-associated infection a potentially zoonotic and foodborne disease? *Clinical Microbiology and Infection* 13: 457–459.

Schaffner, C.P., K. Mosbach, V.C. Bibit, and C.H. Watson. 1967. Coconut and *Salmonella* infection. *Applied Microbiology* 15: 471–475.

Scheil, W., S. Cameron, C. Dalton, C. Murray, and D. Wilson. 1998. A South Australian *Salmonella* Mbandaka outbreak investigation using a database to select controls. *Australian and New Zealand Journal of Public Health* 22: 536–539.

Scott, P. 2001. Analysis of agricultural commodities and foods for *Alternaria* mycotoxins. *Journal of AOAC International* 84: 1809–1817.

Shachar, D., and S. Yaron. 2006. Heat tolerance of *Salmonella enterica* serovars Agona, Enteritidis, and Typhimurium in peanut butter. *Journal of Food Protection* 69: 2687–2691.

Shohat, T., M.S. Green, D. Merom, O.N. Gill, A. Reisfeld, A. Matas, D. Blau, N. Gal, and P.E. Slater. 1996. International epidemiological and microbiological study of outbreak of *Salmonella agona* infection from a ready to eat savoury snack – II: Israel. *British Medical Journal* 313: 1107–1109.

Siro, I., F. Devlieghere, L. Jacxsens, M. Uyttendaele, and J. Debevere. 2006. The microbial safety of strawberry and raspberry fruits packaged in high-oxygen and equilibrium-modified atmospheres compared to air storage. *International Journal of Food Science and Technology* 41: 93–103.

Sivapalasingam, S., C. Friedman, L. Cohen, and R. Tauxe. 2004. Fresh produce: A growing cause of outbreaks of foodborne illness in the United States, 1973 through 1997. *Journal of Food Protection* 67: 2342–2353.

St. Clair, V.J., and M.M. Klenk. 1990. Performance of three methods for the rapid identification of *Salmonella* in naturally contaminated foods and feeds. *Journal of Food Protection* 53: 161–164.

Tauxe, R.V., S.J. O'Brian, and M. Kirk. 2008. Outbreaks of food-borne diseases related to international food trade. In *Imported food: Microbiological issues and challenges*, ed. M.P. Doyle and M.C. Erickson, 69–112. Washington, DC: ASM Press.

Taylor, J.L., J. Tuttle, T. Pramukul, K. O'Brien, T.J. Barrett, B. Jolbaito, Y.L. Lim, D.J. Vugia, J.G. Morris Jr., R.V. Tauxe, and D.M. Dwyer. 1993. An outbreak of cholera in Maryland associated with imported commercial frozen fresh coconut milk. *Journal of Infectious Diseases* 167: 1330–1335.

Todd, E., J. Greig, C. Bartleson, and B. Michaels. 2007. Outbreaks where food workers have been implicated in the spread of foodborne disease. Part 2. Description of outbreaks by size, severity, and settings. *Journal of Food Protection* 70: 1975–1993.

Tournas, V., and M. Stack. 2001. Production of alternariol and alternariol methyl ether by *Alternaria alternata* grown on fruits at various temperatures. *Journal of Food Protection* 64: 528–532.

Tournas, V., J. Heeres, and L. Burgess. 2006. Moulds and yeasts in fruit salads and fruit juices. *Food Microbiology* 23: 684–688.

Uesugi, A.R., and L.J. Harris. 2005. Survival of *Salmonella* Enteritidis PT 30 on almonds after exposure to hot water. Proceedings of the 92nd annual meeting of the international association of food protection, Baltimore.

Uesugi, A.R., M.D. Danyluk, and L.J. Harris. 2006. Survival of *Salmonella* Enteritidis phage type 30 on inoculated almonds stored at −20, 4, 23 and 35°C. *Journal of Food Protection* 69: 1851–1857.

Vojdani, J.D., L.R. Beuchat, and R.V. Tauxe. 2008. Juice-associated outbreaks of human illness in the United States, 1995 through 2005. *Journal of Food Protection* 71: 356–364.

Wang, J., W. Cheung, and D. Grant. 2005. Determination of pesticides in apple-based infant foods using liquid chromatography electrospray ionization tandem mass spectrometry. *Journal of Agricultural and Food Chemistry* 53: 528–537.

Ward, B., and L. Trving. 1987. Virus survival on vegetables spray-irrigated with wastewater. *Water Research* 21: 57–63.

Wareing, P.W., L. Nicolaides, and D.R. Twiddy. 2000. Nuts and nut products. In *The microbiological safety and quality of food*, ed. B.M. Lund, T.C. Baird-Parker, and G.W. Gould, 919–940. Gaithersburg: Aspen Publishers.

Wilson, M.M., and E.F. Mackenzie. 1955. Typhoid fever and salmonellosis due to consumption of infected desiccated coconut. *Journal of Applied Bacteriology* 18: 510–521.

Zhao, Y. 2005. Pathogens in fruit. In *Improving the safety of fresh fruit and vegetables*, ed. W. Jongen, 44–88. Cambridge: Woodhead.

Zhao, Y. 2007. *Berry fruit*. Boca Raton: CRC Press.

Safety of Dairy Products

Elliot T. Ryser

1 Introduction

Milk, defined here as the lacteal secretion practically free from *colostrum* obtained by the complete milking of one or more healthy cows, sheep, or goats, is the "life blood" of the dairy industry; approximately 190 billion pounds of milk was produced in the United States during 2008 (NASS-USDA 2009). Before collection, milk within the udder of a healthy animal is sterile. However, the same is not true for animals suffering from mastitis, which is an infection of their mammary glands. Milk from these animals harbors potentially high numbers of bacteria, some of which are well-recognized human milkborne *pathogens*. During milking, a wide range of beneficial lactic acid bacteria as well as numerous spoilage organisms (e.g., Gram-negative *psychrotrophs*) and bacterial pathogens (e.g., *Salmonella, Listeria*) present in the teat canal and on the external teat surface can contaminate the milk. Under certain circumstances, additional microorganisms entering from the udder/teat surface, bedding material, manure, feed, milking equipment, and/or milk handlers substantially increase the numbers of *psychrotrophic* and *mesophilic* bacteria. Because milk is a highly nutritious growth medium containing approximately 5% lactose, 3.5% fat, and 3% protein (primarily casein), it is imperative that the product be rapidly chilled to a temperature of 1–4°C after milking to minimize microbial growth.

Reported dairy-related outbreaks of illness date back to the inception of the dairy industry, as previously reviewed elsewhere by the author (Ryser 1999). Before World War II, a group of diseases now of primarily historical concern – namely, diphtheria, scarlet fever, tuberculosis, and typhoid fever – were frequently linked to the consumption of raw milk contaminated from ill or previously infected milk handlers. Following the adoption of thermal *pasteurization* of milk as a standard commercial practice in the United States, the involvement of dairy products in foodborne and waterborne outbreaks has decreased drastically, with the majority of foodborne outbreaks now traced to vegetables, seafood, multi-ingredient products, fresh produce, poultry, and beef (CSPI 2007). These earlier historical causes of milk-related illness have now been replaced by such widely recognized bacterial foodborne pathogens as *Salmonella, Campylobacter, Escherichia coli* O157:H7, and *Listeria monocytogenes*, all of which are present in about 1–4% of the raw milk produced in the United States and other developed countries. Hence, it is not surprising that most recent dairy-related outbreaks have involved dairy farmers who regularly consume raw milk (Jayarao et al. 2006) and a small but growing group of raw milk proponents, all of whom comprise <1% of all milk drinkers.

E.T. Ryser (✉)
Department of Food Science and Human Nutrition, Michigan State University, East Lansing, MI, USA
e-mail: ryser@msu.edu

2 Trends in Dairy-Related Outbreaks

The microbiological causes of milkborne diseases have undergone dramatic changes over time. However, more than 90% of all dairy-related illnesses are still of bacterial origin, with 15 of the 21 recognized milkborne bacterial pathogens responsible for most cases of illness (Ryser 1999). Before milk pasteurization became a standard practice in the late 1940s, two bacterial pathogens now primarily of historical concern (*Salmonella typhi* and *Streptococcus pyogenes*, which cause typhoid and scarlet fever, respectively) were responsible for 50–80% and 15–27% of all milkborne cases of illness (Table 1) (Bryan 1983). Due to the unavailability of antibiotics, many of these cases frequently proved fatal, as did more sporadic dairy-related cases of milkborne diphtheria caused by *Corynebacterium diphtheria* and tuberculosis caused by *Mycobacterium tuberculosis* or *Mycobacterium bovis*. The majority of these cases were usually traced back to infected or ill milk handlers who commonly contaminated the milk at the time of collection.

Major improvements in sanitation, milk-handling procedures, and animal health coupled with the adoption of the routine pasteurization of milk and cream in the late 1940s eliminated the threat of milkborne typhoid and scarlet fever in all but a few developing countries. However, during and shortly after World War II, brucellosis, staphylococcal poisoning, and *salmonellosis* emerged as major threats to public health.

During the early 1970s, *Campylobacter* was identified as a major public health concern in the United States among those individuals who still consumed raw milk. This pathogen was responsible for 37 of 51 (72.5%) raw milk-related outbreaks reported from 1990 to 2006 (CSPI 2007). A similarly high incidence of such campylobacteriosis cases has also been reported in England and Scotland (Robinson and Jones 1981). In 1985, 85 primarily Hispanic individuals died of listeriosis after consuming California-made Mexican-style cheese containing *L. monocytogenes*. This rare illness was epidemiologically linked to the consumption of pasteurized milk in Massachusetts 2 years earlier. Since the 1990s, the consumption of raw milk contaminated with *E. coli* O157:H7 – the leading cause of acute kidney failure in children – has led to sporadic outbreaks of illness in the United States, Canada, and England. More recently, *Cronobacter sakazakii* has become a major concern in powdered infant formula. In addition, the possible role of the milkborne pathogen *Mycobacterium avium* subsp. *paratuberculosis* in *Crohn's disease*, a chronic and often severe form of colitis in humans, is still being debated.

While able to cause potentially serious health problems, the *rickettsiae* (e.g., *Coxiella burnetti* – Q fever) and parasites (e.g., *Cryptosporidium*, *Toxoplasma*) have each been most recently responsible for less than 1% of all dairy-related outbreaks. Mycotoxins such as *aflatoxin* M_1 also pose minimal public health concerns in developed countries. More recently, the safety of the milk supply has been questioned due to the emergence of *mad cow disease*. However, no evidence currently exists for shedding of the prion, the infectious protein responsible for this fatal disease, in milk.

Historically, more than 90% of all dairy-related outbreaks of known cause have been of bacterial origin, with the vast majority traced to raw milk and certain soft cheeses illegally prepared from raw milk (Yilmaz et al. 2009). Ice cream, nonfat dry milk, and butter pose far less concern due to their decreased ability to support bacterial growth (Table 2). During 1990–2006, a total of 79 milk-related outbreaks was reported, 51 (65%) of which were traced to the consumption of raw milk. Of these confirmed raw milk outbreaks, *Campylobacter*, enterohemorrhagic *E. coli*, and *Salmonella* were responsible for 72%, 20%, and 8%, respectively (Table 3). During the same period, 31 salmonellosis outbreaks were traced to the consumption of ice cream and frozen custard, most of which was prepared at home with raw eggs added to the mix. Hence, the eggs were the most likely source for *Salmonella*, particularly *Salmonella* serotype Enteritidis. Various cheeses, particularly certain soft varieties such as Brie, Camembert, and Mexican-style cheese prepared illegally from raw milk, were implicated in 62 outbreaks during this same period, 37% of which involved *Salmonella*. Additionally,

Table 1 Dairy-related outbreaks reported in the United States, 1900–1987 and 1990–2006

Cause	1900–1909	1910–1919	1920–1929	1930–1939	1940–1949	1950–1959	1960–1969	1970–1979	1980–1982	1983–1987	1990–2006
Bacillus cereus	–	–	–	–	–	–	–	–	1	–	1
Brucella	–	–	–	1	8	4	9	1	–	3	1
Campylobacter	–	–	–	–	–	–	–	3	40	32	28
Clostridium botulinum	–	1	<1	1	–	1	–	–	–	–	<1
Corynebacterium diphtheriae	8	2	4	1	1	–	–	–	–	–	–
Enterotoxigenic E. coli	–	–	–	–	–	–	–	1	1	8	–
Enterohemorrhagic E. coli	–	–	–	–	–	–	–	–	–	–	8
Listeria monocytogenes	–	–	–	–	–	–	–	–	–	–	3
Mycobacterium tuberculosis	<1	<1	<1	<1	<1	<1	–	–	–	–	–
Salmonella	–	1	3	2	7	21	28	42	50	19	34
Salmonella typhi (typhoid)	78	80	68	50	17	3	–	–	–	–	–
Streptococcus pyogenes	15	15	18	27	8	–	–	–	–	–	–
Shigella	–	–	1	2	4	3	–	–	–	–	2
Staphylococcus aureus	–	–	–	8	26	50	30	5	–	–	5
Yersinia	–	–	–	–	–	–	–	–	–	–	<1
Total rickettsial	–	–	–	–	–	1	–	–	–	–	–
Total parasites	–	–	–	–	–	–	–	1	–	–	<1
Total viral	–	<1	<1	–	1	–	2	1	1	–	14
Total chemical	–	–	–	–	–	–	–	4	<1	11	2
Unknown etiology	<1	1	3	8	26	17	31	41	6	24	<1

Source: Adapted from Bryan (1983), Bean et al. (1990), CSPI (2007), and Olsen et al. (2000)

Table 2 Percentage of outbreaks involving various dairy products in the United States, 1900–2006

Product	1900–1909	1910–1919	1920–1929	1930–1939	1940–1949	1950–1959	1960–1969	1970–1979	1980–1989	1990–2006
Milk	–	–	–	–	–	–	36	–	62	12
Raw milk/cream	100	86	93	90	54	36	–	23	–	36
Certified raw milk	–	–	1	–	–	2	–	4	–	–
Pasteurized milk	–	4	3	3	16	2	–	3	–	–
Nonfat dry milk	–	–	–	–	2	2	–	4	–	1
Cheese	–	1	<1	3	9	34	20	17	19	14
Cheese, raw milk	–	–	–	–	–	–	–	–	–	12
Butter	–	1	–	–	<1	–	–	4	–	1
Ice cream	<1	8	2	4	17	24	44	44	19	24

Source: Adapted from Bean et al. (1996), Bryan (1983), CSPI (2007), MacDonald and Griffin (1986), and Olsen et al. (2000)

Table 3 Number of dairy-related outbreaks by product and cause, 1990–2006

Causative agent	Milk	Ice cream	Cheese	Sour cream	Butter
Bacillus cereus	1	0	1	0	0
Brucella	0	0	2	0	0
Campylobacter	45 (37)[a]	2	5	0	0
Clostridium botulinum	0	0	1	0	0
Enterohemorrhagic *E. coli*	13 (10)	1	2	0	0
Listeria monocytogenes	1	0	4	0	0
Salmonella	10 (4)	31	23	0	0
Shigella	0	0	2	0	1
Staphylococcus aureus	2	4	2	1	0
Yersinia	1	0	0	0	0
Parasites	0	1	0	0	0
Norovirus	1	2	19	0	2
Hepatitis A	0	0	1	0	0
Total chemical	1	2	0	0	0
Unknown etiology	4	0	0	0	0

Source: Adapted from CSPI (2007)
[a]Unpasteurized milk confirmed

Table 4 Number of food safety recalls issued by FDA for dairy products, 1990–2009

| Agent | Product | | | | | |
	Fluid milk and cream	Butter	Ice cream and frozen yogurt	Dried milk and whey	Cheese	Total (%)
L. monocytogenes	2	6	35	–	71	114 (77)
Salmonella	–	–	4	2	5	11 (7)
E. coli	–	–	–	–	3	3 (2)
S. aureus	–	1	–	–	3	4 (3)
C. botulinum	–	–	–	–	7	7 (5)
Cryptosporidium	–	–	–	–	1	1 (0.5)
Aflatoxin	1	–	–	–	–	1 (0.5)
Yeast/mold	–	–	–	–	7	7 (5)

Source: Adapted from FDA (2009)

19 outbreaks involved norovirus-contaminated cheeses that were purchased from restaurants or delicatessens. However, because norovirus is of human fecal origin, these cheeses were most likely contaminated by an infected restaurant or delicatessen worker at the time of purchase. Hence, such outbreaks that are more closely linked to improper food handling than to dairy products will not be considered further.

In accordance with the *Federal Food, Drug and Cosmetic Act* of 1938, an examination of the voluntary recalls issued by dairy manufacturers for pasteurized milk and other dairy products containing pathogenic microorganisms provides additional insight into current public health concerns facing the dairy industry. As seen in Tables 3 and 4, the "zero tolerance" for *L. monocytogenes* in milk and other dairy products, established in 1985 in response to the aforementioned cheese-related outbreak of listeriosis in California, has decreased the number of dairy-related outbreaks, especially those involving fluid milk and cheese (Box 1).

> **Box 1** FDA Recalls of Dairy Products from 1990 to 2009
>
> A total of 148 microbiologically related FDA recalls were issued for dairy products from 1990 to 2009. Of the 114 (78%) recalls due to *L. monocytogenes* contamination, 71 (62%) were associated with cheese and and 35 (31%) with ice cream/frozen yogurt. While certain *Listeria*-contaminated soft cheeses that support the growth of *Listeria* to high numbers constitute a significant health risk to certain segments of the population, including pregnant women, immunocompromised individuals, and the elderly, the same is not true for ice cream and other frozen dairy products. Given the complete absence of any listeriosis cases linked to the consumption of commercially produced ice cream, frozen yogurt, or related frozen dairy desserts, several industry-sponsored petitions have been put forward to relax the current "zero tolerance" policy for *L. monocytogenes* in ice cream and other products in which this pathogen is unable to grow.

3 Development of Pasteurization

The pasteurization process was developed in 1864–1865 by the famous French scientist Louis Pasteur, who found that heating wine at 50–60°C rendered it free of the spoilage microorganisms responsible for abnormal fermentations. He later reported that milk souring from the growth of undesirable microorganisms could be prevented by using a similar heat treatment procedure. In the United States, the thermal processing of milk began in the late 1880s, with interest in pasteurization growing steadily during the early 1900s as a means of inactivating *Salmonella typhi, Streptococcus pyogenes, Mycobacterium tuberculosis*, and *Corynebacterium diphtheriae*, all of which were endemic in raw milk.

The primary purpose of pasteurization is to reduce the numbers of microbial pathogens that could be anticipated in milk to levels that do not constitute a risk to human health. The two most commonly used commercial pasteurization treatments for fluid milk include low-temperature, long-time (LTLT) pasteurization at not less than 62.8°C for no less than 30 min, and continuous high-temperature, short-time (HTST) pasteurization at not less than 71.1°C for a minimum of 15 s. These pasteurization methods will inactivate *Coxiella burnetti* and *M. tuberculosis* (the two most heat-resistant milkborne pathogens) and all other bacterial pathogens of concern in raw milk, including *Salmonella, Campylobacter, L. monocytogenes,* and *E. coli* O157:H7. However, certain *thermoduric* bacteria (e.g., *Lactococcus* and *Streptococcus*) and spore-forming bacteria (e.g., *Bacillus* and *Clostridium*) can survive pasteurization and adversely affect the product quality during cold storage (Ryser 2003). An alternative process known as ultrahigh-temperature (UHT) pasteurization, in which milk is heated to at least 135°C for a minimum of 1 s, has been used in combination with *aseptic processing/packaging* to produce a shelf-stable product that can be kept unrefrigerated until opened. In addition to reducing all human milkborne pathogens to acceptable numbers, all pasteurization procedures minimize any adverse chemical, physical, and organoleptic changes that may occur during heating. Although nonthermal pasteurization methods, including high-pressure processing and ultraviolet light, are now attracting some attention, thermal pasteurization has been used very successfully for more than 120 years and will likely remain the treatment of choice to render milk, cream, and other dairy products safe for consumption.

In 1923, the U.S. Public Health Service began publishing summaries of all suspected outbreaks of milkborne disease. These efforts led to the passage of the first Model Milk Ordinance, of 1924, which was eventually followed by the Grade "A" Pasteurized Milk Ordinance of 1963, both of which stressed the need for the nationwide pasteurization of milk. The latest version of the Grade "A"

> **Box 2** The Risk of Raw Milk Consumption
>
> Of the 79 raw milk-related outbreaks reported from 1990 to 2006, only 11 occurred before 2000. During this same 17-year period, the consumption of tainted raw milk was responsible for 82% of *Campylobacter*, 77% of *Escherichia coli* O157:H7, and 40% of *Salmonella* outbreaks. Hence, the majority of the milk-related outbreaks could be eliminated by avoiding the consumption of raw milk (CSPI 2007).

Pasteurized Milk Ordinance, published in 2009, dictates the requirements for pasteurization and incorporates a series of provisions governing the processing, packaging, and sale of all Grade "A" milk and milk products, including cream, buttermilk, whey, dry milk, and condensed milk (DHHS/PHS/FDA 2009). For more details on the Grade "A" milk ordinance, refer to Chap. 14.

Despite the widespread adoption of milk pasteurization by the late 1940s and the ban on the interstate shipment and sale of raw milk after 1973, 26 U.S. states have legalized the intrastate sale of raw milk, 15 of which require a warning label on the container clearly stating that the product has not been pasteurized and may therefore contain harmful bacteria. This raw milk is usually acquired locally for its supposed enhanced "nutritional and health benefits" (Yilmaz et al. 2009). In response to this growing number of pasteurized milk opponents, who falsely claim that milk pasteurization leads to decreased nutritional value, lactose intolerance, allergic reactions, arthritis, and the growth of bacterial pathogens, raw milk has now become readily available through various "cow share" and "cow leasing" programs. Under these programs, individuals lease or purchase shares in a dairy herd, which entitles them to the raw milk from their animals, which are managed by a dairy farmer (Box 2). In addition, some raw milk is also being legally sold for pets to circumvent the federal trade laws. In reality, the vast majority of dairy-related illnesses among consumers could be eliminated by avoiding raw milk and certain soft cheeses that have been illegally prepared from raw milk in the United States or illegally imported.

4 Major Pathogens of Current Concern

Public health concerns surrounding the safety of dairy products have changed dramatically over the last 100 years and will continue to change in response to ongoing advances in animal husbandry, milk-handling practices, dairy sanitation, outbreak detection strategies, diagnostics, and treatment of disease. The infectious agents of such rampant pre-World War II scourges as typhoid, scarlet fever, and diphtheria have been replaced by an important group of *gastroenteritis*-causing bacterial pathogens, namely, *Salmonella*, *Campylobacter*, *Yersinia enterocolitica*, *Bacillus cereus*, and *Staphylococcus aureus*. However, other less common milkborne bacterial pathogens, including *L. monocytogenes*, *Brucella* spp., *Clostridium botulinum*, and enterohemorrhagic *Escherichia coli* (*E. coli* O157:H7), have become of equal or greater importance due to the severity of their diseases and their high fatality rate (Table 5).

4.1 Salmonella in Dairy Products

Currently, the leading cause of dairy-related illness in the United States and Western Europe is *Salmonella* infections, which are highest among consumers of raw milk and certain raw milk cheeses (Box 3). In addition to humans, dairy cattle and other ruminants are also prone to *Salmonella*

Table 5 Historical, current, uncommon, and emerging public health concerns regarding milk and dairy products

Historical	Current	Uncommon	Emerging
Diphtheria	*Campylobacter*	Aflatoxin	*Cronobacter sakazaxii*
Poliomyelitis	*E. coli*	*Bacillus cereus*	Crohn's disease
Scarlet fever	*Listeria monocytogenes*	Brucellosis	Creutzfeldt–Jakob disease
Septic sore throat	*Salmonella*	*Citrobacter freundii*	Cryptosporidiosis
Tuberculosis		*Clostridium botulinum*	
		Corynebacterium ulcerans	
		Haverhill fever	
		Histamine poisoning	
		Infectious hepatitis	
		Mycotoxins	
		Q fever	
		Shigella	
		Staphylococcus aureus	
		Streptococcus zooepidemicus	
		Tickborne encephalitis	
		Toxoplasmosis	
		Yersinia enterolitica	

Source: Adapted from FDA enforcement reports (1990–2009)

Box 3 Contamination Can Result in Large Outbreaks of Foodborne Pathogens

Two particularly massive outbreaks of *Salmonella* in the United States have been traced to pasteurized milk and ice cream. In the first of these outbreaks in 1985, at least 16,000 culture-confirmed cases of salmonellosis, including 18 fatalities, were traced to one particular brand of pasteurized 2% milk marketed in Chicago. This milk was contaminated with a highly infectious, multi-antibiotic-resistant strain of *Salmonella* serovar Typhimurium. A subsequent follow-up survey estimated the total number of cases at nearly 200,000, making this the second-largest foodborne outbreak of salmonellosis ever recorded (Ryan et al. 1987; Schuman et al. 1989). The source of contamination in this state-of-the-art milk processing facility, which never reopened, was traced to a potential cross-connection between several holding tanks that would have allowed raw milk to contaminate pasteurized skim milk, condensed milk, and cream. The second and largest outbreak ever documented occurred 9 years later in Minnesota, Wisconsin, and South Dakota, with an estimated 240,000 cases of gastroenteritis traced to commercially produced ice cream containing *Salmonella* serovar Enteritidis, an organism typically associated with poultry and eggs (Vought and Tatini 1998; Hennessey et al. 1996). Follow-up investigations revealed that tankers used to haul raw liquid eggs were also used to haul pasteurized ice cream mix to the ice cream factory, where the mix was not repasteurized.

infections, with symptomatic and asymptomatic fecal shedding of this pathogen. However, *Salmonella* is unable to colonize the udder and is therefore seldom associated with mastitis. This pathogen is a frequent contaminant of raw milk, with 2.6–6.1% of such samples from farm bulk tanks testing positive in recent U.S. surveys (Jayarao and Henning 2001; Jayarao et al. 2006; van Kessel et al. 2004). HTST and LTLT pasteurization both provide a wide margin of safety in eliminating expected levels of salmonellae in milk [e.g., <100 *colony-forming units (CFU)*/ml], with populations of *Salmonella* decreasing in milk, cheese, butter, yogurt, and other dairy products during refrigerated storage at temperatures less than 7°C (Ryser 1999).

In addition to many smaller localized outbreaks traced to the consumption of raw milk and homemade frozen custard prepared from contaminated eggs, *Salmonella* has also been a public health concern in nonfat dry milk and various cheeses, including Cheddar, Mozzarella, Mexican-style, and goat's milk cheese. In 1966, 29 cases of salmonellosis, among children predominantly, were diagnosed in 17 states over a 10-month period. These cases were linked to nonfat dry milk containing *Salmonella* serovar Newbrunswick, a serovar seldom seen in the United States (Collins et al. 1968). Incomplete pasteurization of the milk before spray-drying and post-processing contamination were cited as likely causes because *Salmonella* can reportedly survive in heavily inoculated nonfat dry milk during storage for up to 2 months (LiCari and Potter 1970).

Among the wide range of fermented dairy products produced, cheese has been most often implicated in outbreaks of illness involving *Salmonella*, enterohemorrhagic *E. coli*, and *L. monocytogenes*, with relatively few safety concerns associated with sour cream, buttermilk, yogurt, or other fermented products. Most of the cheeses implicated in foodborne illnesses in the United States have been soft varieties (e.g., farmstead and Mexican-style) that were illegally prepared from raw milk (Ryser 1999); however, similar outbreaks have been traced to soft cheeses legally prepared from raw milk in Europe (Dominguez et al. 2009). Cheddar cheese containing very low numbers of *Salmonella* was responsible for three notable North American outbreaks over the past 30 years. In the first of these outbreaks, the consumption of Cheddar cheese produced in Kansas from pasteurized milk was responsible for 339 cases of gastroenteritis (Fontaine et al. 1980), with the remaining two outbreaks involving more than 2,000 cases of illness traced to Cheddar cheese prepared from raw or subpasteurized milk in Canada (Bezanson et al. 1985). Several earlier studies have demonstrated the long-term persistence of *Salmonella* in Cheddar cheese, with this pathogen surviving 84 and 300 days in Cheddar cheese aged at 13 and 7°C, respectively (Goepfert et al. 1968; Park et al. 1970). While these findings clearly indicate the ability of *Salmonella* to survive well beyond the required 60-day aging period at ≥1.7°C for Cheddar and other hard cheeses that can be legally prepared from raw milk, very few outbreaks of salmonellosis or other dairy-related diseases have been linked to such aged cheeses, suggesting that the current 60-day aging rule is still satisfactory.

4.2 Importance of Contamination with Campylobacter jejuni

Long recognized as an important cause of abortion in cattle, *Campylobacter jejuni* is now the leading cause of foodborne bacterial gastroenteritis in Canada and Europe, and the third most common bacterial foodborne pathogen in the United States (Scallon et al. 2011). However, *Campylobacter* remained a relatively obscure foodborne pathogen until the late 1970s due to difficulties in culturing this bacterium from milk. Typical *attack rates* of 50% or greater in raw milk-related outbreaks indicate that the oral infectious dose is extremely low, further supporting the importance of this organism as a major milkborne pathogen.

Campylobacter spp. are easily inactivated in fluid dairy products during pasteurization and are unable to survive in Cheddar (Ehlers et al. 1982; Brodsky 1984) and other cheeses during manufacture and 60 days of ripening (Ryser 1999). The incidence of *Campylobacter* in the raw milk supply continues to be very low, with *C. jejuni* generally recovered from only 0.4–2.0% of the raw milk bulk tank samples examined in the United States (Doyle and Roman 1982; Jayarao et al. 2006; Lovett et al. 1983; McManus and Lanier 1987). The natural shedding of this pathogen in raw milk as a result of mastitis is considered rare, with fecal contamination of the milk during or after milking now regarded as the primary route of contamination.

Dairy-related outbreaks involving *Campylobacter*, which date back to 1976, have almost invariably been traced to the consumption of raw or underpasteurized cow's milk. According to Wood et al. (1992), 14 of 20 raw milk-associated campylobacteriosis outbreaks from 1981 to 1990 involved elementary school children who became ill after participating in class-sponsored field trips to dairy farms. While children are no longer given the opportunity to consume raw milk during such visits, raw milk can still be legally purchased in 26 states and is clearly available to dairy farmers as well as others who participate in cow sharing/leasing programs. Hence, these raw milk outbreaks are likely to continue in the future.

4.3 *Escherichia coli* in Dairy Products

Escherichia coli is another common raw milk contaminant that frequently enters pasteurized milk and other finished products as a post-pasteurization contaminant. The vast majority of *E. coli* strains reside as harmless commensals in the gastrointestinal tract of humans and animals. However, three groups of *E. coli* – enterotoxigenic *E. coli* (ETEC), enteroinvasive *E. coli* (EIEC) and enterohemorrhagic *E. coli* (EHEC) – have been responsible for dairy-related illnesses, with EHEC being the leading cause of *hemolytic-uremic syndrome (HUS)* in children. ETEC strains produce one or more *enterotoxins* responsible for "traveler's diarrhea" (refer to Chap.2). The involvement of ETEC in dairy-associated illnesses is limited to one multistate/multinational outbreak of more than 169 cases that was traced to imported French Brie cheese (MacDonald et al. 1985). Unlike ETEC, strains classified as EIEC can penetrate the mucosal lining of the colon to produce a severe bloody diarrhea resembling shigellosis. Dairy-related EIEC infections are also confined to a single report (Marier et al. 1973) in which the consumption of imported French Brie cheese led to 347 cases of illness in and around Washington, DC.

EHEC, which includes *E. coli* O157:H7, is now the most serious *E. coli* threat to the food industry, including those manufacturing dairy products. Dairy cattle are a major reservoir for EHEC, with fecal shedding at rates of 1.9–3.4% in the United States (Doane et al. 2007; Murinda et al. 2002), and raw milk contamination rates of 0.1–3.2% during or after milking (Cobbold et al. 2008; Jayarao et al. 2006,;Karns et al. 2007; Murinda et al. 2002). Most dairy-related outbreaks involving EHEC have been traced to the consumption of raw milk that was legally purchased (Kenne et al. 1997) or obtained through cow share programs (Denny et al. 2008), with one earlier outbreak involving 46 kindergarten children in Canada who consumed raw milk during a dairy farm visit (Borczyk et al. 1987). Cheese has been less frequently involved, with one outbreak linked to Wisconsin-produced cheese curds that were aged less than 60 days (Durch et al. 1998) and a second outbreak in France in which three cases of illness (including two cases of HUS) resulted from consuming a fresh unpasteurized cheese prepared from goat's milk (Espié et al. 2006). In several studies that assessed the fate of *E. coli* O157:H7 during the manufacture and ripening of various cheeses, this pathogen survived 138–210 days in Cheddar cheese prepared from pasteurized milk inoculated at 1–10 CFU/ml (Reitsma and Henning 1996; Schlesser et al. 2006), and persisted in both feta and Camembert during 65–70 days of ripening (Ramsaran et al. 1998). However, *E. coli* O157:H7 was generally unable to grow in Mexican-type cheeses that were inoculated with this pathogen after manufacture (Kasrazadeh and Genigeorgis 1995). While most cheese surveys for *E. coli* O157:H7 have yielded negative results, the consumption of such soft and surface-ripened cheeses that are sometimes illegally prepared from raw milk still poses a potential health risk.

4.4 *Importance of Listeria monocytogenes*

Known as a cause of human illness since 1929, *L. monocytogenes* did not emerge as a major foodborne pathogen until 1985 when a large outbreak of listeriosis in southern California was directly linked to the consumption of contaminated Mexican-style cheese (Ryser 2007a). Although an estimated 1,600 cases of listeriosis, including 250 fatalities, occur annually in the United States (Scallon et al. 2011), most of these cases are confined to pregnant women, newborn infants, the elderly, and immunocompromised individuals, with healthy adults rarely affected. Unlike most other forms of foodborne illnesses, *L. monocytogenes* infections can lead to *meningitis*, abortion, and *perinatal septicemia*. The 15–20% mortality rate for listeriosis, combined with this pathogen's acid and salt tolerance and ability to grow to high numbers in milk and certain varieties of soft and surface-ripened cheeses during refrigerated storage, prompted the United States to adopt its current policy of "zero tolerance" (absence in a 25-g sample) for *L. monocytogenes* in dairy products and all other cooked *ready-to-eat (RTE) foods*. As a result of this policy, *L. monocytogenes* accounted for 78% of all dairy-related recalls (primarily cheese and ice cream) from 1990 to 2009 (Table 4). Since 2000, most of these recalls have involved Mexican-style and soft surface-ripened cheeses (both domestic and imported) in which *Listeria* can grow to high numbers, with no cases of listeriosis linked to the consumption of commercially prepared ice cream.

Dairy farm environments are an important reservoir for *Listeria*, where cows, sheep, and goats can ingest this pathogen and become infected. Both symptomatic and asymptomatic animals may intermittently shed *L. monocytogenes* in their milk for long periods as a result of *Listeria*-related mastitis (an infection of the udder), *encephalitis*, or abortion. Based on numerous surveys of raw milk bulk tanks on the farm, approximately 2.5–5% of the raw milk supply can be expected to contain very low numbers of *L. monocytogenes* (e.g., <10 CFU/ml) at any given time (Ryser 2007b). Despite being one of the most heat-resistant milkborne bacterial pathogens, LTLT and HTST pasteurization are both sufficient to inactivate the expected numbers of *Listeria* in raw milk, with most dairy-related outbreaks traced to products that were either prepared from raw milk or contaminated after manufacture.

As a postprocessing contaminant, *L. monocytogenes* can grow to high numbers in fluid dairy products, including milk, chocolate milk, and cream during refrigerated storage. However, as a result of improved sanitation practices that now minimize the presence of *Listeria* in the dairy plant environment, this pathogen is now seldom isolated from pasteurized fluid milk. Even the very low numbers of *Listeria* occasionally found in ice cream are unable to increase to hazardous levels. Based on the number of product recalls issued (Table 4), *L. monocytogenes* is most frequently found in certain Mexican-style cheeses (e.g., Queso fresco) and soft surface-ripened cheeses (e.g., Brie and Camembert). During the ripening of these cheeses, *Listeria* populations can increase to $\geq 10^6$ CFU/g as the cheese surface becomes less acidic (pH > 6.0). Although growth in cheese generally ceases at a pH < 5.2, *L. monocytogenes* can survive for weeks or months in various cheeses experimentally produced from inoculated milk. In one report this pathogen persisted for 434 days in Cheddar cheese (Ryser and Marth 1987).

Despite its being frequently contaminated, the consumption of raw milk has only resulted in sporadic cases of listeriosis. However, in 1983 one commercial brand of pasteurized milk was epidemiologically linked to 49 cases of listeriosis in Massachusetts. This milk originated from a dairy farm on which veterinarians diagnosed several cases of listeriosis during the outbreak (Fleming et al. 1985). A similar but far smaller outbreak in Massachusetts involved five cases, including two fatalities, and was documented in 2007. In this case the pathogen again contaminated the milk after pasteurization (CDC 2008). The only other milk-related outbreak of listeriosis was traced to the consumption of temperature-abused chocolate milk in which *L. monocytogenes* had grown to levels of 10^8–10^9 CFU/ml (Dalton et al. 1997). However, in contrast to previous outbreaks, the primary symptom was gastroenteritis with no fatalities reported among the 66 victims.

Various cheeses, most importantly the soft and surface-ripened varieties, remain the leading cause of dairy-related listeriosis cases, with seven major outbreaks and numerous sporadic cases reported in both North America and Europe. The first and largest of these outbreaks occurred in the Los Angeles area in mid-1985, when 142 cases, including 42 fatalities, were linked to the consumption of Mexican-style cheese that was prepared illegally from partially pasteurizal milk and where environmental contamination within the factory was also found as a contributing factor (Linnan et al. 1988). Two years later in Switzerland, a 5-year-long listeriosis outbreak involving 122 cases and 34 fatalities was traced to the consumption of Vacherin Mont d'Or soft surface-ripened cheese that was ripened on contaminated wooden shelves in various cheese cellars (Bula et al. 1995). During the mid-1990s, two types of French-produced soft cheese (Brie de Meaux and Pont l'Évêque) were responsible for at least 34 cases of listeriosis in France, with two more recent outbreaks in Canada traced to soft and semihard cheeses prepared from raw milk. In the United States, homemade domestic and imported Queso fresco and other Mexican-style soft cheeses, illegally prepared from raw milk or illegally imported and then sold door-to-door or at open-air markets, continue to pose a risk among the rapidly growing Hispanic population (Ryser 2007a). Hence, pregnant women and others who are at an increased risk of contracting listeriosis are now being advised to avoid these cheeses.

4.5 Bacillus cereus

A common contaminant of raw milk and the dairy farm environment, *B. cereus* is of occasional public health and spoilage concern. *B. cereus* is responsible for both gastroenteritis and the "sweet curdling" of milk (Ryser 1999). Heat-resistant spores produced by this psychrotrophic pathogen survive thermal pasteurization and germinate as a result of the heating, with outgrowth of this organism at abusive temperatures leading to potential cases of illness. Several dairy-related *B. cereus* outbreaks have been associated with pasteurized milk, in which *B. cereus* grew to populations of 10^6 CFU/ml or greater during extended refrigerated storage. However, powdered milk and powdered infant formula pose the greatest risk since both pasteurization and spray-drying will induce germination and outgrowth of the spores in reconstituted products. Hence, dairy-related outbreaks involving this psychrotrophic pathogen are best prevented by storing both fluid and reconstituted milk at <4°C.

4.6 Brucella Species

Brucellosis, a classic *zoonotic* disease, is primarily acquired through direct or indirect contact with infected animals harboring three of six bacterial species belonging to the genus *Brucella*. Two of these species, *Brucella abortus,* which infects cattle, and *Brucella melitensis,* which infects sheep and goats, are of the greatest concern to the dairy industry (Ryser 1999). *Brucella* typically enters raw milk from infected animals that can shed the pathogen for several months as a result of mastitis. However, the incidence of *Brucella* contamination in raw milk in the United States, Canada, and other developed countries is extremely low due to active brucellosis eradication programs in dairy cattle. This pathogen is also readily inactivated in milk by LTLT and HTST pasteurization. In humans, symptoms of brucellosis occurring after an incubation period of 3–21 days may range from a mild flulike illness to undulant fever, which is characterized by sweating, chills, chest and joint pain, weight loss, and anorexia. Dairy-related cases of brucellosis have remained a rare occurrence in most well-developed countries. However, Mexico and some Mediterranean and Middle Eastern countries are seeing a resurgence in human cases due to ongoing infections in domestic livestock. Most dairy-related cases in the United States have been associated with Hispanic people who consumed various soft, unripened cheeses that were produced in Mexico from raw milk.

4.7 Clostridium botulinum

Botulism, among the rarest of the dairy-related diseases, results from ingesting minute amounts of a preformed neurotoxin produced by *Clostridium botulinum*. Spores generated by this *anaerobic* Gram-positive organism are widely distributed in the environment, with soil serving as a primary reservoir. These spores, which frequently contaminate raw milk, will survive both LTLT and HTST pasteurization. However, milk and other fluid dairy products are insufficiently anaerobic for *C. botulinum* growth and subsequent toxin production. Since 1899, only 13 dairy-related outbreaks of botulism, 12 traced to various anaerobically packaged cheeses or cheese spreads and one to yogurt, have been documented worldwide, with fewer than 20 cases recorded in the United States (Ryser 1999). The threat of growth and toxin production by *C. botulinum* in cheese and cheese spreads can be easily eliminated by controlling the pH, moisture content, water activity, phosphate level, and *nisin* content of the final product.

4.8 *Staphylococcus aureus*

Staphylococci are frequent contaminants of raw milk, with *S. aureus* widely recognized as a cause of clinical and subclinical *mastitis* in dairy cattle, sheep, and goats. While the mammary gland of ruminant animals represents an important reservoir for *S. aureus*, this pathogen is also commonly found on human skin, with both of these sources contributing to the contamination of raw milk and other dairy products, including cheese and ice cream, during manufacture. The ingestion of a preformed enterotoxin produced during the growth of *S. aureus* in temperature-abused (>15°C) food leads to nausea, vomiting, and diarrhea, with hospitalization seldom required. Milk, cheese, ice cream, butter, and nonfat dry milk are historically well-known vehicles for staphylococcal poisoning, which accounted for up to 50% all dairy-related illnesses during the 1950s (Bryan 1983). Subsequent improvements in milk pasteurization and dairy sanitation practices have now made such outbreaks and product recalls a rare occurrence, and *S. aureus* accounted for less than 5% of all dairy-related outbreaks reported in the United States from 1990 to 2006 (Table 4). However, in 2000, a massive outbreak of staphylococcal poisoning in Japan involving more than 10,000 cases of illness was traced to the consumption of reconstituted skim milk (Ikeda et al. 2005).

4.9 *Yersinia enterocolitica*

Not widely recognized as a cause of foodborne illness until the 1970s, *Y. enterocolitica* infections have been historically linked to pork, particularly chitterlings, with hogs being the primary reservoir for human pathogenic strains. In children less than 7 years of age, this Gram-negative enteric bacterial pathogen produces gastroenteritis characterized by low fever, diarrhea, and severe abdominal pain that, like *Campylobacter*, can mimic appendicitis and lead to unnecessary appendectomies. However, *septicemic infections*, accompanied by various secondary complications including *reactive arthritis* and skin infections, also have been reported, particularly in elderly patients with underlying conditions such as alcoholism, liver disease, *hemolytic anemia*, and other immunosuppressive disorders. Ruminant animals are frequent shedders of this Gram-negative enteric organism, with one study from Scotland identifying *Y. enterocolitica* in 50% of the fecal samples from healthy dairy cows (Davey et al. 1983). However, only a few of these isolates belonged to pathogenic serotypes associated with milkborne yersiniosis. Based on surveys conducted in the United States, the incidence of *Y. enterocolitica* in raw milk ranged from 1.2% to 48%, with none of these isolates being virulent (Jayarao et al. 2006; McManus and Lanier 1987; Moustafa et al. 1983; Rohrbach et al. 1992). Despite these findings and the inability of *Y. enterocolitica* to cause mastitis in dairy cows, raw milk remains the primary vehicle of infection for dairy-related cases of yersiniosis. Since 1972, seven yersiniosis outbreaks involving raw milk, powdered milk, chocolate milk, and pasteurized milk have been documented in the United States and Canada. In one of these outbreaks, 138 Canadian schoolchildren developed gastroenteritis after a class field trip to a dairy farm (Kasatiya 1976). In 1978, 217 yersiniosis cases in a school cafeteria in the state of New York were traced to the consumption of chocolate milk that most likely became contaminated when chocolate syrup was hand-mixed into the pasteurized milk in an open vat (Black et al. 1978). Two additional outbreaks involved crates of unsold pasteurized milk that were delivered to pig farmers as feed and then returned to the dairy, thereby reinforcing pigs as a source for pathogenic *Y. enterocolitica* strains (Ackers et al. 2000; Tacket et al. 1984). Even though this psychrotrophic pathogen can grow to high numbers in pasteurized milk during refrigerated storage, *Y. enterocolitica* remains an unusual cause of milkborne illness due to the low incidence of human pathogenic strains in raw milk and the organism's high susceptibility to pasteurization.

4.10 Uncommon Milkborne Pathogens

As discussed above, *Salmonella, Campylobacter, E. coli,* 0157:H7 and *L. monocytogenes* are currently responsible for over 95% of all dairy-related outbreaks, with the remaining 5% of these outbreaks caused by a diverse group of microorganisms (Table 5). The most important of these dairy-associated diseases that were just discussed included *Bacillus cereus* poisoning, botulism, brucellosis, staphylococcal poisoning, and yersiniosis, all of which are caused by bacteria. Several uncommon diseases, all of which are described in greater detail elsewhere (Ryser 1999), are produced by other bacterial agents, including *Citrobacter freundii, Corynebacterium ulcerans, Corynebacterium ulcerans, Coxiella burnetii* (Q fever), *Shigella* (shigellosis), *Streptobacillus moniliformis* (Haverhill fever), *Streptococcus zooepidemicus,* and certain lactic acid bacteria (histamine poisoning), all of which are infrequently encountered in dairy products. Aflatoxin M1 and other *mycotoxins* produced by a wide range of molds can also enter the milk supply through contaminated animal feed, with many of these toxins being carcinogenic and/or mutagenic. Finally, several cases of infectious hepatitis caused by hepatitis A virus and toxoplasmosis, caused by the protozoan parasite *Toxoplasma gondi*, also have been documented.

5 Emerging Milkborne Concerns

The list of dairy-associated pathogens has continued to evolve, with both *L. monocytogenes* and *E. coli* O157:H7 not identified as a serious threat to the dairy industry until the 1980s. In response to recent improvements in foodborne outbreak investigations, new causes of foodborne illness are continually being identified. Hence, the number of "emerging" milkborne pathogens has also continued to grow and now includes at least four previously unrecognized concerns: *Creutzfeldt–Jakob disease,* cryptosporidiosis, *Crohn's disease,* and *Cronobacter* (formerly *Enterobacter*) *sakazakii* (Table 5).

5.1 Creutzfeldt–Jakob Disease

The main form of *human spongiform encephalopathy,* Creutzfeldt–Jakob disease (CJD) is an extremely rare and fatal degenerative brain disorder in older adults that is caused by an infectious protein called a *prion*. In 1996, a slightly different form of CJD, called new variant Creutzfeldt–Jakob Disease (nv-CJD), was identified in younger adults in Great Britain. Upon further investigation, the emergence of nv-CJD was found to be related to earlier cases of *bovine spongiform encephalopathy* (BSE), known today as "mad cow disease." The transmission of nv-CJD has been epidemiologically linked to the consumption of ground beef containing an infectious *prion* that resides in the nervous tissue of infected animals. No cases of nv-CJD have been directly linked to the consumption of BSE-infected beef or dairy products since the disease symptoms do not appear until long after exposure to the prion. However, cows suffering from BSE can reportedly shed a noninfectious form of the prion in their milk (Everest et al. 2006), with other noninfectious prions identified in milk from sheep (Maddison et al. 2009) suffering from a related prion disease called *scrapie*. Thus, while these observations have heightened concerns in the dairy industry, the actual risk of contracting nv-CJD from dairy products appears to be extremely low.

5.2 Cryptosporidiosis

One of the most common parasitic forms of gastroenteritis in healthy individuals, cryptosporidiosis is primarily a waterborne illness caused by ingesting protozoan *parasites* of the genus *Cryptosporidium*. Cows, sheep, and goats are a major source for *Cryptospordium*, with infected animals capable of shedding high numbers of *oocysts* in feces, which can lead to the subsequent contamination of the milk. Evidence for involvement of *Cryptosporidium* in milkborne illness is growing, with at least five major outbreaks outside the United States traced to the consumption of raw or improperly pasteurized milk (Ryser 1999). Although HTST pasteurization is sufficient to render any oocysts in milk noninfectious, the potential exposure of cheese curd to *Cryptosporidium*-contaminated water during cheese manufacture raised some concerns following one massive outbreak of waterborne cryptosporidiosis in Milwoutee, Wisconsin.

5.3 Johne's Disease

An economically devastating disease in cattle, Johne's disease is characterized by *ileocolitis*, which eventually leads to diarrhea, weight loss, debilitation, and death. High numbers of the causative bacterium *Mycobacterium avium* ssp. *paratuberculosis* (MAP) are shed in feces, with up to 35% of clinically infected cows and 12% of symptom-free carriers yielding positive milk samples. This pathogen has generated considerable interest given its possible association with Crohn's disease – a nearly identical form of ileocolitis in humans (Grant 2006). Unlike most other milkborne pathogens, MAP is presumably capable of surviving HTST pasteurization in low numbers and can also enter milk as a postpasteurization contaminant. Cheese constitutes one additional source for MAP. However, the importance of this organism as a milkborne pathogen remains unclear given the lack of a definitive link between Johne's disease in ruminants and Crohn's disease in humans.

5.4 Cronobacter sakazakii

Cronobacter sakazakii, a Gram-negative enteric bacterium formerly known as *Enterobacter sakazakii*, was recently identified as a rare cause of *sepsis*, meningitis, and *necrotizing enterocolitis* in neonates and infants who have consumed rehydrated powdered infant formula (Chenu and Cox 2009). Given a mortality rate of 20–80% along with the irreversible neurological disorders encountered by surviving neonates, *C. sakazakii* is now a leading concern among manufacturers of infant formulas. Although typically absent in raw milk and easily inactivated by HTST and UHT pasteurization,

C. sakazakii is widespread in many dairy processing environments, with this opportunistic pathogen entering milk and powdered infant formula as a postpasteurization contaminant. Hence, appropriate control strategies during the production of powdered infant formula, including adherence to strict hygienic practices and a sound environmental monitoring program, are needed to minimize contamination of the final product.

6 Summary

Milk in the udder of healthy animals is sterile. However, during its collection, storage, and transport, milk can be contaminated with different pathogens. Most of these pathogens come from the hide of the cows, bedding material, manure, feed, milking equipment, and/or milk handlers. More than 90% of all dairy-related outbreaks have been traditionally caused by a few bacterial pathogens, such as *Salmonella* and *Campylobacter*, but since the 1980s, *E. coli* O157:H7 and *L. monocytogenes* have been frequently associated with milk and milk-derived products (soft cheese). Other, emerging concerns include cryptosporidiosis, *Cronobacter sakazakii,* and the very rare Creutzfeldt–Jakob disease. Therefore, aseptic techniques have been developed for the safe collection of milk. Scarlet fever, tuberculosis, and typhoid fever have been widely controlled by the pasteurization of fluid milk. However, 26 states in the United States have legalized the intrastate sale of raw milk in response to a growing number of opponents to the pasteurization of milk. Most current dairy-related illnesses could be eliminated by avoiding the consumption of raw milk and some soft cheeses made with raw milk that are illegally prepared or imported into the United States.

References

Ackers, M.L., S. Schoenfeld, J. Markman, M.G. Smith, M.A. Nicholson, W. DeWitt, D.N. Cameron, P.M. Griffin, and L. Slutsker. 2000. An outbreak of *Yersinia enterocolitica* O:8 infections associated with pasteurized milk. *Journal of Infectious Diseases* 181: 1834–1837.

Bean, N.H., P.M. Griffin, J.S. Goulding, and C.B. Ivey. 1990. Foodborne disease outbreaks, 5-year summary, 1983–1987. *Morbidity and Mortality Weekly Report* 39(SS-1): 15–23.

Bean, N.H., J.S. Goulding, C. Lao, and F.J. Angulo. 1996. Surveillance of foodborne disease outbreaks – US, 1988–1992. *Morbidity and Mortality Weekly Report* 45(SS-5): 1–55.

Bezanson, G.S., R. Khakhria, D. Duck, and H. Lior. 1985. Molecular analysis confirms food source and simultaneous involvement of two distinct but related subgroups of *Salmonella typhimurium* bacteriophage type 10 in major interprovincial *Salmonella* outbreak. *Applied and Environmental Microbiology* 50: 1279–1284.

Black, R.E., R.J. Jackson, T. Tsai, M. Medvesky, M. Shaayegani, J.C. Feeley, K.I.E. MacLeod, and A.M. Wakelee. 1978. Epidemic *Yersinia enterocolitica* infection due to contaminated chocolate milk. *The New England Journal of Medicine* 298: 76–79.

Borczyk, A.A., M.A. Karmail, H. Lior, and L.M.C. Duncan. 1987. Bovine reservoir for verotoxin-producing *Escherichia coli* O157:H7. *The Lancet* 329(8524): 98.

Brodsky, M.H. 1984. Bacteriological survey of freshly formed cheddar cheese. *Journal of Food Protection* 47: 546–548.

Bryan, F.L. 1983. Epidemiology of milk-borne disease. *Journal of Food Protection* 46: 637–649.

Bula, C.J., J. Bille, and M.P. Glauser. 1995. An epidemic of food-borne listeriosis in western Switzerland: description of 57 cases involving adults. *Clinical Infectious Diseases* 20: 66–72.

Center for Science in the Public Interest. 2007. *Outbreak alert!* 9th ed. Washington, DC: Center for Science in the Public Interest.

Centers for Disease Control and Prevention. 2008. Outbreak of *Listeria monocytogenes* infections associated with pasteurized milk from a local dairy – Massachusetts, 2007. *Morbidity and Mortality Weekly Report* 57: 1097–1100.

Chenu, J.W., and J.M. Cox. 2009. *Cronobacter* ("*Enterobacter sakazakii*"): current status and future prospects. *Letters in Applied Microbiology* 49: 153–159.

Cobbold, R.N., M.A. Davis, D.H. Rice, M. Szymanski, P.I. Tarr, T.E. Besser, and D.D. Hancock. 2008. Associations between bovine, human, and raw milk and beef isolates of non-O157 shiga toxigenic *Escherichia coli* within a restricted geographic area of the US. *Journal of Food Protection* 71: 1023–1027.

Collins, R.N., M.D. Treger, J.B. Goldsby, J.R. Boring III, D.B. Coohon, and R.N. Barr. 1968. Interstate outbreak of *Salmonella newbrunswick* infection traced to powdered milk. *Journal of the American Medical Association* 203: 838–844.

Dalton, C.B., C.C. Austin, J. Sobel, P.S. Hayes, W.F. Bibb, L.M. Graves, B. Swaminathan, M.E. Proctor, and P.M. Griffin. 1997. An outbreak of gastroenteritis and fever due to *Listeria monocytogenes* in milk. *The New England Journal of Medicine* 336: 100–105.

Davey, G.M., J. Bruce, and E.M. Drysdale. 1983. Isolation of *Yersinia enterocolitica* and related species from faeces of cows. *Journal of Applied Bacteriology* 55: 439–443.

Denny, J., M. Bhat, and K. Eckmann. 2008. Outbreak of *Escherichia coli* O157:H7 associated with raw milk in the Pacific Northwest. *Foodborne Pathogens and Disease* 5: 321–328.

DHHS, PHS, and FDA. 2009. Grade "A" pasteurized milk ordinance. Available at http://www.fda.gov/downloads/Food/FoodSafety/Product-SpecificInformation/MilkSafety/NationalConferenceonInterstateMilkShipmentsNCIMSModelDocuments/UCM209789.pdf. Accessed 4 Feb 2011.

Doane, C.A., P. Pangloli, H.A. Richards, J.R. Mount, D.A. Golden, and F.A. Draughon. 2007. Occurrence of *Escherichia coli* O157:H7 in diverse farm environments. *Journal of Food Protection* 70: 6–10.

Dominguez, M., N.J. da Silva, V. Vaillant, N. Pihier, C. Kermin, F.X. Weill, G. Delmas, A. Kerouanton, A. Brisabois, and H. de Valk. 2009. Outbreak of *Salmonella* enteric serotype Montevideo infections in France linked to consumption of cheese made from raw milk. *Foodborne Pathogens and Disease* 6: 121–128.

Doyle, M.P., and D.J. Roman. 1982. Prevalence and survival of *Campylobacter jejuni* in unpasteurized milk. *Applied and Environmental Microbiology* 44: 1154–1158.

Durch, J., T. Ringhanr, K. Manner, M. Barnett, M. Proctor, S. Ahrabi-Fard, J. Davis, and D. Boxrud. 1998. Outbreak of *Escherichia coli* O1157:H7 infection associated with eating fresh cheese curds – Wisconsin, June 1998. *Morbidity and Mortality Weekly Report* 49: 911–913.

Ehlers, J.G., M. Chapparo-Serrano, R.I. Richter, and C. Vanderzant. 1982. Survival of *Campylobacter fetus* subsp. *jejuni* in cheddar and cottage cheese. *Journal of Food Protection* 45: 1018–1021.

Espié, E., V. Vaillant, P. Mariani-Kurkdjian, F. Grimont, R. Martin-Schaller, H. De Valk, and C. Vernozay-Rozand. 2006. *Escherichia coli* O157 outbreak associated with fresh unpasteurized goats' cheese. *Epidemiology and Infection* 134: 143–146.

Everest, S.J., L.T. Thorne, J.A. Hawthorn, R. Jenkins, C. Hammersley, A.M. Ramsay, S.A. Hawkins, L. Venables, L. Flynn, R. Sayers, J. Kilpatrick, A. Sach, J. Hope, and R. Jackman. 2006. No abnormal prion protein detected in the milk of cattle infected with the bovine spongiform encephalopathy agent. *Journal of General Virology* 87: 2433–2441.

FDA. 2009. FDA Enforcement reports. Available at http://www.fda.gov/safety/recalls/enforcementreports/default.htm. Accessed 4 Feb 2011.

Fleming, D.W., S.L. Cochi, K.L. MacDonald, J. Brondum, P.S. Hayes, B.D. Plikaytis, M.B. Homes, A. Audurier, C.V. Broome, and A.L. Reingold. 1985. Pasteurized milk as a vehicle of infection in an outbreak of listeriosis. *The New England Journal of Medicine* 312: 404–407.

Fontaine, R.E., M.L. Cohen, W.T. Martin, and T.M. Vernon. 1980. Epidemic salmonellosis from cheddar cheese – surveillance and prevention. *American Journal of Epidemiology* 111: 247–253.

Goepfert, J.M., N.F. Olson, and E.H. Marth. 1968. Behavior of *Salmonella* during manufacture and curing of Cheddar cheese. *Applied Microbiology* 16: 862–866.

Grant, I.R. 2006. *Mycobacterium avium* ssp. *paratuberculosis* in foods: current evidence and potential consequences. *International Journal of Dairy Technology* 59: 112–117.

Hennessy, T.W., C.W. Hedberg, L. Slutsker, K.E. White, J.M. Besset-Wiek, M.E. Moen, J. Feldman, W.W. Coleman, L.M. Edmonson, K.L. MacDonald, M.T. Osterholm, and The Investigation Team. 1996. A national outbreak of *Salmonella enteritidis* infections from ice cream. *The New England Journal of Medicine* 334: 1281–1286.

Ikeda, T., N. Tomate, K. Yamaguchi, and S.-I. Makino. 2005. Mass outbreak of food poisoning disease caused by small amounts of staphylococcal enterotoxins A and H. *Applied and Environmental Microbiology* 71: 2793–2795.

Jayarao, B.M., and D.R. Henning. 2001. Prevalence of foodborne pathogens in bulk tank milk. *Journal of Dairy Science* 84: 2157–2162.

Jayarao, B.M., S.C. Donaldson, B.A. Straley, A.A. Sawant, N.V. Hegde, and J.L. Brown. 2006. A survey of foodborne pathogens in bulk tank milk and raw milk consumption among farm families in Pennsylvania. *Journal of Dairy Science* 89: 2451–2458.

Karns, J.S., J.S. van Kessel, B.J. McClusky, and M.L. Perdue. 2007. Incidence of *Escherichia coli* O157:H7 and *E. coli* virulence factors in US bulk tank milk as determined by polymerase chain reaction. *Journal of Dairy Science* 90: 3212–3219.

Kasatiya, S.S. 1976. *Yersinia enterocolitica* outbreak – Montreal. *Canada Diseases Weekly Report* 2: 73–74.

Kasrazadeh, M., and C. Genegeorgis. 1995. Potential growth and control of *Escherichia coli* O157:H7 in soft hispanic type cheese. *International Journal of Food Microbiology* 25: 289–300.

Kenne, W.E., K. Hedberg, D.E. Herriott, D.D. Hancock, R.W. McKay, T.J. Barrett, and D.W. Fleming. 1997. A prolonged outbreak of *Escherichia coli* O157:H7 infections caused by commercially distributed raw milk. *Journal of Infectious Diseases* 176: 815–818.

LiCari, J.J., and N.N. Potter. 1970. *Salmonella* survival during spray drying and subsequent handling of skim milk powder. III. Effects of storage temperature on *Salmonella* and dried milk products. *Journal of Dairy Science* 53: 877–882.

Linnan, M.J., L. Mascola, X.D. Lou, V. Goulet, S. May, C. Salminen, D.W. Hird, M.L. Yonkura, P. Hayes, R. Weaver, A. Audurier, B.D. Plikaytis, S.L. Fannin, A. Kleks, and C.V. Broome. 1988. Epidemic listeriosis associated with Mexican-style cheese. *The New England Journal of Medicine* 319: 823–828.

Lovett, J., D.W. Francis, and J.M. Hunt. 1983. Isolation of *Campylobacter jejuni* from raw milk. *Applied and Environmental Microbiology* 46: 459–462.

MacDonald, K.L., and P.M. Griffin. 1986. Foodborne disease outbreaks, annual summary, 1982. *Morbidity and Mortality Weekly Report* 35(SS-1): 7ss–10ss.

MacDonald, K.L., M. Edison, C. strohmeyer, M.E. Levy, J.G. Wells, N.D. Puhr, K. Wachsmuth, N.T. Hargett, and M.L. Cohen. 1985. A multistate outbreak of gastrointestinal illness caused by enterotoxigenic *Escherichia Coli* in imported semisoft cheese. *Journal of Infectious Diseases* 151: 716–720.

Maddison, B.C., C.A. Baker, H.C. Rees, L.A. Terry, L. Thorne, S.J. Bellworthy, G.C. Whitelam, and K.C. Gough. 2009. Prions are secreted in milk from clinically normal Scrapie-exposed sheep. *Journal of Virology* 83: 8293–8296.

Marier, R., J.G. Wells, R.C. Swanson, W. Callahan, and I.J. Mehlman. 1973. An outbreak of enteropathogenic Escherichia coli foodborne disease traced to imported French cheese. *The Lancet* 2: 1376–1378.

McManus, C., and J.M. Lanier. 1987. *Salmonella, Campylobactrer jejuni* and *Yersinia enterocolitica* in raw milk. *Journal of Food Protection* 50: 51–55.

Moustafa, M.K., A.A.-H. Ahmed, and E.H. Marth. 1983. Occurrence of *Yersinia enterocolitica* in raw and pasteurized milk. *Journal of Food Protection* 46: 276–278.

Murinda, S.E., L.T. Nguyen, S.J. Ivey, B.E. Gillespie, R.A. Almeida, F.A. Draughon, and S.P. Oliver. 2002. Prevalence and molecular characterization of *Escherichia coli* O157:H7 in bulk tank milk and fecal samples from cull cows: a 12-month survey of dairy farms in Tennessee. *Journal of Food Protection* 65: 752–759.

NASS-USDA. 2009. Annual milk production. Available at http://www.nass.usda.gov/Statistics_by_State/Wisconsin/Publications/Dairy/anmkpd.pdf. Accessed 15 Dec 2010.

Olsen, S.J., C.L. MacKinon, J.S. Goulding, N.H. Bean, and L. Slutsker. 2000. Surveillance for foodborne disease outbreaks – US, 1993–1997. *Morbidity and Mortality Weekly Report* 49(SS01): 1–51.

Park, H.S., E.H. Marth, J.M. Goepfert, and N.F. Olson. 1970. The fate of *Salmonella typhimurium* in the manufacture and ripening of low-acid Cheddar cheese. *Journal of Milk and Food Technology* 33: 280–284.

Raamsaran, H., J. Chen, B. Brunke, A. Hill, and M.W. Griffiths. 1998. Survival of bioluminescent *Listeria monocytogenes* and *Escherichia coli* O157:H7 in soft cheeses. *Journal of Dairy Science* 81: 1810–1817.

Reitsma, C.J., and D.R. Henning. 1996. Survival of enteropathogenic *Escherichia coli* O157:H7 during the manufacture and curing of Cheddar cheese. *Journal of Food Protection* 59: 460–464.

Robinson, D.A., and D.M. Jones. 1981. Milk-borne *Campylobacter* infection. *British Medical Journal* 282: 1374–1376.

Rohrbach, B.W., F.A. Draughon, P.M. Davidson, and S.P. Oliver. 1992. Prevalence of *Listeria monocytogenes*, *Campylobacter jejuni*, *Yersinia enterocolitica*, and *Salmonella* in bulk tank milk: risk factors and risk of human exposure. *Journal of Food Protection* 55: 93–97.

Ryan, C.A., M.K. Nickels, N.T. Hargrett-Bean, M.E. Potter, T. Endo, L. Mayer, C.W. Langkop, C. Gibson, R.C. MacDonald, R.T. Kenney, N.D. Puhr, P.J. McDonnell, R.J. Martin, M.L. Cohen, and P.A. Blake. 1987. Massive outbreak of antimicrobial resistant salmonellosis traced to pasteurized milk. *Journal of the American Medical Association* 258: 3269–3274.

Ryser, E.T. 1999. Public health concerns. In *Applied dairy microbiology*, 2nd ed, ed. E.H. Marth and J.K. Steele, 397–545. New York: Marcel Dekker.

Ryser, E.T. 2003. Pasteurisation of liquid milk products: principles, public health aspects. In *Encyclopedia of dairy sciences*, ed. H. Roginski, P.F. Fox, and J.W. Fuqyay, 2232–2237. London: Academic.

Ryser, E.T. 2007a. Incidence and behavior of *Listeria monocytogenes* in cheese and other fermented dairy products. In *Listeria, listeriosis and food safety*, 3rd ed, ed. E.T. Ryser and E.H. Marth, 405–501. Boca Raton: CRC Press.

Ryser, E.T. 2007b. Incidence and behavior of *Listeria monocytogenes* in unfermented dairy products. In *Listeria, listeriosis and food safety*, 3rd ed, ed. E.T. Ryser and E.H. Marth, 357–403. Boca Raton: CRC Press.

Ryser, E.T., and E.H. Marth. 1987. Behavior of *Listeria monocytogenes* during manufacture and ripening of Cheddar cheese. *Journal of Food Protection* 50: 7–13.

Scallan, E., R.M. Hoeskstra, F.J. Angulo, R.V. Tauxe, M.-A. Widdowson, S.L. Roy, J.L. Jones, and P.M. Griffin. 2011. Foodborne illness acquired in the United States - Major pathogens. *Emerging Infectious Diseases* 17: 7–15.

Schlesser, J.E., R. Gerdes, S. Ravishankar, K. Madsen, J. Mowbray, and A.Y. Teo. 2006. Survival of a five-strain cocktail of *Escherichia coli* O157:H7 during the 60-day aging period of Cheddar cheese made from unpasteurized milk. *Journal of Food Protection* 69: 990–998.

Schuman, J.D., E.A. Zottola, and S.K. Harlander. 1989. Preliminary characterization of a food-borne multiple-antibiotic-resistant *Salmonella typhimurium* strain. *Applied and Environmental Microbiology* 55: 2344–2348.

Tacket, C.O., J.P. Narain, R. Sattin, J.P. Lofgren, C. Konigsberg, R.C. Rendtorff, A. Rausa, B.R. Davis, and M.L. Cohen. 1984. A multistate outbreak of infections caused by *Yersinia enterocolitica* transmitted by pasteurized milk. *Journal of the American Medical Association* 251: 483–486.

van Kessel, J.S., J.S. Karns, L. Gorski, B.J. McCluskey, and M.L. Perdue. 2004. Prevalence of salmonellae, *Listeria monocytogenes*, and fecal coliforms in bulk tank milk in US dairies. *Journal of Dairy Science* 87: 2822–2830.

Vought, K.J., and S.R. Tatini. 1998. *Salmonella* contamination in ice cream associated with a 1994 multistate outbreak. *Journal of Food Protection* 61: 5–10.

Wood, R.C., K.L. MacDonald, and M.T. Osterholm. 1992. *Campylobacter* enteritis outbreaks associated with drinking raw milk during youth activities: a 10-year review of outbreaks in the US. *Journal of the American Medical Association* 268: 3228–3230.

Yilmaz, T., B. Moyer, R.E. MacDonell, M. Cordero-Coma, and M.J. Gallagher. 2009. Outbreaks associated with unpasteurized milk and soft cheese: an overview of consumer safety. *Food Protection Trends* 29: 211–222.

Safety of Meat Products

Paul Whyte and Séamus Fanning

1 Introduction

Despite concerted efforts for several decades, foodborne disease remains a major global public health issue, with substantial *morbidity* and *mortality* associated with the consumption of contaminated foodstuffs (Havelaar et al. 2010; WHO 2007). The ingestion of foodborne *pathogens* or their *toxins* can result in a diverse spectrum of disease, ranging from mild infections to severe disease, sometimes resulting in death, particularly within susceptible subpopulations. In addition to human health consequences, significant economic costs can be associated with foodborne diseases (Whyte et al. 2005; USDA-ERS 2009). Meatborne biological hazards can enter and persist in the food chain at any point in the preharvest, harvest, or postharvest stages, including processing and storage (Fig. 1).

2 Foodborne Disease and the Role of Meat and Meat Products

The availability of accurate data estimating the scale of global foodborne disease is limited, and these data gaps have been acknowledged by the World Health Organization (2009a), which has recently started collaborative initiatives to address these shortfalls. It is accepted that approximately 75% of new communicable diseases over the past 10 years have been caused by pathogens originating from animals or products of animal origin. Many of these *zoonoses* have been transmitted by direct contact with infected animals or through the handling or consumption of contaminated products derived from animals (WHO 2009b). However, data gaps remain in relation to the *attribution* of disease burden caused by various food categories. This kind of information is vital from the perspective of public health protection, so that resources for disease prevention and control along with surveillance can be prioritized and targeted to maximize performance in a cost-effective manner (Doyle and Erickson 2006).

P. Whyte (✉)
Centre for Food Safety & Institute of Food and Health,
School of Veterinary Medicine, University College Dublin, Ireland
e-mail: paul.whyte@ucd.ie

S. Fanning
Centre for Food Safety & Institute of Food and Health,
School of Public Health, Physiotherapy and Population Science,
University College Dublin, Ireland

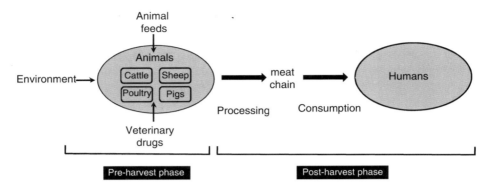

Fig. 1 A simplified schematic showing where pathogens and chemical residues may enter the food chain. The diagram shows the pre- and postharvest phases of the meat production chain

Based on published data, a considerable proportion of foodborne disease in developed countries has been linked to the consumption or handling of contaminated meat or meat products. For example, beef products were associated with 3.4% of foodborne outbreaks in the United States between 1993 and 1997 (CDC 2000), whereas in Europe, 10.3% of reported outbreaks in 2006 could be attributed to contaminated red meat/meat products (EFSA 2007). Greig and Ravel (2009) conducted a large-scale retrospective analysis of international outbreaks from 1988 to 2007 and reported that 12.2% of these outbreaks were associated with the consumption of beef or beef products, with a range of zoonotic pathogens as causative agents. Poultry meat and associated products have also been frequently linked with outbreaks produced by *Campylobacter* and *Salmonella*, the two most frequently identified foodborne agents in these products (Adak et al. 2002; EFSA 2007; Greig and Ravel 2009).

3 Hazards Associated with Meat and Meat Products

Meat and meat products have traditionally been identified as potentially hazardous and frequently contaminated with pathogenic microorganisms. As a result, *organoleptic*-based inspection systems have evolved over the last 100 years or more to assess the suitability of meat for human consumption. Such systems typically involve a brief visual examination of animals *antemortem* to ensure that they are not clinically infected or ill, and *postmortem* examination to detect pathological evidence of parasitic infestation or microbiological infection in various carcass tissues and specimens. Such an approach has achieved considerable success to date in identifying various conditions of relevance to animal and/or public health, for example, bovine tuberculosis or various parasitic infestations, where any affected carcass or meat can be excluded from the human food chain. In addition, monitoring and surveillance based on this organoleptic approach has been effectively used before harvest, for example, to identify high-risk herds, to establish disease prevalences at a national level, or to assess intervention measures such as disease eradication and control programs. However, in recent years it has become apparent, based on human foodborne disease and epidemiological data, that the most significant zoonotic agents in foods of animal origin are those that frequently colonize or infect animals asymptomatically. Examples of these pathogens include *Salmonella* spp., *Campylobacter* spp., and enterohemorrhagic *Escherichia coli* (EHEC). These pathogens can spread easily within and between herds during production, with infected animals appearing clinically normal. As a result, additional control measures are required to complement traditional visual inspection systems and provide satisfactory levels of safety assurance.

3.1 Salmonella spp.

Human disease caused by *Salmonella* can be divided into two categories: typhoidal *salmonellosis*, which is a severe systemic infection, and nontyphoidal or *gastroenteritis* infection of the intestinal epithelium. *Salmonella* spp. causing nontyphoidal symptoms are one of the most common causes of foodborne disease in humans and globally represent a significant public health challenge. In the European Union (EU), the number of reported cases of human salmonellosis was 167,240 in 2006, representing an average of 34.6 cases per 100,000 population. However, the range of reported cases per 100,000 population varied markedly between individual member states: 3.7–235.9 (EFSA 2007). In the United States, it has been estimated that 1.4 million cases of nontyphoidal salmonellosis occur each year, resulting in 15,000 hospitalizations and 580 deaths, with substantial associated economic costs (WHO 2005).

Salmonella infections are transmitted within food animal species via the fecal–oral route and can be either asymptomatic or symptomatic depending on different factors, such as the host species, age, immune status, pathogen *serotype* and virulence, and challenge or dose ingested. Clinical infections can commonly manifest as septicemia, enteritis, and abortion. However, subclinical infection in farm animals plays a important role in this pathogen's epidemiology and can result in variable and sporadic levels of fecal shedding. Animals within herds may be exposed as a result of direct contact with infected animals and feces or indirectly as a result of contact with a contaminated environment where *Salmonella* have been shown to survive for prolonged periods. Livestock may also become infected as a result of transmission via contaminated feed or water, wildlife vectors, contaminated vehicles, or milk from infected animals fed to young.

As with other *enteric* pathogens, *Salmonella* can contaminate the hide or skin of animals, the intestinal tract, and in some cases some internal organs, such as the liver and spleen. Therefore, the burden of pathogens in live animals provides opportunities for the contamination of carcasses and meat during slaughter and dressing and further along the food chain during the subsequent processing of meat products. The prevalence of *Salmonella* in food animals and meat can vary widely depending, for example, on geographical areas, which in turn might influence farming practices and control measures at preharvest and during the slaughter and processing stages (Wegener 2010).

3.2 Enterohemorrhagic E. coli (EHEC)

Since the early 1980s, EHEC have emerged as highly significant enteropathogens linked to large high-profile foodborne outbreaks (Riley et al. 1983; CDC 1997; Keene 1999). EHEC now represents a highly significant food safety challenge given the often severe and potentially life-threatening illnesses in humans associated with this infection. The infectious dose for these organisms is low, with an ingestion of ten organisms or fewer thought to be sufficient to cause infection in some individuals (Willshaw et al. 1994; Coia 1998). Although *E. coli* O157 remains the most prevalent serotype associated with disease, several other EHEC serotypes have also emerged more recently as pathogens of concern, including O26, O103, O111, and O145 (Doyle et al. 2001; Mora et al. 2007). Ruminant animals have been identified as the main reservoir of these enteropathogens, with the gastrointestinal tract becoming colonized resulting in their excretion in feces. Fecal carriage rates in cattle have varied widely among different studies, with prevalences ranging from 0% to 48.8% (Minihan et al. 2003; Hussein and Sakuma 2005). Hides and fleeces can become contaminated with EHECs at the farm level or during transportation of animals to the abattoir, and pathogens can subsequently be transferred to carcasses during slaughter and dressing operations (O'Brien et al. 2005; Lenehan et al. 2007; Carney et al. 2006). The dissemination of EHECs can also occur during further processing of meat. The mincing of beef represents a significant risk due to the mixing of meat from different animals and the potential for distributing contamination throughout the product, which increases the likelihood of organisms surviving internally in such products if not adequately cooked (Beth et al. 1994). Red meat and meat products can be contaminated with these organisms with varying prevalences. In the EU in 2006, for example, the mean isolation rates for EHECs in ground beef, bovine meat, sheep meat, and pig meat were 1.9%, 0.3%, 1.8%, and 0.8%, respectively. No EHECs were detected in poultry meat during this period (EFSA 2007). In the United States, Rangel et al. (2005) analyzed *E. coli* O157 outbreaks that occurred between 1982 and 2002 and reported that 41% and 6% of foodborne outbreaks could be attributed to ground beef and other beef products, respectively. Outbreaks have also been linked to other processed meat products, including salami (CDC 1995).

3.3 Campylobacter spp.

The genus *Campylobacter* consists of more than 18 species, with two predominant species, *Campylobacter jejuni* and *Campylobacter coli*, that are associated with most infections in humans (Gillespie et al. 2002). This bacterium is a leading cause of bacterial gastroenteritis in Europe, North America, and Australia. The numbers of reported confirmed human cases of campylobacteriosis in the EU increased from 175,561 in 2006 to 200,507 in 2007, giving a crude incidence rate (CIR) of 45.2 cases per 100,000 population in 2007 (EFSA 2007). It is estimated that in the United States, campylobacteriosis affects approximately two million people each year (Mead et al. 1999). Most cases are sporadic and typically the infective dose is low. Usually, infections are self-limiting, lasting only a few days. Peak reporting of *Campylobacter*-associated enteritis infections occurs during the warmer months (Sopwith et al. 2003; Miller et al. 2004). In a study of the incidences of *Campylobacter* in broilers and humans, in several European countries, colonization of these birds was related to seasonality and temperature (Jore et al. 2010). A small number of *Campylobacter* cases may result in complications, such as *Guillain–Barré syndrome*, an acute autoimmune polyradiculoneuropathy affecting the peripheral nervous system (Gruenewald et al. 1991). In Finland, a follow-up study of infected individuals identified musculoskeletal symptoms associated with *C. jejuni* enteritis in 39% of cases, pains in the joints in 81% of cases, and the incidence of reactive arthritis was low (Schonberg-Norio et al. 2010). Improvements in understanding the epidemiology of this bacterium may lead to the development of new intervention strategies to control the infectious process (Sherman et al. 2010).

The gastrointestinal tracts of food production animals, such as poultry, cattle, and pigs, are frequently colonized with *Campylobacter* although the highest prevalence is found in poultry species (Moore et al. 2005; Sherman et al. 2010). The bacterium can also be cultured from wild birds (Waldenstrom et al. 2007). In both developed and developing countries, more than half of the raw retail poultry and byproducts is often contaminated with *Campylobacter*, and *C. jejuni* is the most common species isolated in many parts of the world (Suzuki and Yamamoto 2009).

3.4 Listeria monocytogenes

Listeria monocytogenes is recognized as a significant foodborne pathogen that can cause severe infection in susceptible human populations (Gudbjornsdottir et al. 2004). This pathogen has been associated with intrauterine infection, meningitis, and septicemia. Infection in pregnant women can result in severe systemic infection in the unborn child as well as mild influenza-like symptoms in the mother. In adults, the main symptoms are septicemia and/or infection of the central nervous system. The immunosuppressed, elderly persons, or those with underlying conditions such as diabetes or alcoholism are the most susceptible (McLauchlin et al. 2004). Although the numbers of reported infections are quite low, mortality rates in vulnerable groups are high, with estimates ranging from 20% to 40% (Farber and Peterkin 1991; Smerdon et al. 2001). In England and Wales during the period 1996–2000, *Listeria monocytogenes* was confirmed in 221 of 1.7 million episodes of reported foodborne illness but accounted for over 11% of fatalities (Adak et al. 2005). The incubation period in humans can be up to 90 days, making epidemiological and food attribution investigations difficult.

In contrast to most other bacterial foodborne pathogens, *L. monocytogenes* can grow at refrigeration temperatures, which potentially has serious safety implications for foods during storage (Rhoades et al. 2009). *Listeria* spp. are ubiquitously distributed in nature and have been isolated from soil, vegetation, water, animal feed, raw and processed meats, seafood, and the feces of healthy animals and humans (McLauchlin et al. 2004). As a result, numerous sources and opportunities exist for contamination to occur on meat during slaughter and subsequent processing operations. In addition, *Listeria* spp. have the ability to form biofilms on solid surfaces, including equipment, and in the environment within food processing plants. This can allow organisms to become established, persistent, and difficult to control or remove once introduced in processing plants (Gudbjornsdottir et al. 2004).

Most cases of listeriosis have been associated with ready-to-eat (RTE) foods, including cooked and processed meats (Jacquet et al. 1995; Goulet et al. 1998; Health Canada 2009). As adequate cooking of meat products is sufficient to destroy *Listeria*, contamination from the processing environment post-heating is the most likely means of introduction. The reported prevalence of *L. monocytogenes* in raw meats has varied among studies, with 16%, 22%, and 24% of pork, poultry, and beef samples, respectively, found to be positive (Chasseignaux et al. 2002; Gudbjornsdottir et al. 2004; Rhoades et al. 2009). Official statistics compiled within the EU for the year 2007 have shown mean isolation rates of 1.8%, 2.2%, 2.5%, and 2.6% for *Listeria monocytogenes* in RTE processed bovine, pig, red meat/mixed, and poultry meats, respectively (EFSA 2009).

3.5 Parasites and Viruses

Infections by parasites represent a significant cause of morbidity in human populations globally. The burden of disease associated with such organisms is likely to be underreported, can vary considerably, and is often dependent on the geographical region, local agricultural practices, and socioeconomic factors. A number of parasitic species have been identified in the United States as capable of being transmitted to humans via the consumption of contaminated foods. These include *Giardia*, *Cryptosporidium*, and *Cyclospora*, with estimated annual numbers of human cases in the United States of 200,000, 30,000, and 14,000, respectively (Mead et al. 1999; Tauxe 2002). However, most infections associated with these parasitic species are transmitted via water, via direct contact with animals, or as a result of consuming RTE foods (mainly fresh fruit and vegetables), which may have come in contact with animal feces either directly or indirectly or through infected human food handlers (Shield et al. 1990; Furness et al. 2000; Dawson 2005).

The most common meatborne parasites include *Toxoplasma gondii*, *Taenia* spp. (*T. saginata*, *T. saginata asiatica*, and *T. solium*), and *Trichinella spiralis*. *Toxoplasma* may be transmitted to humans via contact with feline feces or by the consumption of raw or undercooked meat contaminated with tissue *oocysts*. These parasites have been detected in meat from game, sheep, goat, horse, chicken, and pigs but not from bovines (Tenter et al. 2000). Intensive indoor rearing of animals can substantially reduce prevalences compared to extensively reared animals, including organic production systems (Kijlstra et al. 2004; Dorny et al. 2009). The seroprevalence of *T. gondii* in pigs within the EU is estimated at 6% (EFSA 2009) and cysts were detected in 0.38% of retail pork meat in the United States (Dubey et al. 2005).

Cysticercosis and *taeniosis* are terms used to describe larval and adult taeniid tapeworms, respectively (Dorny et al. 2009). *T. saginata* can be transmitted to humans via cattle, while *T. saginata asiatica* and *T. solium* are disseminated by pigs. Humans usually become infected with *T. saginata* following the ingestion of raw or undercooked beef containing *cysticerci*. Symptoms include abdominal pain, nausea, weight loss, and anal pruritus; however, more severe infection may occur, resulting in intestinal perforation and peritonitis. Reported prevalences in food animals (cattle and pigs) and wild boar at the time of slaughter in the EU are low, ranging from 0% to 0.2% (EFSA 2007). Infection with *Taenia solium* can also be transmitted to humans at the cysticercus stage through the ingestion of eggs that have been shed by a tapeworm-carrying host and can lead to the development of neurocysticercosis in the brain. *Taenia solium* has been eradicated in most developed countries, where sanitation standards are high and intensive pig production systems are used. However, human infection levels in many poorer regions still remain high (Phiri et al. 2003; Dorny et al. 2009).

The genus *Trichinella* are nematodes and comprise eight species and four genotypes within encapsulated and nonencapsulated groups (Dorny et al. 2009), with most human infections attributed to *Trichinella spiralis*. They are widely distributed in wild and domestic animals on all continents with the exception of Antarctica (Pozio and Murrell 2006). Human trichinellosis occurs following the ingestion of larvae distributed in the muscle tissue of infected wild or domestic animals. Pigs are the most significant source of *Trichinella*, although outbreaks of human disease have also been associated with wild boar and horse meat (Gottstein et al. 2009). The incidence of *Trichinella* in modern intensively produced pig herds is extremely low in most developed countries, and human infection in these regions is often the result of consuming raw or undercooked wild game (boar) or horse meat. It has been estimated that there are approximately 10,000 cases globally each year (Pozio 2007); symptoms include abdominal pain, nausea, and diarrhea, followed by muscle aches, itching, fever, chills, and joint pains.

In developed countries, risks associated with human exposure to parasitic organisms through the consumption of infected meat and meat products is relatively low as a result of high standards in sanitation, modern farm management and husbandry practices, along with specific interventions used to detect, prevent, and control infection in domestic livestock. However, these parasites remain a significant meat safety and public health issue in many developing countries, where control measures are less evolved. Furthermore, the waterborne transmission of many parasitic species remains the most likely route of exposure for human populations in a global context.

Foodborne viruses have emerged in recent years as a significant cause of morbidity in humans, and their incidence is thought to be increasing (Koopmans et al. 2002). It is widely accepted that the burden of disease caused by foodborne viruses is grossly underestimated by routine surveillance (De Wit et al. 2001) and disease can be associated with outbreaks or sporadic cases. A number of viruses have been identified that can be transmitted through the food chain, including, for example, norovirus, astrovirus, adenovirus, rotavirus, and hepatitis A. Of these, norovirus and hepatitis A are the most significant in terms of the numbers of human cases in developed countries (Koopmans and Duizer 2004). In the United States, noroviruses have been identified as the single biggest cause of foodborne disease in humans, with an estimated 9.2 million cases annually (Mead et al. 1999), and the associated economic costs are high due to their frequent occurrence and high transmissibility (Noda et al. 2008). Clinical manifestations associated with norovirus infection tend to be mild, and many are asymptomatic, which can contribute to the spread of infection. Once introduced in a community, additional spread can occur quite rapidly due to the highly infectious nature of these pathogens, resulting in a significant number of secondary infections in approximately 50% of contacts (Koopmans and Duizer 2004). The transmission of these viruses is predominantly human to human, but indirect exposure via contaminated foods is increasingly recognized. Foodborne transmission has been linked to a wide variety of RTE foods that have become contaminated by infected handlers during processing or at the catering level. Shellfish harvested from waters contaminated with human waste have also been recognized as a significant risk associated with the transmission of these viruses

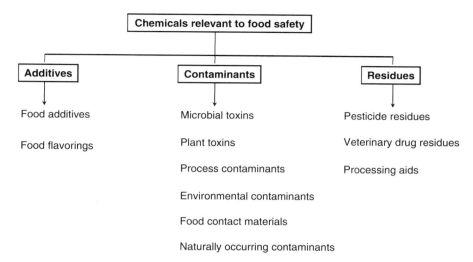

Fig. 2 Summary of the major chemical compounds affecting food safety

due to the concentration effect of filter feeding (Noda et al. 2008). More recently, data have been published suggesting a possible role of animals in the epidemiology of foodborne viruses. A debate has focused on the zoonotic potential of certain noroviruses and their circulation between animal and human populations following the detection of caliciviruses indistinguishable from noroviruses in pigs in Japan and cattle in Great Britain, Germany, and the Netherlands (Liu et al. 1995, 1999; Sugieda et al. 1998; van der Poel et al. 2000). Other authors have indicated that animal noroviruses have not been detected in humans to date, and their role in human health remains unclear (Bank-Wolf et al. 2010) although human strains can be carried in animals. The zoonotic potential of rotavirus has also been examined and indicates that human infection with animal strains can occur, with infection via direct or indirect contact with carrier animals possible. Foodborne transmission has also been identified as a possible exposure route through the contact of RTE foods with feces from infected animals, contaminated water or raw meats, or infected food handlers (Cook et al. 2004). For more information about foodborne viruses, refer to Chap. 6.

3.6 *Microbial Toxins*

Microbial toxins occur as a natural consequence of bacterial and fungal metabolism. Bacteria commonly associated with preformed toxins include *Bacillus cereus, Clostridium botulinum,* and *Staphylococcus aureus.* These preformed toxins are ingested in contaminated food and absorbed into the bloodstream, with the resulting illness arising from intoxication rather than infection. Several species of fungi produce toxic substances; of these, ochratoxin A and sterigmatocystin have been linked with contaminated animal feeds (Rosa et al. 2009) and meat products (Martin et al. 2004; Sorensen et al. 2008; Lacumin et al. 2009). Acute toxicity due to mycotoxins is associated with kidney and liver damage. The important fungi associated with the production of mycotoxins include the genera *Aspergillus, Fusarium,* and *Penicillium.*

Pesticides are used to control harmful or undesired organisms, such as the poultry red mite, *Dermanyssus gallinae* (Sparagano et al. 2009). This group of chemicals includes fungicides and insecticides, among others. Most are highly selective and their levels are controlled by setting maximum residue limit (MRL) levels for each compound (Fig. 2).

4 Controls in the Food Chain

The colonization of livestock by a range of zoonotic pathogens and their dissemination through the food chain remain a global public health challenge. These issues are compounded by the fact that many of the principal zoonoses are asymptomatically carried in animal populations and effective control measures at the farm level are frequently unavailable or impractical. However, it is essential that appropriate measures are taken, including the implementation of good farming practices, feed and waste management, and biosecurity to minimize the possibility of introducing or spreading these organisms. The absence of reliable risk reduction options at the primary production level for pathogens such as *E. coli* O157, *Campylobacter* spp., and *Salmonella* spp. has also meant that control measures are necessary at subsequent stages along the food chain (Madsen 2007). For meat-producing animals, controls must be directed at ensuring that clean and healthy animals are presented for slaughter and good hygienic practices are maintained during carcass dressing and processing of meat products to prevent opportunities for microbial contamination to occur. It is also essential that product chilling is carried out quickly and efficiently to prevent or retard pathogen growth rates (Buncic 2006).

It is widely accepted that effective control of meatborne hazards will most likely be achieved through the development and implementation of integrated, sequential risk reduction–based approaches at all levels of the food chain. The implementation of integrated food safety management programs using Hazard Analysis and Critical Control Points (HACCP) is now universally recommended and has been mandated in food safety legislation in many countries around the world (USDA 1996; European Union 2004). It is also imperative that responsibility for meat safety be shared between all stakeholders along the food chain, including farmers, animal transporters, managers of abattoirs and processing plants, distributors and retailers, consumers, and regulatory agencies.

The monitoring and surveillance of zoonotic pathogens and levels of antimicrobial resistance, drug residues, and other foodborne hazards in food chains is another essential component of an effective public health protection strategy. Robust systems must be implemented so that prevalences can be established, associated public health risks assessed, and trends over time reviewed so that timely interventions can be made as and where necessary. An integrated approach must be taken in the development of such monitoring programs so that risks from existing and potential/emerging hazards can be considered in animal populations, farm inputs (e.g., feed, water, and the farm environment), foods of animal origin, and human populations.

The adoption of molecular techniques in testing laboratories is also recommended, as recent advances have provided methods capable of rapidly and reliably detecting zoonotic hazards and characterizing the pathogenicity and virulence of these agents. Furthermore, molecular subtyping has greatly assisted epidemiological studies at various stages of the food chain as well as investigations of human disease outbreaks. As a consequence, the dissemination and integration of such techniques in veterinary, food, and public health laboratories should be actively encouraged.

Finally, the search for novel approaches to control zoonotic agents at all stages of the food chain must continue. These should include strategies to improve herd health and prevent or reduce pathogen burdens at the farm level (e.g., through the development and implementation of disease prevention strategies using improved animal husbandry practices and developments in vaccines, probiotics, and bacteriophages) and at the processing level, by incorporating chemical decontamination strategies or novel nonthermal physical technologies such as high pressure, ultrasonication, pulsed electric fields, or others. Such initiatives may result in reducing the prevalence of infected animals and/or the concentration of pathogen in animal populations and subsequently in meat products, thereby reducing the risk of exposure and the burden of foodborne disease in humans.

5 Summary

Meat and meat products can get contaminated with different pathogens, such as *Salmonella*, enterohemorrhagic *Escherichia coli*, *Listeria monocytogenes*, or parasites. The reservoirs of these pathogens are live animals, and their environment and meat products are contaminated during the processing of these animals into human food. Microbial hazards can be introduced along the food chain, and their presence varies throughout the processing, storage, distribution, and preparation of foods. Therefore, an integrated food safety management based in part on sequential risk reduction procedures is required to control the risks associated with these hazards and increase the safety of meat products.

References

Adak, G.K., S.M. Meakins, H. Yip, B.A. Lopman, and S.J. O'Brien. 2002. Trends in indigenous foodborne disease and deaths, England and Wales: 1992–2000. *Gut* 51: 832–841.

Adak, G.K., S.M. Meakins, H. Yip, B.A. Lopman, and S.J. O'Brien. 2005. Disease risks from foods, England and Wales, 1996–2000. *Emerging Infectious Diseases* 11: 365–372.

Bank-Wolf, B.R., M. Konig, and H.-J. Thiel. 2010. Zoonotic aspects of infections with noroviruses and sapoviruses. *Veterinary Microbiology* 140: 204–212.

Beth, P., B.P. Bell, M. Goldoft, P.M. Griffin, M.A. Davis, D.C. Gordon, P.I. Tarr, C.A. Bartleson, J.H. Lewis, T.J. Barrett, J.G. Wells, R. Baron, and J. Kobayashi. 1994. A multistate outbreak of *Escherichia coli* O157:H7– associated bloody diarrhea and hemolytic uremic syndrome from hamburgers: the Washington experience. *Journal of the American Medical Association* 272: 1349–1353.

Buncic, S. 2006. *Integrated food safety and veterinary public health*, 139–178. Oxfordshire: CABI.

Carney, E., S.B. O'Brien, J.J. Sheridan, D.A. McDowell, I.S. Blair, and G. Duffy. 2006. Prevalence and level of *Escherichia coli* O157 on beef trimmings, carcasses and boned head meat at a beef slaughter plant. *Food Microbiology* 23: 52–59.

CDC. 1995. *Escherichia coli* O157:H7 outbreak linked to commercially distributed dry-cured salami – Washington and California, 1994. *Morbidity and Mortality Weekly Report* 44: 157–160.

CDC. 1997. *Escherichia coli* O157:H7 infections associated with eating a nationally distributed commercial brand of frozen ground beef patties and burgers – Colorado, 1997. *Morbidity and Mortality Weekly Report* 46: 777–778.

CDC. 2000. Surveillance for foodborne outbreaks – US, 1993–1997. *Morbidity and Mortality Weekly Report* 49(ss-1): 1–66.

Chasseignaux, E., P. Gerault, M.T. Toquin, G. Salvat, P. Colin, and G. Ermel. 2002. Ecology of *Listeria monocytogenes* in the environment of raw poultry meat and raw pork meat processing plants. *FEMS Microbiology Letters* 210: 271–275.

Coia, J. 1998. Clinical, microbiological and epidemiological aspects of *Escherichia coli* O157 infection. *FEMS Immunology and Medical Microbiology* 20: 1–9.

Cook, N., J. Bridger, K. Kendall, M.I. Gomara, L. El-Attar, and J. Gray. 2004. The zoonotic potential of rotavirus. *Journal of Infection* 48: 289–302.

Dawson, D. 2005. Foodborne protozoan parasites. *International Journal of Food Microbiology* 103: 207–227.

De Wit, M., M. Koopmans, T. Kortbeek, W. van Leeuwen, A. Bartelds, and Y. van Duynhoven. 2001. c Sensor: a population-based cohort study on gastroenteritis in the Netherlands: incidence and etiology. *American Journal of Public Health* 154: 666–674.

Dorny, P., N. Praet, N. Deckers, and S. Gabriel. 2009. Emerging food-borne parasites. *Veterinary Parasitology* 163: 196–206.

Doyle, M.P., and M.C. Erickson. 2006. Emerging microbiological food safety issues related to meat. *Meat Science* 74: 98–112.

Doyle, M.P., L.R. Beuchat, and T.J. Montville. 2001. *Food microbiology: fundamentals and frontiers*. Washington, DC: ASM Press.

Dubey, J.P., D.E. Hill, J.L. Jones, A.W. Hightower, E. Kirkland, J.M. Roberts, P.L. Marcet, T. Lehmann, M.C. Vianna, K. Miska, C. Sreekumar, O.C. Kwok, S.K. Shen, and H.R. Gamble. 2005. Prevalence of viable *Toxoplasma gondii* in beef chicken and pork from retail meat stores in the US: risk assessment to consumers. *Journal for Parasitology* 91: 1082–1093.

EFSA (European Food Safety Authority). 2007. The community summary report on trends and sources of zoonoses, zoonotic agents, antimicrobial resistance and foodborne outbreaks in the European Union in 2006. *European Food Safety Authority Journal* 130: 3–352.

EFSA. 2009. The community summary report on trends and sources of zoonoses, zoonotic agents in the European Union in 2007. *European Food Safety Authority Journal* 223: 3–312.

EU. 2004. European Union. Regulation (EC) no. 852/2004 of the European Parliament and of the Council of 29 April 2004 on the hygiene of foodstuffs. Official Journal of the European Union L139/1.

Farber, J.M., and P.I. Peterkin. 1991. *Listeria monocytogenes*, a food-borne pathogen. *Microbiology Reviews* 55: 752–811.

Furness, B.W., M.J. Beach, and J.M. Roberts. 2000. Giardiasis surveillance – US, 1992–1997. *Morbidity and Mortality Weekly Report* 49: 1–13.

Gillespie, I.A., S.J. O'Brien, J.A. Frost, G.K. Adak, P. Horby, A.V. Swan, M.J. Painter, and K.R. Neal. 2002. A case-case comparison of *Campylobacter coli* and *Campylobacter jejuni* infection: a tool for generating hypotheses. *Emerging Infectious Diseases* 8: 937–942.

Gottstein, B., E. Pozio, and K. Nockler. 2009. Epidemiology, diagnosis, treatment, and control of trichinellosis. *Clinical Microbiology Reviews* 22: 127–145.

Goulet, V., J. Rocourt, I. Rebiere, C. Jacquet, C. Moyse, P. Dehaumont, G. Salvat, and P. Veit. 1998. Listeriosis outbreak associated with the consumption of rillettes in France in 1993. *Journal of Infection* 177: 155–160.

Greig, J.D., and A. Ravel. 2009. Analysis of foodborne outbreak data reported internationally for source attribution. *International Journal of Food Microbiology* 130: 77–87.

Gruenewald, R., A.H. Ropper, H. Lior, J. Chan, R. Lee, and V.S. Molinaro. 1991. Serologic evidence of *Campylobacter jejuni/coli* enteritis in patients with Guillain–Barré syndrome. *Archives of Neurology* 48: 1080–1082.

Gudbjornsdottir, B., M.L. Suihko, P. Gustavsson, G. Thorkelsson, S. Salo, A.M. Sjoberg, O. Niclasen, and S. Bredholt. 2004. The incidence of *Listeria monocytogenes* in meat, poultry and seafood plants in the Nordic countries. *Food Microbiology* 21: 217–225.

Havelaar, A.H., S. Brul, A. de Jong, R. de Jonge, M.H. Zwietering, and B.H. Ter Kuile. 2010. Future challenges to microbial food safety. *International Journal of Food Microbiology* 139(Suppl 1): S79–S94.

Health Canada. 2009. Lessons learned report – health Canada's response to the 2008 Listeriosis outbreak. http://www.hc-sc.gc.ca/fn-an/alt_formats/hpfb-dgpsa/pdf/securit/listeriosis-eng.pdf. Accessed 15 Dec 2010.

Hussein, H.S., and T. Sakuma. 2005. Invited review: prevalence of Shiga toxin-producing *Escherichia coli* in dairy cattle and their products. *Journal of Dairy Science* 88: 450–465.

Jacquet, C., B. Catimel, R. Brosch, C. Buchrieser, P. Dehaumont, V. Goulet, A. Lepoutre, P. Veit, and J. Rocourt. 1995. Investigations related to the epidemic strain involved in the French listeriosis outbreak in 1992. *Applied and Environmental Microbiology* 61: 2242–2246.

Jore, S., H. Viljugrein, E. Brun, B.T. Heier, B. Borck, S. Ethelberg, M. Hakkinen, M. Kuusi, J. Reiersen, I. Hansson, E. Olsson Engvall, M. Lofdahl, J.A. Wagenaar, W. van Pelt, and M. Hofshagen. 2010. Trends in *Campylobacter* incidence in broilers and humans in six European countries, 1997–2007. *Preventive Veterinary Medicine* 93: 33–41.

Keene, W.E. 1999. Lessons from investigations of foodborne disease outbreaks. *Journal of the American Medical Association* 281: 1845–1847.

Kijlstra, A., O. Eissen, A. Cornelissen, J. Munniksma, K. Eijck, and I.T. Kortbeek. 2004. *Toxoplasma gondii* infection in animal-friendly pig production systems. *Investigative Ophthalmology & Visual Science* 45: 3165–3169.

Koopmans, M., and E. Duizer. 2004. Foodborne viruses: an emerging problem. *International Journal of Food Microbiology* 90: 23–41.

Koopmans, M., C.H. von Bonsdorff, J. Vinje, D. de Medici, and S. Monroe. 2002. Foodborne viruses. *FEMS Microbiology Reviews* 26: 187–205.

Lacumin, L., L. Chiesa, D. Boscolo, M. Manzano, C. Cantoni, S. Orlic, and G. Comi. 2009. Moulds and ochratoxin A on surfaces of artisanal and industrial dry sausages. *Food Microbiology* 26: 65–70.

Lenehan, M., S. O'Brien, K. Kinsella, T. Sweeney, and J.J. Sheridan. 2007. Prevalence and molecular characterization of *E. coli* O157:H7 on Irish lamb carcasses, fleece and in feces samples. *Journal of Applied Microbiology* 103: 2401–2409.

Liu, B.L., I.N. Clarke, E.O. Caul, and P.R. Lambden. 1995. Human enteric caliciviruses have a unique genome structure and are distinct from the Norwalk-like viruses. *Archives of Virology* 140: 1345–1356.

Liu, B.L., P.R. Lambden, H. Gunther, P. Otto, M. Elschner, and I.N. Clarke. 1999. Molecular characterization of a bovine enteric calicivirus: relationship to the Norwalk-like viruses. *Journal of Virology* 73: 819–825.

Madsen, M. 2007. *Campylobacter*. In *Food safety handbook – microbiological challenges*, ed. M. Storrs, 20–37. Paris: BioMérieux Editions.

Martin, A., M. Jurado, M. Rodriguez, F. Nunez, and J.J. Cordoba. 2004. Characterization of molds from dry-cured meat products and their metabolites by micellar electrokinetic capillary electrophoresis and random amplified polymorphic DNA PCR. *Journal of Food Protection* 67: 2234–2239.

McLauchlin, J., R.T. Mitchell, W.J. Smerdon, and K. Jewell. 2004. *Listeria monocytogenes* and listeriosis: a review of hazard characterisation for use in microbiological risk assessment of foods. *International Journal of Food Microbiology* 92: 15–33.

Mead, P., L. Slutsker, V. Dietz, L. McCaig, J. Bresee, C. Shapiro, P. Griffin, and R. Tauxe. 1999. Food-related illness and death in the US. *Emerging Infectious Diseases* 5: 607–625.

Miller, G., G.M. Dunn, A. Smith-Palmer, I.D. Ogden, and N.J. Strachan. 2004. Human campylobacteriosis in Scotland: seasonality, regional trends and bursts of infection. *Epidemiology and Infection* 132: 585–593.

Minihan, D., P. Whyte, M. O'Mahony, T. Clegg, and J.D. Collins. 2003. *Escherichia coli* O157 in Irish feedlot cattle: a longitudinal study involving pre-harvest and harvest phases of the food chain. *Journal of Food Safety* 23: 167–178.

Moore, J.E., D. Corcoran, J.S. Dooley, S. Fanning, B. Lucey, M. Matsuda, D.A. McDowell, F. Megraud, B.C. Millar, R. O'Mahony, L. O'Riordan, M. O'Rourke, J.R. Rao, P.J. Rooney, A. Sails, and P. Whyte. 2005. *Campylobacter*. *Veterinary Research* 36: 351–382.

Mora, A., M. Blanco, J.E. Blanco, G. Dahbi, C. Lopez, P. Justel, M.P. Alonso, A. Echeita, M.I. Bernardez, E.A. Gonzalez, and J. Blanco. 2007. Serotypes, virulence genes and intimin types of Shiga toxin (verocytotoxin)-producing *Escherichia coli* isolates from minced beef in Lugo (Spain) from 1995 through 2003. *BMC Microbiology* 7: 13.

Noda, M., S. Fukuda, and O. Nishio. 2008. Statistical analysis of attack rate in norovirus foodborne outbreaks. *International Journal of Food Microbiology* 122: 216–220.

O'Brien, S.B., G. Duffy, E. Carney, J.J. Sheridan, D.A. McDowell, and I.S. Blair. 2005. Prevalence and numbers of *Escherichia coli* O157 on bovine hide at a beef slaughter plant. *Journal of Food Protection* 68: 660–665.

Phiri, I.K., H. Ngowi, S. Afonso, E. Matenga, M. Boa, S. Mukaratirwa, S. Githigia, M. Saimo, C. Sikasunge, N. Maingi, G.W. Lubega, A. Kassuku, L. Michael, S. Siziya, R.C. Krecek, E. Noormahomed, M. Vilhena, P. Dorny, and A.L. Willingham. 2003. The emergence of *Taenia solium* cysticercosis in Eastern and Southern Africa as a serious agricultural problem and public health risk. *Acta Tropica* 87: 13–23.

Pozio, E. 2007. Taxonomy, biology and epidemiology of *Trichinella* parasites. In *Guidelines for the surveillance, management, prevention and control of trichinellosis, FAO/WHO/OIE*, ed. J. Dupouy-Camet and K.D. Murrell, 1–35. Paris: OIE.

Pozio, E., and K.D. Murrell. 2006. Systematics and epidemiology of *Trichinella*. *Advances in Parasitology* 63: 367–439.

Rangel, J.M., P.H. Sparling, C. Crowe, P.M. Griffin, and D.L. Swerdlow. 2005. Epidemiology of *Escherichia coli* O157:H7 outbreaks, US, 1982–2002. *Emerging Infectious Diseases* 11: 603–609.

Rhoades, J.R., G. Duffy, and K. Koutsoumanis. 2009. Prevalence and concentration of verocytotoxigenic *Escherichia coli*, *Salmonella enterica* and *Listeria monocytogenes* in the beef production chain: a review. *Food Microbiology* 26: 357–376.

Riley, L.W., R.S. Remis, S.D. Helgerson, H.B. McGee, J.G. Wells, B.R. Davis, R.J. Herbert, E.S. Olcott, L.M. Johnson, N.T. Hargrett, P.A. Blake, R.A. Chordash, and M.L. Cohen. 1983. Hemorrhagic colitis associated with a rare *Escherichia coli* serotype. *The New England Journal of Medicine* 308: 681–685.

Rosa, C.A., K.M. Keller, L.A. Keller, M.L. Gonzalez Pereyra, C.M. Pereyra, A.M. Dalcero, L.R. Cavaglieri, and C.W. Lopes. 2009. Mycological survey and ochratoxin A natural contamination of swine feedstuffs in Rio de Janeiro state, Brazil. *Toxicon* 53: 283–288.

Schonberg-Norio, D., L. Mattila, A. Lauhio, M.L. Katila, S.S. Kaukoranta, M. Koskela, S. Pajarre, J. Uksila, E. Eerola, S. Sarna, and H. Rautelin. 2010. Patient-reported complications associated with *Campylobacter jejuni* infection. *Epidemiology and Infection* 138: 1004–1011.

Sherman, P.M., J.C. Ossa, and E. Wine. 2010. Bacterial infections: new and emerging enteric pathogens. *Current Opinion in Gastroenterology* 26: 1–4.

Shield, J., J.H. Baumer, J.A. Dawson, and P.J. Wilkinson. 1990. Cryptosporidiosis – an educational experience. *Journal of Infection* 21: 297–301.

Smerdon, W.J., R. Jones, J. McLauchlin, and M. Reacher. 2001. Surveillance of listeriosis in England and Wales 1995–1999. *Communicable Disease and Public Health* 4: 188–193.

Sopwith, W., M. Ashton, J.A. Frost, K. Tocque, S. O'Brien, M. Regan, and Q. Syed. 2003. Enhanced surveillance of *Campylobacter* infection in the Northwest of England 1997–1999. *Journal of Infectious* 46: 35–45.

Sorensen, L.M., T. Jacobsen, P.V. Nielsen, J.C. Frisvad, and A.G. Koch. 2008. Mycobiota in the processing areas of two different meat products. *International Journal of Food Microbiology* 124: 58–64.

Sparagano, O., A. Pavlicevic, T. Murano, A. Camarda, H. Sahibi, O. Kilpinen, M. Mul, R. van Emous, S. le Bouquin, K. Hoel, and M.A. Cafiero. 2009. Prevalence and key figures for the poultry red mite *Dermanyssus gallinae* infections in poultry farm systems. *Experimental & Applied Acarology* 48: 3–10.

Sugieda, M., H. Nagaoka, Y. Kakishima, T. Ohshita, S. Nakamura, and S. Nakajima. 1998. Detection of Norwalk-like virus genes in the caecum contents of pigs. *Archives of Virology* 143: 1215–1221.

Suzuki, H., and S. Yamamoto. 2009. *Campylobacter* contamination in retail poultry meats and by-products in the world: a literature survey. *The Journal of Veterinary Medical Science* 71: 255–261.

Tauxe, R.V. 2002. Emerging foodborne pathogens. *International Journal of Food Microbiology* 78: 31–41.

Tenter, A.M., A.R. Heckeroth, and L.M. Weiss. 2000. *Toxoplasma gondii*: from animals to humans. *International Journal for Parasitology* 30: 1217–1258.

USDA. 1996. Pathogen reduction: hazard analysis critical control point (HACCP) systems: final rule. *Federal Register* 61: 38806–38988.

USDA-ERS. 2009. Food safety: economic costs of foodborne illness. http://www.ers.usda.gov/briefing/foodsafety/economic.htm. Accessed 15 Dec 2010.

van der Poel, P., J. Vinje, R. van der Heide, I. Herrera, A. Vivo, and M. Koopmans. 2000. Norwalk-like calicivirus genes in farm animals. *Emerging Infectious Diseases* 6: 36–41.

Waldenstrom, J., S.L. On, R. Ottvall, D. Hasselquist, and B. Olsen. 2007. Species diversity of campylobacteria in a wild bird community in Sweden. *Journal of Applied Microbiology* 102: 424–432.

Wegener, H.C. 2010. Danish initiatives to improve the safety of meat products. *Meat Science* 84: 276–283.

WHO. 2005. Drug resistant *Salmonella*. Fact Sheet no. 139.

WHO. 2007. The world health report, 2007. http://www.who.int/whr/2007/en/index.html.

WHO. 2009a. Initiative to estimate the global burden of foodborne diseases. http://www.who.int/foodsafety/foodborne_disease/ferg/en/print.html. Accessed 15 Dec. 2010.

WHO. 2009b. Zoonoses and veterinary public health. http://www.who.int/zoonoses/vph/en/. Accessed 15 Dec 2010.

Whyte, P., S. Fanning, M. O'Mahony, R. O'Mahony, and D. Drudy. 2005. Public health and economic burden of infectious food-borne diseases in Ireland. *Irish Veterinary Journal* 58: 279–283.

Willshaw, G.A., J. Thirlwell, A.P. Jones, S. Parry, R.L. Salmon, and M. Hickey. 1994. Verocytotoxin-producing *Escherichia coli* O157 in beefburgers linked to an outbreak of diarrhoea, haemorrhagic colitis and haemolytic uraemic syndrome in Britain. *Letters in Applied Microbiology* 19: 304–307.

Safety of Fish and Seafood Products

Kenneth Lum

1 Introduction

There are more than 800 species of vertebrate and invertebrate animals that comprise the vast selection of seafood on the market today. Multiplied by the various forms of finished products and packaging available to the consumer, it is readily apparent that seafood represents the most diverse protein food source available. This diversity provides not only greater market opportunity, but also a greater number of food safety considerations related to production, packaging, storage, and distribution. This chapter describes in general many of the food safety aspects that must be considered for the development of effective and comprehensive food safety programs for some common seafood product forms.

2 Regulatory Requirements for Seafood Safety

Regulations enforced by the U.S. Food and Drug Administration (FDA) provide the essential elements of seafood safety through good manufacturing practices (GMPs) and sanitation by *Hazard Analysis Critical Control Point (HACCP)* requirements.

Fish and fishery products processed in the United States are subject to the HACCP and sanitation requirements published in the FDA Fish and Fishery Products regulation, Title 21 of the Code of Federal Regulations, Part 123 (21 CFR 123). Much of the information presented in this chapter is based on the requirements of this regulation, as well as information provided in the FDA Fish and Fishery Products Hazards and Controls Guidance (Hazard Guide).

3 Core Programs for Seafood Safety

As with all food manufacturing operations, seafood producers must implement core programs to ensure the production of safe, wholesome seafood products. These core programs are considered GMPs. GMPs describe the methods, equipment, facilities, personnel practices, and controls for

K. Lum (✉)
Seafood Products Association, Seattle, WA, USA
e-mail: klum@spa-food.org

producing safe food. The basis for initially developing GMPs was to provide a set of regulatory requirements that could be used to enforce provisions of the Federal Food Drug and Cosmetic Act (FDCA), specifically the two sections of the FDCA that are directly related to conditions in a facility where food has been manufactured.

- Section 402(a)(3) specifies that food has been manufactured under such conditions that it is unfit for consumption.
- Section 402(a)(4) considers that food may be adulterated if it is prepared, packed, or held under insanitary conditions whereby it may have become contaminated with filth or rendered injurious to health.

Regulatory requirements for GMPs are published in 21 CFR 110. As prerequisite sanitary and processing requirements for producing safe and wholesome food, GMPs are essential industry practices that also provide regulators with a tool for applying control over the safety of the nation's food supply. Though it is beyond the scope of detailed discussion for this chapter, it is nevertheless important to emphasize the key role that GMPs play in seafood safety. It is important for food safety professionals to be familiar with the regulatory requirements for GMPs prescribed in 21 CFR 110 as well as the specific requirements for monitoring GMPs for seafood processors in the FDA Fish and Fishery Products regulation in 21 CFR 123. GMPs are also occasionally referred to as "Current" GMPs (CGMPs) to recognize the value of reviewing current science and technologies to ensure the most effective practices are in place.

4 Species-Related Considerations

There are more than 600 species of vertebrates and more than 200 species of invertebrates listed in the FDA Hazard Guide. Vertebrate species are primarily any type of finfish, and invertebrates include molluscan and crustacean shellfish. Potential hazards have been identified for each species listed, and each potential hazard must be evaluated to determine if controls are necessary for the process to ensure end product safety. Food safety considerations are unique to vertebrates (species with a backbone) and invertebrates (species without a backbone). Vertebrate species are represented primarily by finfish marketed in such forms as whole fish, headed and gutted, fillets, portions, and steaks and as the primary ingredient in many minced and *value-added products,* such as fish sticks and breaded and battered portions. Invertebrate species commonly harvested and processed as seafood are divided into molluscan shellfish and crustacean shellfish. Examples of molluscan shellfish include animals such as clams, oysters, mussels, snails, and octopus. Crustacean shellfish are animals that include crab, shrimp, lobster, and crayfish. The distinction between molluscan shellfish and crustacean shellfish is important to keep in mind when considering GMPs and controls for allergens, as mentioned later in this chapter.

Species-related seafood safety considerations include the following:

- Bacterial *pathogens*
- Parasites
- Natural *toxins,* both intrinsic sources and extrinsic sources from environmental exposure
- Histamine, combined intrinsic properties and environmental temperature exposure
- Chemical contaminants
- Aquaculture drugs
- Allergens

4.1 Bacterial Pathogens

Generally, raw finfish have some natural microbial load from the environment from which they are harvested. Wild caught finfish is typically chilled rapidly and handled and stored under conditions that are not likely to result in the contamination or growth of pathogens. The natural flora of microorganisms on raw finfish are usually predominantly nonpathogenic spoilage organisms that provide competitive growth barriers to pathogen growth. Maintaining sanitary conditions and practices generally prevent pathogens from becoming significant food safety concerns for raw finfish. Most raw finfish is also intended to be cooked prior to consumption. Typical home cooking temperatures, as well as retail/restaurant cooking procedures, will destroy microorganisms of public health significance.

The FDA *Food Code* (FDA 2009) recommends a cook process that results in

- Whole muscle fish products reaching 145°F for at least 15 s or
- Comminuted (minced) fish products reaching 155°F for at least 15 s

The Food Code is a model code jointly developed by the U.S. Food and Drug Administration (FDA) and the Centers for Disease Control and Prevention (CDC) of the U.S. Department of Health and Human Services (HHS) and the Food Safety and Inspection Service of the U.S. Department of Agriculture (USDA). The code is used as a reference document for state, city, county, and tribal agencies that regulate restaurants, retail food stores, vending operations, and foodservice operations in institutions such as schools, hospitals, nursing homes, and childcare centers.

The Food Code establishes practical, science-based guidance and enforceable provisions for mitigating risk factors known to cause foodborne illness. The FDA Hazard Guide does not identify pathogens as a potential hazard for vertebrate species of fish. While pathogens may not be considered inherent food safety hazards specifically associated with particular species of raw finfish, they must be considered once the finfish is processed and packaged into other product forms that may increase the risk of foodborne illness due to pathogens. These product- and package-related hazards will be discussed later in this chapter.

Many invertebrate species of fish (primarily molluscan shellfish) are harvested in near-shore waters closer to potential sources of human and industrial waste that may lead to waters contaminated with pathogens. Many of these molluscan shellfish (clams, mussels, oysters, etc.) can accumulate the pathogens. In addition to these, there are also a number of naturally occurring pathogens, particularly from the genus *Vibrio,* that can cause severe illness and in some cases even death. If consumed raw or partially cooked, pathogens will survive and may be present at levels sufficient to cause illness. This is a particular concern for shellfish like oysters, clams, and scallops, where some cuisines consider raw consumption a delicacy. Cooking molluscan shellfish is usually sufficient to kill pathogens.

The primary control of pathogens for molluscan shellfish is source control. Source control is typically achieved by the closure of waters for shellfish harvesting where certain levels of pathogens have been detected. Elevated levels of pathogens are often correlated with environmental water temperatures with peak populations coinciding with times of the year when the water temperature is highest. Shellfish control authorities (typically state agencies) monitor the water as well as marine animals and have the authority to close harvest areas when high levels of pathogens are detected.

4.2 Parasites

Parasites naturally occur in many finfish species, as well as in a few molluscan shellfish (octopus, squid, and snails). Many of these parasites, if ingested alive, may present a human health hazard. Human illnesses have occurred when fish products are consumed raw or partially cooked, because they have not been exposed to sufficient heat or freezing-time/temperature combinations that kill the parasites. The most common parasites of human health concern are roundworms, tapeworms, and flukes. Typical symptoms are nausea, diarrhea, and abdominal swelling and pain.

Similar to pathogens, parasites of human health concern are destroyed by the heat applied by typical home cooking or restaurant/foodservice cooking procedures.

The Food Code and the FDA Hazards Guide recommend freezing procedures that have demonstrated effectiveness in killing parasites, as follows:

Before service or sale in *ready-to-eat (RTE)* form, raw, raw-marinated, partially cooked, or marinated-partially cooked fish shall be

- Frozen and stored at a temperature of −20°C (−4°F) or below for a minimum of 168 h (7 days) in a freezer or
- Frozen at −35°C (−31°F) or below until solid and stored at −35°C (−31°F) or below for a minimum of 15 h or
- Frozen at −35°C (−31°F) or below until solid and stored at −20°C (−4°F) or below for a minimum of 24 h

4.3 Natural Toxins

There are several categories of "natural toxins" that may affect the safety of many species of finfish and molluscan and crustacean shellfish. The natural toxins of concern in finfish are

- Ciguatoxin: Ciguatera fish poisoning (CFP)
- Planktonic toxins:
  - Domoic acid causes amnesic shellfish poisoning (ASP)
  - Azaspiracid shellfish poisoning (AZP)
  - Okadaic acid and dinophysistoxins cause diarrhetic shellfish poisoning (DSP)
  - Brevetoxin causes neurotoxic shellfish poisoning (NSP)
  - Saxitoxin causes paralytic shellfish poisoning (PSP)
- Gempylotoxin
- Tetrodotoxin

4.3.1 Ciguatoxin

Consuming fish containing certain levels of ciguatoxin can result in an illness referred to as ciguatera fish poisoning (CFP). CFP can occur when food is prepared from fish harvested in subtropical and tropical waters around the world where toxic algae are consumed by small reef fish, which are eaten by larger reef fish that are then consumed by humans. CFP cannot reliably be inactivated by heat. Symptoms include numbness and tingling around the mouth and possibly other extremities, nausea, vomiting, diarrhea, joint and muscle pain and weakness, headache, reversal of hot/cold sensation, vertigo, irregular heartbeat, and reduced blood pressure. The primary prevention is source control accomplished by harvesting fish from waters that are not subject to state or other authority advisories, as well as harvester and processor monitoring and knowledge of reported CFP cases in the areas where fish are harvested.

Table 1 Planktonic toxins: animals and geographical areas typically affected

Toxin	Illness	Animal(s) affected	Location of implicated animal harvest, and reported outbreaks
Domoic acid	Amnesic shellfish poisoning	Molluscan and crustacean shellfish, anchovies	NE and NW coasts of North America
Azaspiracid	Azaspiracid shellfish poisoning	Mussels	Ireland, the Netherlands, Europe
Okadaic acid, dinophysistoxin	Diarrhetic shellfish poisoning	Molluscan shellfish	Japan, Southeast Asia, Scandinavia, Western Europe, Chile, New Zealand, Canada
Brevetoxin	Neurotoxic shellfish poisoning	Molluscan shellfish	Gulf of Mexico, southern Atlantic coast of the US, Gulf of Florida, New Zealand
Saxitoxin	Paralytic shellfish poisoning	Molluscan and crustacean shellfish, some finfish (mackerel and puffer fish)	Global distribution in temperate to tropical waters

Source: Adapted from Botana (2000)

4.3.2 Planktonic Toxins

Planktonic fish poisoning compounds that cause amnesic shellfish poisoning (ASP), azaspiracid shellfish poisoning (AZP), diarrhetic shellfish poisoning (DSP), neurotoxic shellfish poisoning (NSP), and paralytic shellfish poisoning (PSP) are caused by planktonic algae or metabolic products of the algae. These toxins typically coincide with algae blooms, and individual toxins are often associated with particular species and geographical regions of the world. The toxins can be found in the viscera of marine animals that inhabit areas of the ocean where these algae blooms occur. Table 1 summarizes the animals typically affected by these toxins and the geographical areas where implicated animals have been harvested and illnesses have been reported. Table 1 is not all-inclusive but reflects historical illness data. The movement of algae types to other geographical areas is dynamic, and seafood safety professionals must be vigilant to track outbreaks and conduct epidemiological analysis to prevent or minimize the continuation of outbreaks.

The planktonic toxins cause illnesses that result in a variety of gastrointestinal and neurological symptoms ranging in severity from tingling sensations to death. Other reported symptoms include diarrhea, vomiting, nausea, abdominal pain, burning sensation, reversal of hot/cold sensation, numbness, drowsiness, headache, short-term memory loss, incoherence, and coma. PSP and ASP have a history of more severe illness, both resulting in fatality cases, with PSP having the highest mortality rate among the planktonic toxins.

The most effective control for preventing illnesses from planktonic toxins is source controls administered by state and local agencies in the United States and other foreign government agencies referred to as shellfish control authorities. Their function typically includes monitoring levels of toxins in shellfish meats and/or levels of implicated plankton where molluscan shellfish in particular are harvested. When predetermined levels are exceeded, the shellfish control authorities can restrict harvesting in affected areas. Monitoring and enforcing the harvesting restrictions are achieved through requirements for harvesters and processors to document and certify the date and location of the harvest of all shellfish being handled. In some areas where persistent levels of some of these planktonic algae occur at lower levels, the processing of shellfish harvested from the area may be allowed at the discretion of the regulatory authority, but only in the eviscerated form when this product form eliminates the presence of the toxins.

4.3.3 Gempylotoxin

Gempylotoxin is not derived from marine algae but is a natural constituent of two particular species of fish that have a very high oil content, escolar and oilfish. In addition to these two species, other white flesh high-oil fish such as butterfish, black cod, and seabass have also been implicated, but it is unclear whether or not species substitution resulted in the misidentification of the actual causative species.

Symptoms of gempylotoxin are primarily gastrointestinal, causing severe diarrhea due to the purgative effect of the toxin. While the sale of the fish is not prohibited, the FDA advises against it and recommends adequate consumer notification of the purgative effect when such fish is offered for sale. The FDA statement on escolar and oilfish follows:

[EXT]There are naturally occurring toxins in some species that do not involve marine algae. Escolar (Scientific Name *Lepidocybium flavobrunneum*) and its relative Oilfish or Cocco (Scientific Name *Ruvettus pretiosus*) contain a strong purgative oil that when consumed can cause diarrhea known as Gempylid Fish Poisoning or Gempylotoxism. FDA advises against the sale of the fish in intrastate/interstate commerce, and recommends that seafood manufacturers/processors inform potential buyers/sellers, etc. of the purgative effect associated with the consumption of these fish (Anonymous 2011).

4.3.4 Tetrodotoxin

Illness caused by tetrodotoxin (anhydrotetrodotoxin 4-epitetrodotoxin, tetrodonic acid) is commonly referred to pufferfish poisoning, fugu poisoning, or tetradon poisoning. As the name implies, the illness has been associated primarily with the consumption of pufferfish. The gonads, liver, intestines, and skin of pufferfish can contain levels of tetrodotoxin sufficient to produce rapid and violent death. Pufferfish is considered a delicacy in many cultures, and chefs must receive special training to ensure the edible parts of the fish flesh are not contaminated by the toxin. Fortunately, the incidence of poisoning attributed to tetrodotoxin is very small.

Tetrodotoxin has also been detected in blue-ringed octopus, starfish, marine angelfish, xanthid crabs, and even some terrestrial amphibians, but illnesses have been predominantly attributed to the consumption of improperly prepared pufferfish.

There is no evidence that tetrodotoxin originates from planktonic algae; however, recent reports of the production of tetrodotoxin by several bacterial species, including strains of the family *Vibrionaceae*, *Pseudomonas sp.*, and *Photobacterium phosphoreum*, point toward a bacterial origin.

The key to the prevention of tetrodotoxin poisoning is to ensure preparers are well trained and experienced in handling the susceptible fish species. In the United States, importation is only allowed under special circumstances, which greatly reduces exposure and risk.

4.4 Histamine and Scombrotoxin

Scombrotoxin can be produced in certain fish that contain high levels of the amino acid histidine. When certain strains of bacteria grow on/in the fish, they produce enzymes that convert histidine to histamine. Histamine is the primary compound in scombrotoxin that causes illness referred to as scombroid poisoning, or histamine poisoning. The fish most commonly associated with scombroid poisoning are tuna (e.g., skipjack and yellowfin), mahi mahi, bluefish, sardines, mackerel, amberjack, and abalone. These fish typically have naturally higher levels of free histidine. Other foods and beverages such as Swiss cheese and wine can also have levels of histamine that can cause illness. Histamine can form even after frozen products have been thawed and prior to consumption. Scombrotoxin is a stable compound and, once formed, can continue to cause illness even after fish with high levels of scombrotoxin have been canned and thermally processed.

Symptoms of scombrotoxin poisoning may include a tingling or burning sensation in the mouth, a rash on the upper body, itching of the skin, headaches, and a drop in blood pressure. More severe symptoms may include nausea, vomiting, and diarrhea. The severity of symptoms is related to individual sensitivity, levels of histamine, and the health of the consumer at the time of consumption. Scombrotoxin poisoning is still a leading cause of illness related to the consumption of seafood.

Since scombrotoxin is an end result of bacterial growth that transforms histidine into histamine, it is imperative that fish be handled, stored, and processed as swiftly as possible under chilled (<40°F) conditions to limit bacterial growth. Maintaining appropriate temperatures from harvest through processing to a finished product is the primary control for preventing histamine formation. The FDA Hazard Guide includes several strategies for preventing and/or detecting levels of histamine in scombroid fish species. Since bacterial growth can result in decomposition as well as histamine formation, sensory testing provides a good indication of histamine formation in those fish that exhibit decomposition. Sensory testing alone or in combination with histamine testing are effective indicators of fish that may contain scombrotoxin. FDA Compliance Policy Guide 540.525 provides the administration's opinion that product represented by fish that were tested and confirmed to have a histamine level equal to or greater than 50 ppm, or fish that fail the sensory test due to decomposition, would be considered adulterated under the Food Drug and Cosmetic Act.

4.5 *Chemical Contaminants*

Fish may be exposed to chemical contaminants from environments to which they are exposed throughout their life cycle and/or from harvest area waters. These contaminants may accumulate in fish at levels that can cause illness. Illnesses can range in symptoms and severity and may include neurological as well as physical ailments. Chemical contamination is most commonly associated with bioaccumulation in the food animal from long-term exposure to these contaminants. Aquatic animals are generally able to naturally cleanse or purify (depurate) tissues from contaminants when there is a single or very short-term exposure, given sufficient time prior to harvest. Concern for these contaminants primarily focuses on fish harvested from fresh water, estuaries, and near-shore coastal waters (e.g., areas subject to shoreside contaminant discharges), rather than from the open ocean.

The presence of *methylmercury* in fish has been widely discussed and is of great interest to consumers. Generally, larger, longer-lived species of fish tend to have higher levels of methylmercury due to bioaccumulation over time and predation on smaller fish. The effect of methylmercury exposure from the consumption of fish is not thoroughly understood, and recent studies have focused on the balance of both the benefits and the risks associated with the consumption of fish that may contain methylmercury. These studies generally indicate the benefits outweigh the risks for the general population. While demonstrated health benefits for the general population have been shown, federal, state, or local authorities may occasionally issue consumption advisories when certain species of fish are found to have relatively higher amounts of methylmercury or when at-risk subpopulations are identified.

Table 2 provides a list of contaminants, regulatory limits, the associated aquatic animals affected, and regulatory references. Regulatory limits are "Tolerances" or "Action Levels" at or above which the FDA may take legal action. The action levels are described and enforced in accordance with Compliance Policy Guides (CPG), are revised according to criteria specified in Title 21, Code of Federal Regulations, Parts 109 and 509, and are revoked when a regulation establishing a tolerance

Table 2 Environmental chemical contaminants: regulatory tolerances and action levels

Contaminant	Level in edible tissue (ppm[a])	Fish to which limit is applied	Reference
PCBs	2.00	All fish	21 CFR 109.30
Carbaryl	0.25	Oysters	40 CFR 180.169
Diquat	2.00	Fish	40 CFR 180.226
Diquat	20.00	Shellfish	40 CFR 180.226
Diuron and its metabolites	2.00	Farm-raised Freshwater fish	40 CFR 180.106
Endothall and its monomethyly ester	0.10	All fish	40 CFR 180.293
Fluridone	0.50	Finfish and crayfish	40 CFR 180.420
Glyphosphate	0.25	Fish	40 CFR 180.364
Glyphosphate	3.00	Shellfish	40 CFR 180.364
2,4-D	0.10	Fish	40 CFR 180.142
2-4-D	1.00	Shellfish	40 CFR 180.142
Aldrin and dieldrin	0.30[b]	All fish	CPG Sec. 575.100
Benzene hexachloride	0.30	Frog legs	CPG Sec. 575.100
Chlordane	0.30	All fish	CPG Sec. 575.100
Chlordecone	0.30	All fish	CPG Sec. 575.100
Chlordecone	0.40	Crabmeat	CPG Sec. 575.100
DDT, TDE, and DDE	5.00[b]	All fish	CPG Sec. 575.100
Methylmercury	1.00	All fish	CPG Sec. 575.600
Heptachlor and heptachlorepoxide	0.30[c]	All fish	CPG Sec. 575.100
Mirex	0.10	All fish	CPG Sec. 575.100

Source: Adapted from the Code of Federal Regulations (CFR), Title 21, Chap. I, Food and Drug Administration, Department of Health and Human Services, Part 109, Unavoidable Contaminants in Food for Human Consumption and Food – Packaging Material; Title 40, Protection of Environment, Chap. I, Environmental Protection Agency, Subchapter E, Pesticide Programs, Part 180, Tolerances and Exemptions for Pesticide Chemical Residues in Food; and from the Compliance Policy Guides (CPG), Section 575, Pesticide Residues in Food and Feed – Enforcement Criteria

[a]ppm = parts per million
[c]Action level for residues individually or in combination

for the same substance and use becomes effective. Where no established action level or tolerance exists, the FDA may take legal action against the product at the minimal detectable level of the contaminant. Identified contaminants and regulatory limits are subject to change as additional science and risk assessment become available.

There are several options for controlling chemical contaminants. Depending on the risk, a producer may choose to select one or a combination of controls to ensure the product is not contaminated. Commonly employed controls are

- On-farm visits: including sample collection and testing of farmed (aquaculture) fish
- Supplier certification: supplier certification of compliance with contaminant limits
- Testing and monitoring records: reports of analysis of harvest or growing waters
- Chemical contaminant testing: testing raw material/ingredients for chemical contaminants
- Source controls: monitoring to ensure raw materials are harvested from open waters

4.6 Aquaculture Drugs

Aquaculture drugs are used for disease control and the treatment of cultured finfish and shellfish. The use of unapproved aquaculture drugs can affect the safety of seafood. Their use for aquaculture finfish or shellfish could result in human toxicity, allergic reactions, build-up of pathogen resistance, and/or an increased risk of cancer. Most governments, including the United States', have stringent approval protocols for aquaculture drugs. However, not all governments share the same list of approved drugs. Therefore, it is important to ensure that specifications for sourced aquaculture finfish or shellfish prohibit aquaculture drugs that are not approved in the country of consumption. Specifications should not only limit which drugs may be used, but also require appropriate application as specified by label instructions provided by the approving authority. In the United States, drugs used for aquaculture are approved by the Center for Veterinary Medicine, while the FDA regulates the use of aquacultured fish and shellfish as food.

There are several controls that may be applied to prevent and/or detect the use of unapproved aquaculture drugs. These include

- On-farm visits: observation to ensure the appropriate use of aquaculture drugs
- Supplier certifications: suppliers provide certification of proper drug use
- Supplier drug use records: records of drug use provided by supplier
- Drug residue testing: testing to ensure no unapproved drugs, or levels of approved drugs that exceed maximum limits, are present in the edible portion of the fish

4.7 Allergens

Fish and crustacean shellfish are identified as two of the eight major food allergens requiring consumer notification through labeling in accordance with the Food Allergen Labeling and Consumer Protection Act (FALCPA). Other countries also recognize molluscan shellfish as allergens as well. The notification requirement for compliance with FALCPA is satisfied by the declaration of the fish or crustacean in the ingredient statement. The type of fish (e.g., bass, flounder, cod) and the type of crustacean shellfish (e.g., crab, lobster, shrimp) must be declared. While the ingredient declaration is required by other labeling regulation, the FALCPA requirement can also be satisfied by placing the word "Contains" followed by the name of the food source from which the major food allergen is derived, immediately after or adjacent to the list of ingredients, in type size that is no smaller than the type size used for the list of ingredients, for example, "Contains fish."

5 Product- and Package-Related Considerations

In addition to the species-related considerations that have been described, it is important to consider how the finished product characteristics and package type may affect the safety of the seafood product during manufacture, storage, distribution, retail handling, and eventual consumer handling.

Finished product categories range from highly perishable raw products to shelf-stable low-acid canned foods that are thermally processed to commercial sterility.

For the purpose of evaluating food safety considerations, there are two primary categories of packaging:

1. Reduced-oxygen packaging (ROP)
2. Other than reduced-oxygen packaging

The primary reason for distinguishing between these two package forms is to identify products that require controls for preventing botulism, an illness caused by the growth and toxin formation of *Clostridium botulinum*, which produces an extremely potent neurotoxin that can cause severe illness. Data reflect a mortality rate of approximately 10% for foodborne cases of botulism (Anonymous 2002). Product/package characteristics that have a direct effect on the growth and toxin formation are the temperature of storage and reduced-oxygen conditions, which are conducive to growth. Therefore, measures for controlling *C. botulinum* must be considered for products packed in ROP. The FDA considers vacuum and modified atmosphere packaging (MAP), hermetically sealed containers, packing in deep containers from which the air is expressed, and packing in oil to be forms of ROP.

Strains of *C. botulinum* are classified into four groups and seven types depending on the chemical and physiological characteristics and toxin susceptibility in animal species. There are generally two categories of *C. botulinum* that are of food safety concern:

1. Type E and nonproteolytic strains of Types B and F. This category can grow and produce toxin at or above 38°F. Spores of this category are generally more heat-labile than spores of the proteolytic category, but recommended lethality values ($F_{194°F} = 10.0$ min) are still high relative to vegetative cells.
2. Type A and proteolytic strains of Types B and F. This category can grow and produce toxin at or above 50°F. Spores of this category are very heat-resistant, and seafood products, other than canned seafood processed to commercial sterility, are generally not processed sufficiently to kill spores of this category of *C. botulinum*.

The following are examples of different product categories, with each category further broken down to the packaging type – either ROP, other than ROP, or all packaging. Food safety considerations and control measures are identified for each combination of product and package type.

5.1 Product Type: Cooked or Pasteurized (e.g., Shrimp, Crab, Finfish, Surimi Seafood)

5.1.1 In Reduced-Oxygen Packaging

- Proper temperature control to prevent the growth of pathogens during preparation and processing steps
- Use of time temperature integrators (TTI) for products that are cooked but not pasteurized, to ensure time/temperature exposure during storage and distribution does not present an opportunity for *C. botulinum* growth and toxin formation
- If frozen, appropriate handling and thawing instructions on the label
- Pasteurization sufficient to control nonproteolytic *C. botulinum* spores for product intended to be distributed and stored unfrozen
- Prevention of postpasteurization contamination for product intended to be distributed and stored unfrozen
- Allergen/additive labeling, to comply with FALCPA and requirements for labeling other food sensitivity additives such as sulfites
- Foreign material control: prevention of metal and glass inclusion through appropriate GMPs or prerequisite programs

5.1.2 In Other Than Reduced-Oxygen Packaging

- Proper temperature control to prevent the growth of pathogens
- Cooking sufficient to control pathogens
- Allergen/additive labeling, to comply with FALCPA and requirements for labeling other food sensitivity additives such as sulfites
- Foreign material control: prevention of metal and glass inclusion through appropriate GMPs or prerequisite programs

5.2 Product Type: Smoked (e.g., Fish and Shellfish)

5.2.1 In Reduced-Oxygen Packaging

- Proper temperature control to prevent the growth of pathogens during preparation and processing steps
- Ensure brining, drying, and smoking procedures that result in water-phase salt (WPS) levels sufficient to control nonproteolytic *C. botulinum* spores for product intended to be distributed and stored unfrozen; generally 3.5% WPS
- Prevention of postprocessing contamination prior to packaging
- If frozen, appropriate handling and thawing instructions on the label
- Allergen/additive labeling, to comply with FALCPA and requirements for labeling other food sensitivity additives such as sulfites
- Foreign material control: prevention of metal and glass inclusion through appropriate GMPs or prerequisite programs

5.2.2 In Other Than Reduced-Oxygen Packaging

- Proper temperature control to prevent the growth of pathogens during preparation and processing steps
- Allergen/additive labeling, to comply with FALCPA and requirements for labeling other food sensitivity additives such as sulfites
- Foreign material control: prevention of metal and glass inclusion through appropriate GMPs or prerequisite programs

5.3 Product Type: Battered/Breaded (e.g., Raw Shrimp, Finfish, Shellfish)

5.3.1 All Packaging Types

- GMP and time/temperature controls during preparation of batter to minimize the risk of contamination, growth, and toxin production by *Staphylococcus aureus*
- Allergen/additive labeling, to comply with FALCPA and requirements for labeling other food sensitivity additives such as sulfites
- Foreign material control: prevention of metal and glass inclusion through appropriate GMPs or prerequisite programs

5.4 Product Type: Stuffed Seafood (e.g., Stuffed Finfish, Shellfish)

5.4.1 All Packaging Types

- GMP and time/temperature controls during processing to minimize the risk of contamination and growth of pathogens
- Allergen/additive labeling, to comply with FALCPA and requirements for labeling other food sensitivity additives such as sulfites
- Foreign material control: prevention of metal and glass inclusion through appropriate GMPs or prerequisite programs

5.5 Product Type: Dried, Cured, and Salted Fish

5.5.1 All Packaging Types

- GMP and time/temperature controls during processing to minimize the risk of contamination, growth, and toxin production by pathogens, including *Staphylococcus aureus*
- Ensuring proper salt, water-phase salt, and/or water activity (a_w) levels to prevent growth and toxin formation by *C. botulinum*
- Proper evisceration of fish to prevent growth and toxin formation by *C. botulinum*
- Allergen/additive labeling, to comply with FALCPA and requirements for labeling other food sensitivity additives such as sulfites.
- Foreign material control: prevention of metal and glass inclusion through appropriate GMPs or prerequisite programs

5.6 Product Type: Raw Shellfish (Products That May Be Consumed Raw)

5.6.1 In Reduced-Oxygen Packaging

- Proper temperature control to prevent the growth of pathogens during preparation and processing steps
- Use of time temperature integrators (TTI) for products that are cooked but not pasteurized, to ensure time/temperature exposure during storage and distribution does not present an opportunity for *C. botulinum* growth and toxin formation
- Use of a validated pathogen-reduction process
- If frozen, appropriate handling and thawing instructions on the label
- Allergen/additive labeling, to comply with FALCPA and requirements for labeling other food sensitivity additives such as sulfites
- Foreign material control: prevention of metal and glass inclusion through appropriate GMPs or prerequisite programs

5.6.2 In Other Than Reduced-Oxygen Packaging

- Proper temperature control to prevent the growth of pathogens during preparation and processing steps
- Use of a validated pathogen-reduction process
- Allergen/additive labeling, to comply with FALCPA and requirements for labeling other food sensitivity additives such as sulfites
- Foreign material control: prevention of metal and glass inclusion through appropriate GMPs or prerequisite programs

5.7 Product Type: Raw or Partially Cooked Finfish (Products May Be Consumed Raw)

5.7.1 In Reduced-Oxygen Packaging

- Proper temperature control to prevent the growth of pathogens during preparation and processing steps

Safety of Fish and Seafood Products

- Use of time temperature integrators (TTI) for products that are cooked but not pasteurized, to ensure time/temperature exposure during storage and distribution does not present an opportunity for *C. botulinum* growth and toxin formation
- Use of appropriate freezing process to control parasites in certain species if intended to be consumed raw
- If frozen, appropriate handling and thawing instructions on the label
- Allergen/additive labeling, to comply with FALCPA and requirements for labeling other food sensitivity additives such as sulfites
- Foreign material control: prevention of metal and glass inclusion through appropriate GMPs or prerequisite programs

5.7.2 In Other Than Reduced-Oxygen Packaging

- Proper temperature control to prevent the growth of pathogens during preparation and processing steps
- Allergen/additive labeling, to comply with FALCPA and requirements for labeling other food sensitivity additives such as sulfites
- Foreign material control: prevention of metal and glass inclusion through appropriate GMPs or prerequisite programs

5.8 Product Type: Shelf-Stable Low-Acid or Acidified Canned Seafood

5.8.1 All Packaging Types

- Compliance with the low-acid canned food (LACF) regulation 21 CFR 113 for foods with a finished equilibrium pH greater than 4.6 and a water activity (a_w) greater than 0.85.
- Compliance with the acidified food regulation 21 CFR 114 for foods that are low-acid foods to which acid(s) or acid food(s) are added. These foods have a water activity (a_w) greater than 0.85 and have a finished equilibrium pH of 4.6 or below.
- Both the low-acid canned food and the acidified food regulations are based on requirements that destroy or prevent the growth of microorganisms of public health significance that would otherwise be capable of reproducing in the food under normal nonrefrigerated conditions of storage and distribution.

6 Summary

There is a large variety of seafood species and products available on the marketplace, and as the demand grows, processors are challenged to develop new product forms maintaining high quality and strict safety. This requires an understanding of the potential hazards associated with each species of finfish or shellfish and the procedures necessary to control these hazards. The traditional use of refrigeration and freezing temperatures, as well as high cooking temperatures, are appropriate controls for most of the parasites and bacterial foodborne hazards. Finfish and molluscan and crustacean shellfish may contain "natural toxins" that may affect the safety of these species. In addition, fish and crustacean shellfish are identified as two of the eight major food allergens requiring consumer notification through labeling in accordance with the Food Allergen Labeling and Consumer Protection Act.

Appropriate regulations addressing the safety of seafood products by federal agencies and strong and effective seafood safety programs driven by seafood companies ensure that safe products are delivered to the consumer.

References

Anonymous 2002. Botulism. Fact Sheet no. 270. World Health Organization. http://www.who.int/mediacentre/factsheets/fs270/en/. Accessed 3 Feb 2011.

Anonymous 2011. Bad bug book: foodborne pathogenic microorganisms and natural toxins handbook – Gempylotoxin. http://www.fda.gov/food/foodsafety/foodborneillness/foodborneillnessfoodbornepathogensnaturaltoxins/badbugbook/ucm071191.htm. Accessed 3 Feb 2011.

Botana, L.M. 2000. *Seafood and freshwater toxins: pharmacology, physiology, and detection*. New York: Marcel Dekker.

FDA. 2009. Food code. http://www.fda.gov/Food/FoodSafety/RetailFoodProtection/FoodCode/FoodCode2009/. Accessed 3 Feb 2011.

Part III
Risk Analysis, Interventions and Regulations

Food Risk Analysis

Thomas P. Oscar

1 Introduction

Risk analysis is a holistic approach to food safety that involves three interactive processes: risk assessment, risk management, and risk communication (Dennis et al. 2008). Risk assessment is the scientific process of predicting the risk posed by a hazard, whereas risk management is the process of combining the scientific risk assessment with nonscientific aspects of the problem, such as social, political, cultural, and economic considerations, to arrive at a decision of how to manage the risk. Risk communication is the process of informing stakeholders about the risk and how it is being managed (Hallman 2008).

Risk assessment modeling is the foundation of risk analysis and is most often accomplished using Monte Carlo simulation methods to combine existing knowledge and data into a prediction of risk (Vose 1996). The prediction of risk is relative rather than absolute because of knowledge, data, and model uncertainty. However, through the process of scenario analysis, relative risk can be assessed and used to help inform risk management decisions. The purpose of this chapter is to introduce the reader to the Monte Carlo simulation methods used in risk assessment modeling and then to illustrate how risk assessment modeling can be used to provide a relative assessment of risk for risk management decisions.

2 Risk Assessment Modeling Concepts

Risk assessment modeling consists of four steps: (1) hazard identification; (2) exposure assessment; (3) hazard characterization; and (4) risk characterization. The scope and design of a risk assessment model depend on the purpose of the risk assessment and on the food and hazard being evaluated.

T.P. Oscar (✉)
Microbial Food Safety Research Unit, U. S. Department of Agriculture,
University of Maryland Eastern Shore, Princess Anne, MD, USA
thomas.oscar@ars.usda.gov

2.1 Hazard Identification

The first step in building a risk assessment model is to determine the distribution of *hazards* in a given food product at some point in the risk pathway. The quantification of a hazard in food is expensive and time-consuming; therefore, it is only likely to be performed at one point in the risk pathway. The point in the risk pathway where hazard identification is performed depends on the design and purpose of the risk assessment.

If the hazard is not uniformly distributed in the food, the initial distribution of the hazard will depend on the size of the food sample used in hazard identification and will increase in a nonlinear fashion as a function of sample size (Oscar 2004c). Therefore, when the hazard is not uniformly distributed in the food, it is recommended that the food unit used in the risk assessment model be the same as the size of the food sample used in hazard identification because this will provide the most accurate assessment of relative risk. Alternatively, Monte Carlo simulation methods can be used to extrapolate hazard identification results obtained with one sample size to larger sample sizes (Oscar 2004c).

2.2 Exposure Assessment

Exposure assessment is the second step in building a risk assessment model and focuses on how the distribution of hazards in food changes from hazard identification to consumer exposure. After hazard identification, risk assessors develop a Monte Carlo simulation model of the risk pathway from hazard identification to consumption and then use the model to predict consumer exposure (Nauta 2008). In microbial risk assessment, the risk pathway consists of a series of process steps and associated hazard events, such as growth and inactivation (Whiting and Buchanan 1997).

Hazard events in food are uncertain (i.e., random), variable (i.e., stochastic), and rare (i.e., occur less than 100% of the time). The modeling of random, stochastic, and rare events is accomplished by linking a discrete distribution for the incidence of the event with a continuous distribution for the extent of the event (Winston 1996). The extent of most hazard events in a microbial risk assessment is either a normal or log-normal distribution, and thus, a good distribution to use is the Pert distribution (see below), which can vary in shape from a normal to a log-normal distribution. In addition, the Pert distribution is robust and easy to use, as it only takes three values (minimum, most likely, maximum) to define it.

Predictive microbiology models that calculate the growth and inactivation of the microbial hazard as a function of conditions in the risk pathway can be configured to define input distributions (i.e., discrete and Pert) for hazard events in a microbial risk assessment model (Oscar 2004a). To keep the risk assessment model simple and to reduce the simulation time, it is recommended that predictive microbiology models be used outside the risk assessment model.

2.3 Hazard Characterization

This term predicts how the interaction among the hazard, food, and host affects the response of consumers to hazard exposure (Coleman and Marks 1998). Knowledge of the dose of the hazard that causes a response is obtained from epidemiological investigations and, in the past, from controlled feeding trials with humans (McCullough and Eisele 1951). The response can be acute toxicity, infection, illness, chronic disability (e.g., reactive arthritis), cancer, or death.

In most feeding trials, a fairly uniform group of humans (e.g., healthy adults) are fed log doses of the hazard in a standard food vehicle and then the percentage of individuals that express the response is determined (McCullough and Eisele 1951). A graph of the percentage of the population expressing the response versus the log dose is usually sigmoid in shape (Teunis et al. 1999). The log dose that causes 50% of the population to exhibit the response is the response dose 50, or RD_{50}.

In reality, contaminated food is consumed by a nonuniform population of hosts that includes individuals at high risk for a response to the hazard. In addition, multiple forms of the hazard, of different potency, are often present in the food, which is consumed with a variety of food and beverage items that can alter the dose response. In fact, when multiple forms of a hazard with different RD_{50} values are present in food, the population dose-response curve changes from a sigmoid shape to a nonsigmoid shape that reflects the incidence and potency of the hazard forms present (Oscar 2004b).

To model the dose response in a food risk assessment, the random interaction of hazard, food, and host is simulated using Monte Carlo methods. Existing data are used to define an array of probability distributions for the effects of hazard, food, and host factors on the response dose (RD):

$$RD_i = \text{Discrete}\left[(f_1, f_2, \ldots, f_j), (x_1, x_2, \ldots, x_k)\right] \quad (1)$$

$$x_k = \text{Pert}(RD_{min}, RD_{50}, RD_{max}) \quad (2)$$

where the response dose of the ith food serving (RDi) is a function of the jth frequency of occurrence (f_j) of the kth class of hazard, food, and host factors (xk), and where RD_{min} is the minimum response dose, RD_{50} is the median response dose, and RD_{max} is the maximum response dose. During simulation of the model, a response dose is randomly determined from Eqs. 1 and 2 for each food unit.

2.4 Risk Characterization

This is the fourth step in building a risk assessment model and deals with the prediction of the impact of the food hazard on public health. The risk of an adverse health outcome from the consumption of a food unit depends on rare and random events in the risk pathway. Random chance determines which unit of food is contaminated at packaging, kept at an improper temperature during distribution, washed during meal preparation, cooked improperly, contaminated with a virulent hazard strain, consumed with a permissive meal, and consumed by someone from the high-risk population. In the final step of risk assessment, the hazard identification, exposure assessment, and hazard characterization are combined to characterize the risk by predicting the number of responses (R) in the host population:

$$R_i = \text{IF}(HD_i / RD_i < 1, 0, 1) \quad (3)$$

where HDi is hazard dose and RDi is response dose for the ith food unit; when the ratio of HD to RD is less than 1, no response ("0") occurs; otherwise, a response occurs ("1").

3 Risk Assessment Modeling Methods

With the advent of computer software applications for risk assessment, it is now possible to develop Monte Carlo simulation models for the evaluation and management of food safety risks (Cassin et al. 1998). A Monte Carlo simulation model consists of input distributions, a formula, and output distributions (McNab 1998). During simulation of the model, input distributions are randomly sampled and the values generated are used in a formula to generate output distributions. Each calculation or iteration of the model simulates movement of one food unit through the risk pathway.

In the example presented below, a risk assessment model for a generic food that is contaminated with a generic microbial hazard was developed in an Excel[1] workbook and was simulated using @Risk (Palisade Corp., Newfield, NY), a spreadsheet add-in computer program. The model consists of two spreadsheets: one for hazard identification and exposure assessment (Fig. 1) and one for hazard characterization and risk characterization (Fig. 2). The formulas used in the model are shown in Tables 1 and 2.

3.1 Discrete Distribution

The discrete distribution is used to model the incidence of events in the model. The discrete distribution is defined such that an output of "0" indicates that the event did not occur, whereas an output of "1" indicates that the event occurred. In addition, the discrete distribution is used in hazard characterization to model the frequency of occurrence of the eight combinations of hazard, food, and host factors.

3.2 Pert Distribution

The Pert distribution is used to model the extent of events in the model. Log-transformed values are used to define Pert distributions so that sampled values more closely reflect the original data; using untransformed values for data that vary by several orders of magnitude, a common occurrence in microbial risk assessment, results in sampled distributions that do not closely reflect the original data.

3.3 Logical Function

Modeling rare events involves linking discrete and Pert distributions using a logical "IF" statement. When the output from the discrete distribution is "0," the output of the associated Pert distribution is not used in the calculation of results, whereas when the output from the discrete distribution is "1," the output of the associated Pert distribution is used in the calculation of results. Using this logical function nullifies the sensitivity analysis provided by @Risk because it uncouples the correlation between the input and output distributions.

---

[1] Mention of trade names or commercial products in this publication is solely for providing specific information and does not imply recommendation or endorsement by the U.S. Department of Agriculture.

	A	B	C	D	E	F	G	H
1	Hazard Identification and Exposure Assessment							
2			Output					
4	Process Step	Hazard Event	Number	Discrete	Pert			
5	Packaging	Contamination	79	1	79			
6	Distribution	Growth	1,055	1	13.36			
7	Washing	Removal	27	1	0.026			
8	Cooking	Survival	0	0	0.001098			
9	Serving	Contamination	12	1	0.0113			
10								
11					Input			
12			Incidence			Extent		
13	Process Step	Hazard Event	0	1	Minimum	Most Likely	Maximum	Units
14	Packaging	Contamination	75%	25%	0	1	4	log Δ
15	Distribution	Growth	80%	20%	0.1	1	3	log Δ
16	Washing	Removal	85%	15%	-0.1	-1	-3	log Δ
17	Cooking	Survival	90%	10%	-0.1	-5	-7	log Δ
18	Serving	Contamination	85%	15%	-3	-2	-1	log rate

Fig. 1 Excel spreadsheet model for hazard identification and exposure assessment. The food unit in this iteration was initially contaminated with 79 hazard units at packaging. During food distribution, the food unit was temperature-abused, resulting in a 1.1-log increase in the hazard number to 1,055 hazard units. The food unit was washed before cooking, which resulted in a 1.6-log reduction of the hazard number to 27 hazard units. The food unit was properly cooked, resulting in the death of all 27 hazard units. However, during serving, the cooked food unit was contaminated with 12 hazard units that were removed during washing of the uncooked food unit. The transfer rate was 1.1%

	A	B	C	D	E	F	G	H	I
1	Hazard Characterization			Output		Risk Characterization			
2	Class			8		Response	0	No	
3	Response Dose			1,662					
4									
5	Class				Input			Output	
6	Hazard	Food	Host	%	RD$_{min}$	RD$_{50}$	RD$_{max}$	Class	RD
7	Normal	Normal	Normal	70	4.0	6.0	8.0	1	1,473,783
8	High	Normal	Normal	6	3.0	5.0	7.0	2	109,237
9	Normal	High	Normal	2	3.5	5.5	7.5	3	10,910,247
10	High	High	Normal	2	2.5	4.5	6.5	4	247,677
11	Normal	Normal	High	5	20	4.0	6.0	5	1,510
12	High	Normal	High	9	1.0	3.0	5.0	6	4,776
13	Normal	High	High	3	1.5	3.5	5.5	7	13,842
14	High	High	High	3	0.5	2.5	4.5	8	1,662

Fig. 2 Excel spreadsheet model for hazard characterization and risk characterization. The food unit in this iteration was contaminated with a high-risk strain of the hazard, was consumed with high-risk food or beverage items, and was consumed by someone from the high-risk population (i.e., class 8: hazard_food_host). The response dose was 1,662, whereas the hazard dose was 12 (Fig. 1), and thus, the consumer did not have an adverse health response from consumption of the food unit

Table 1 Formula used in the hazard identification and exposure assessment spreadsheet A

Column/row	Formula
C5	=ROUND(IF(D5=0,0,E5),0)
C6	=ROUND(IF(D6=0,C5,(E6*C5)),0)
C7	=ROUND(IF(D7=0,C6,(E7*C6)),0)
C8	=ROUND(IF(D8=0,0,(E8*C7)),0)
C9	=ROUND((IF(D9=0,C8,E9*(C6-C7)+C8)),0)
D5	=RiskDiscrete({0,1},C15:D15)
D6	=RiskDiscrete({0,1},C16:D16)
D7	=RiskDiscrete({0,1},C17:D17)
D8	=RiskDiscrete({0,1},C18:D18)
D9	=D7
E5	=POWER(10,RiskPert(E15,F15,G15))
E6	=POWER(10,RiskPert(E16,F16,G16))
E7	=POWER(10,RiskPert(G17,F17,E17))
E8	=POWER(10,RiskPert(G18,F18,E18))
E9	=POWER(10,RiskPert(E19,F19,G19))

Table 2 Formula used in the hazard characterization and risk characterization spreadsheet B

Column/row	Formula
D2	=RiskDiscrete({1,2,3,4,5,6,7,8},D7:D14)
D3	=ROUND(LOOKUP(D2,H7:I14),0)
G2	=RiskOutput("Response")+IF(A!C9/D3<1,0,1)
H2	=IF(G2=0,"No," "Yes")
I7	=POWER(10,RiskPert(E7,F7,G7))
I8	=POWER(10,RiskPert(E8,F8,G8))
I9	=POWER(10,RiskPert(E9,F9,G9))
I10	=POWER(10,RiskPert(E10,F10,G10))
I11	=POWER(10,RiskPert(E11,F11,G11))
I12	=POWER(10,RiskPert(E12,F12,G12))
I13	=POWER(10,RiskPert(E13,F13,G13))
I14	=POWER(10,RiskPert(E14,F14,G14))

3.4 "POWER" Function

To simulate uncontaminated servings, the "POWER" function is used to convert log-transformed values from Pert distributions back to untransformed numbers for the calculation of results since it is not possible to take the log of 0. By simulating uncontaminated food units, changes in hazard incidence are predicted and a more realistic simulation of the risk pathway is obtained.

3.5 "ROUND" Function

Although chemical hazards can be present in fractional amounts, it is not possible to have a fraction of a microbe. Thus, the "ROUND" function is used to convert output results in microbial risk assessment models into whole numbers. This function also allows contaminated food units to become hazard-free following an inactivation step, such as cooking.

Food Risk Analysis

3.6 "LOOKUP" Function

In hazard characterization, hazard, food, and host factors are classified as normal or high risk and then Pert distributions for the response dose are defined for each combination of hazard, food, and host factors. A discrete distribution is used to model the frequency of occurrence of these combinations. During simulation of the model, the output of the discrete distribution is used in the "LOOKUP" function to determine which output of the Pert distributions is used in risk characterization.

3.7 @Risk Functions

@Risk is a spreadsheet add-in program that allows the use of probability distributions in Excel to describe model variables. In addition, the @Risk function "Risk Output" allows the designation of cells for the collection of data for output distributions. After model simulation, results for individual iterations as well as summary statistics are generated by @Risk. These results can be filtered to remove uncontaminated servings; this is a recommended practice because it facilitates the calculation of the hazard incidence as a function of process step.

3.8 Scenario Analysis

A scenario in risk assessment modeling is defined as a unique set of input variables. The relative risk is obtained by comparing scenarios. The scenario uncertainty is assessed by using a specified or randomly selected set of random number generator seeds to initiate simulations. The random number generator seed is a number that initiates the random sampling of input distributions. Each random number generator seed generates a unique outcome of the model.

3.9 Transparency

The more complex a risk assessment model is, the more difficult it is to make it transparent. By keeping the model simple, it is easier to describe the model, assumptions, and data used to build it. However, the model should not be so simple that it does not properly simulate the risk pathway.

4 Risk Analysis Example: Design and Input Settings

In this example, a food company has two processing plants that are located in different regions of the country but produce the same food product. End product testing indicated that the food product produced by both plants is contaminated with a single species of a specific microbial hazard. However, while the hazard incidence was higher for food from Plant A, only food from Plant B has caused a foodborne illness outbreak in consumers. These observations caused the risk managers in the food company to pose the following risk question: Why did food from Plant B cause an outbreak when food from Plant B has a lower incidence of the hazard than food from Plant A? To answer this question, the food company decided to conduct a risk assessment.

4.1 Hazard Identification

As a first step, the food company determined the initial distribution of the hazard in single units of the food at packaging. They found that the incidence of contamination was 25% in Plant A and 10% in Plant B, whereas the extent of contamination in both plants ranged from 0–4 log, with a most likely level of 1 log per food unit.

4.2 Exposure Assessment

After hazard identification, the food company developed a Monte Carlo simulation model to predict consumer exposure. After packaging, the first event in the model was hazard growth during food distribution. A predictive microbiology model that predicted hazard growth in and on the food as a function of storage conditions during product shipping, retail display, and consumer transport was used to determine that the incidence of hazard growth in and on the food during distribution was 20% for Plant A and 40% for Plant B, and the extent of hazard growth, when it occurred, ranged from 0.1 to 3 log, with a most likely value of 1 log per food unit for both plants.

The next event in the model was removal of the hazard by washing during meal preparation. A predictive microbiology model that predicted hazard removal from the food during washing as a function of washing conditions was used to determine that the incidence of hazard removal during washing was 15% for Plant A and 30% for Plant B, and the extent of removal of the hazard, when it occurred, ranged from 0.1 to 3 log, with a most likely value of 1 log per food unit for both plants.

The next event in the model was cooking of the food. A predictive microbiology model that predicted hazard inactivation during cooking was used to determine that the potential for hazard survival during cooking was 10% for Plant A and 5% for Plant B, and the extent of inactivation when the hazard had a chance to survive was the same for both plants and ranged from 0.1 to 7 log, with a most likely value of 5 log per food unit.

Cooking conditions that resulted in less than a 7-log reduction in hazard level were considered to present an opportunity for hazard survival. A value of a 0.1-log reduction was used to define the minimum extent of inactivation or maximum potential for survival to include the scenario of close to 100% survival for those rare occasions where a portion of the food unit did not receive heat treatment and was consumed raw.

The next event in the model was cross-contamination of the cooked food with hazard removed from the uncooked food by washing. Based on in-house research, the food company determined that when a food unit was washed, regardless of the plant of origin, 0.1–10% of the removed hazard was transferred to the cooked food and consumed, with a most likely transfer rate of 1%. To model cross-contamination during serving and calculate consumer exposure, the food company calculated the amount of hazard removed by washing, multiplied this number by the transfer rate, and then added the result to the amount of hazard that survived cooking.

4.3 Hazard Characterization

The food company collected data for hazard characterization that included hazard (bacterium) strain, eating habits, and consumer demographics. This information was used to classify the hazard, food, and host effects on the response dose as normal or high risk. The hazard was classified as high risk when the hazard strain was known to cause the response in humans. The food was classified as high risk when the meal consumed with the food was high in fat or when an anti-acid pill was taken with the food. Host effects were classified as high risk when the consumer was from a high-risk group, such as very young, very old, diabetic, cancer patient, pregnant, etc.

When hazard, food, and host effects were classified as normal risk, RD_{min}, RD_{50}, and RD_{max} were set at 4, 6, and 8 log, respectively. When hazard, food, or host effects went from normal to high risk, RD_{min}, RD_{50}, and RD_{max} were decreased by 1, 0.5, and 2 log, respectively. When more than one hazard, food, or host effect went from normal to high risk, the effect on RD_{min}, RD_{50}, and RD_{max} was additive. The incidence of high-risk events for hazard, food, and host effects for Plant A were 20%, 10%, and 20%, respectively, whereas for Plant B they were 60%, 10%, and 30%, respectively.

4.4 Risk Characterization

In this last step of risk assessment, the randomly determined hazard dose at consumption for a food unit was combined with a randomly selected response dose, and the number of responses per 100,000 food units was determined. The model for each scenario (i.e., Plant A and Plant B) was simulated 200 times with @Risk settings of Latin Hypercube sampling, 100,000 iterations, and randomly selected and different random number generator seeds for each replicated simulation.

5 Risk Analysis Example: Results

To assist the risk managers, the risk assessors summarized the results of the risk assessment in a series of tables and graphs.

5.1 Hazard Identification and Exposure Assessment

The first graph prepared by the risk assessors for the risk managers showed hazard incidence as a function of process step (Fig. 3). As expected, the hazard incidence was 25% for Plant A and 10% for Plant B at packaging and did not change during food distribution because microbial growth only changes the hazard number and not the hazard incidence among food units. Washing during meal preparation caused a slight reduction in hazard incidence to 24% for Plant A and 9.4% for Plant B, whereas cooking caused a dramatic reduction in hazard incidence to 0.11% for Plant A and 0.055% for Plant B. After cooking, the hazard incidence among food units increased to 1.55% for Plant A and 1.47% for Plant B due to cross-contamination of the cooked food during serving. Thus, although hazard incidence among food units was 1.5-fold higher in Plant A than Plant B at packaging, the hazard incidence at consumption was similar.

The second graph prepared by the risk assessors for the risk managers showed the total hazard number per 100,000 food units as a function of the process step (Fig. 4). At packaging, the total hazard number was 6.37 log for Plant A and 5.97 log for Plant B. However, after distribution, the total hazard number increased to 7.27 log for Plant A and 7.16 log for Plant B. Washing before cooking reduced the total hazard number to 7.19 log for Plant A and 7.02 log for Plant B, whereas cooking further reduced total hazard number to 3.72 log for Plant A and 2.50 log for Plant B. Cross-contamination of the cooked food during serving increased the total hazard number to 4.64 log for Plant A and to 4.72 log for Plant B. Thus, although the total hazard number per 100,000 food units was 1.4-fold higher for Plant A than Plant B at packaging, consumers of food from Plant B were exposed to more hazard units than consumers of food from Plant A because of events that occurred after packaging.

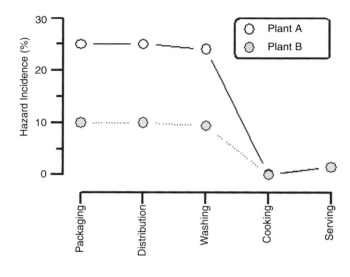

Fig. 3 Hazard incidence results for food produced by Plants A and B. The results were obtained with a single simulation of each scenario using @Risk settings of Latin Hypercube sampling, 100,000 iterations, and a random number generator seed of 3

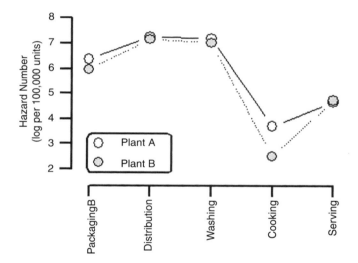

Fig. 4 Hazard number results for food produced by Plants A and B. The results are from a single simulation of each scenario using @Risk settings of Latin Hypercube sampling, 100,000 iterations, and a random number generator seed of 3. The hazard number refers to the total number of hazard units among the 100,000 food units in the simulation

5.2 Hazard Characterization

The third graph showed the population dose response for Plants A and B (Fig. 5). The RD_{50} for Plant B was lower than the RD_{50} for Plant A because the food from Plant B was more often contaminated with a highly virulent hazard strain and because the consumer population for Plant B had a higher percentage of high-risk individuals. Thus, the consumers of food from Plant B were not only exposed to more hazard units, but they were also more likely to have a response to hazard exposure than consumers of food from Plant A.

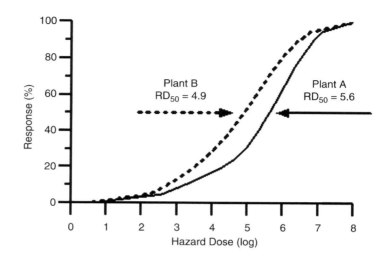

Fig. 5 Hazard characterization results for Plants A and B, where RD_{50} is the hazard dose at which 50% of the population exhibits a response to the hazard. The results are from a single simulation of each scenario using @Risk settings of Latin Hypercube sampling, 100,000 iterations, and a random number generator seed of 3

Table 3 Simulation results for iterations that resulted in an adverse health response

Plant	Iteration	Hazard number per food unit					Class	Response dose
		Packaging	Distribution	Washing	Cooking	Serving		
A	9,904	1,451	300,130	36,260	0	5,945	7	3,065
A	64,376	202	4,539	27	0	23	8	22
A	69,075	316	5,714	3,214	0	20	8	7
A	93,039	100	1,191	173	0	28	8	15
B	53,459	52	4,045	509	0	81	6	67
B	65,115	62	5,188	343	0	111	6	111
B	69,075	316	5,714	3,214	0	20	8	7
B	71,656	668	92,965	13,705	0	948	6	347
B	87,865	153	3,446	611	0	110	6	106
B	93,039	100	1,191	173	0	28	8	15

5.3 Risk Characterization

In fact, results of the simulation indicated that the predicted number of responses per 100,000 food units was 4 for Plant A and 6 for Plant B. To help the risk managers understand the latter result, the risk assessors prepared a summary (Table 3) of those iterations that resulted in a response. The identified risk factors from this summary included the following: (1) contamination of the food at packaging with >50 hazard units; (2) growth of the hazard by >1 log during distribution; (3) cross-contamination of the cooked food with the hazard; and (4) contamination of the cooked food with a high-risk hazard strain and consumption of the food unit by a high-risk host without (class 6) or with (class 8) a high-risk food factor.

Finally, to assess the uncertainty of the public health impact due to rare and random events in the risk pathway, the food company performed 200 replicated simulations of the scenarios for Plants A and B using randomly selected and different random number generator seeds to initiate each replicated simulation. The results of this analysis confirmed that the risk from food produced by Plant B was higher than the risk posed by food from Plant A (Fig. 6). The median level of risk was 3.0

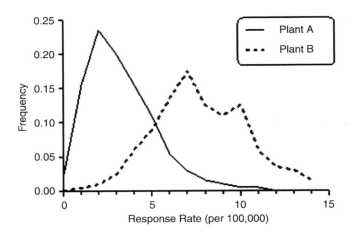

Fig. 6 Risk characterization results for Plants A and B showing the frequency distribution of the response rate for 200 replicated simulations generated using randomly selected random number generator seeds

cases per 100,000 (range: 0–11) for Plant A and 7.5 cases per 100,000 (range: 1–14) for Plant B. The overlap in the distributions for the response rate indicated that by random chance, the risk posed by food from Plant B could be less than the risk posed by food from Plant A.

5.4 Risk Management and Communication

These results indicated to the risk managers that they needed to focus more attention on Plant B and that they needed to take a multihurdle approach to control the food safety risk, which included interventions during production and processing to reduce the level of hazard in and on the food at packaging and interventions to reduce the growth of the hazard during distribution to consumers.

A dilemma for the risk managers was the finding that washing the food before cooking, which was previously believed to be a good food safety practice and was promoted by the company, actually increased the food safety risk. Thus, they had to effectively communicate to consumers that to reduce the food safety risk, they should not wash the food before cooking, which was in sharp contrast to what they had been told before. They were not sure how this would be received by their customers and what effect it might have on consumer confidence in their company and product.

Nonetheless, the food company was pleased with the risk assessment because it demonstrated to them the importance of taking a holistic look at their food safety risk. Had they only looked at the hazard incidence at packaging, the food company would not have detected and been able to properly manage the higher relative risk posed by the food produced by Plant B.

6 Summary

Food risk analysis is a holistic approach to food safety because it considers all aspects of the problem. Risk assessment modeling is the foundation of food risk analysis. The proper design and simulation of the risk assessment model is important to properly predict and control risk. Because of knowledge, data, and model uncertainty, risk assessment models provide a relative rather than an absolute assessment of risk. Nonetheless, through the process of scenario analysis, risk management options can be evaluated and risk management decisions aimed at improving public health can be made. Good risk management and communication involve active listening and sound decision making to address not only risk problems but also risk perceptions.

References

Cassin, M.H., A.M. Lammerding, E.C.D. Todd, W. Ross, and R.S. McColl. 1998. Quantitative risk assessment for *Escherichia coli* O157:H7 in ground beef hamburgers. *International Journal of Food Microbiology* 41: 21–44.

Coleman, M.E., and H.M. Marks. 1998. Topics in dose-response modeling. *Journal of Food Protection* 61: 1550–1559.

Dennis, S.B., J. Kause, M. Losikoff, D.L. Engeljohn, and R.L. Buchanan. 2008. Using risk analysis for microbial food safety regulatory decision making. In *Microbial risk analysis of foods*, ed. D.W. Schaffner, 137–175. Washington, DC: ASM Press.

Hallman, W.K. 2008. Communicating about the microbial risks in foods. In *Microbial risk analysis of foods*, ed. D.W. Schaffner, 205–262. Washington, DC: ASM Press.

McCullough, N.B., and C.W. Eisele. 1951. Experimental human salmonellosis. I. Pathogenicity of strains of *Salmonella meleagridis* and *Salmonella anatum* obtained from spray-dried whole egg. *Journal of Infectious Diseases* 88: 278–279.

McNab, W.B. 1998. A general framework illustrating an approach to quantitative microbial food safety risk assessment. *Journal of Food Protection* 61: 1216–1228.

Nauta, M.J. 2008. The modular process risk model (MPRM): a structured approach to food chain exposure assessment. In *Microbial risk analysis of foods*, ed. D.W. Schaffner, 99–137. Washington, DC: ASM Press.

Oscar, T.P. 2004a. A quantitative risk assessment model for *Salmonella* and whole chickens. *International Journal of Food Microbiology* 93: 231–247.

Oscar, T.P. 2004b. Dose-response model for 13 strains of *Salmonella*. *Risk Analysis* 24: 41–49.

Oscar, T.P. 2004c. Simulation model for enumeration of *Salmonella* on chicken as a function of PCR detection time score and sample size: implications for risk assessment. *Journal of Food Protection* 67: 1201–1208.

Teunis, P.F.M., N.J.D. Nagelkerke, and C.N. Haas. 1999. Dose response models for infectious gastroenteritis. *Risk Analysis* 19: 1251–1260.

Vose, D. 1996. *Quantitative risk analysis. A guide to Monte Carlo simulation modeling*. West Sussex: Wiley.

Whiting, R.C., and R.L. Buchanan. 1997. Development of a quantitative risk assessment model for *Salmonella enteritidis* in pasteurized liquid eggs. *International Journal of Food Microbiology* 36: 111–125.

Winston, W.L. 1996. *Simulation modeling using @Risk*. Belmont: Wadsworth.

Interventions to Inhibit or Inactivate Bacterial Pathogens in Foods

P. Michael Davidson and Faith M. Critzer

1 Introduction

The characteristics of a food product that make it acceptable for consumption vary among the world's populations. One characteristic that is universally accepted by all populations is that foods be free of, or nearly free of, human *pathogens*. To control pathogenic microorganisms, humans have devised several intervention methods that either inhibit the growth of or inactivate microorganisms. The effectiveness of the intervention method is not only dependent upon the method but also the target microorganism, the environment, and the characteristics of the food product.

2 Microbial Factors

One of the factors affecting the susceptibility of microorganisms to inhibition or intervention methods is the characteristics of the microorganisms themselves. For example, there are five major groups of microorganisms that are of concern in foods: bacteria, molds, yeasts, protozoa, and viruses. A major difference among the groups is that bacteria, molds, and yeasts can reproduce in a given food if conditions are conducive for growth. In contrast, protozoa and viruses generally do not reproduce in foods. Protozoa and viruses must infect a living host, replicate, and become excreted into an environment where they can then contaminate a foodstuff and then reinfect a host. Therefore, though protozoa and viruses do not replicate in food, they still must be inactivated to prevent illness.

Some Gram-positive bacteria of the genera *Bacillus*, *Clostridium*, *Alicyclobacillus*, *Desulfotomaculum*, *Geobacillus*, and *Sporolactobacillus* are capable of producing spores. *Spores* are a dormant stage of a microorganism and are more resistant than *vegetative cells* to inactivation treatments such as heat, antimicrobials, and irradiation. The outer surface of spores is composed of two distinct layers: a peptidoglycan layer that makes up the spore cortex and an outer coat that is made up of proteins that aid in the exclusion of compounds that may degrade the coat. The fatty acid profiles have also been found to differ in vegetative and spore states, with endospore-forming bacilli

P.M. Davidson (✉) • F.M. Critzer
Department of Food Science and Technology, University of Tennessee, Knoxville, TN 37996-4591, USA
e-mail: pmdavidson@utk.edu

having higher proportions of saturated branched-chain fatty acids in the spores (Song et al. 2000). The spore also has a lower water content, which aids in survival during heat treatments (Setlow 1995). These properties have given spore-forming bacteria the ability to survive many harsh conditions and germinate to a vegetative state when nutrient or other environmental conditions improve.

3 Environmental Factors

Environmental or storage conditions experienced by microorganisms and their response to those conditions can influence the susceptibility of the microorganisms to the intervention process. These environmental factors are known as extrinsic factors, and the two most important are temperature and atmosphere.

Microorganisms can be classified by the optimum temperatures at which they can grow: psychrophiles, psychrotrophs, mesophiles, and thermophiles. *Psychrophiles* are organisms that thrive in cold environments and have a growth range of −5°C to 35°C. Their minimum growth range is −5°C to 5°C, optimum 12–15°C, and maximum 15–20°C. These species are primarily of marine origin, and very few are of importance to foods except possibly to the *spoilage* of marine foods. The term *psychrotroph* (from "psychro" meaning "cold" and "trophic" meaning "food" or "nourishment") is used to describe those microorganisms that grow relatively well at 7°C and below and have optimum temperatures of 20–30°C (Jay et al. 2005). Examples of these organisms are the foodborne pathogens *Yersinia enterocolitica* and *Listeria monocytogenes*, the spoilage bacterium *Pseudomonas*, as well as yeasts and molds. The category in which most human and animal pathogens and most food spoilage microorganisms are found is the *mesophiles* ("meso" meaning "middle" and "phile" meaning "love"). Their minimum growth range is 5–15°C, optimum 30–45°C, and maximum 35–47°C. Examples of these pathogens in the group are *Escherichia coli*, *Salmonella*, *Clostridium botulinum*, and *Staphylococcus aureus*. *Thermophiles* ("thermo" meaning "hot" and "phile" meaning "love") are the microorganisms that can grow at the highest temperatures. Their minimum growth range is 40–45°C, optimum 55–75°C, and maximum 60–90°C. Thermophiles' primary importance to food microbiology is that they can be a source of spoilage and are often spore formers. These organisms are of interest in the canning industry; some examples include *Thermoanaerobacterium thermosaccharolyticum* and *Geobacillus stearothermophilus*.

Microorganisms can be classified by the atmospheric requirements for their growth. Microorganisms that require oxygen are called *strict aerobes*. Some bacteria (e.g., *Pseudomonas*) and most molds are strict aerobes. This is important because removing air will prevent their growth. Thus, vacuum packaging of foods or using hermetically sealed containers (e.g., metal cans) will prevent the growth of strict aerobes. *Facultative anaerobes* comprise the largest group of food-related microorganisms. They are capable of growing with or without oxygen present. *Strict anaerobes* cannot grow in the presence of oxygen. These organisms do not require the use of oxygen as a final electron acceptor in the electron transport chain during cellular respiration. Therefore, anaerobic atmospheres will allow the growth of facultative and strict anaerobes. Some foodborne pathogens as well as spoilage microorganisms can survive in anaerobic atmospheres, such as *Clostridium*. Carbon dioxide as an atmosphere has variable effects on microbial growth and survival and may stimulate, inhibit, or have no impact on growth (Daniels et al. 1985; Taniwaki et al. 2009).

4 Intrinsic Factors

The properties influencing the susceptibility contributed by food properties are termed *intrinsic factors*. The most important to growth inhibition and inactivation are pH, oxidation-reduction potential, water activity, and inhibitory substances.

Table 1 Ranges of pH for growth of microorganisms

	Minimum	Optimum	Maximum
Bacteria	4.5	6.5–7.5	9.0
Molds	1.5–3.5	4.0–6.8	8.0–11.0
Yeasts	1.5–3.5	4.0–6.5	8.0–8.5

Source: Banwart (1979)

Table 2 Oxidation-reduction potential ranges required for microbial growth

Category	Oxidation-reduction potential (mV)
Aerobes	+350 to +500
Facultative anaerobes	−300 to +350
Aerotolerant (*Clostridium perfringens*)	−200 to +200
Anaerobes	−100
Obligate anaerobes	−300

Source: Banwart (1979) and Jay et al. (2005)

The pH of a food product influences both microbial growth and survival. pH is expressed on a $\log_{10}$ scale and is defined as pH = −log[H$^+$]. pH is the measure of the acidity or alkalinity in an environment. General pH ranges for growth of various types of microorganisms are shown in Table 1. However, there are organisms that can tolerate a pH outside these ranges. For example, lower pH ranges can be tolerated by *Lactobacillus* (3.0–4.4) and *Acetobacter* (2.8). Some proteolytic bacteria can grow at a higher maximum pH because during growth they produce amines to buffer the high pH.

The pH of a food system will often determine the type of microorganism that grows in a food. Neutral-pH foods will allow bacteria to reproduce faster, whereas acidic foods will allow for the faster proliferation of yeasts and molds. The growth of microorganisms is influenced by both the initial pH and the buffering capacity of the food. The *buffering capacity* can be defined as the resistance of a food to changes in pH. The higher the buffering capacity, the more the food product resists changes in pH with the addition of an acid or base. The buffering capacity in foods with a high carbohydrate concentration is low, while in foods with high proteins, such as meat or milk, the buffering capacity is high.

Oxidation is the loss of electrons and reduction is the gaining of electrons. Generally, the *oxidation-reduction potential* (*Eh*) is the ability of a substrate to gain or lose electrons and is measured in millivolts (mV). *Eh* can be measured with a platinum electrode using a pH meter and is calculated by the following equation: $Eh = E_0 + RT/nF \ln[ox]/[red]$ (E_0 = standard electrode potential, n = number of electrons in half-reaction, R = universal gas constant, F = Faraday constant, and T = K). Foods that are more highly oxidized will have a positive electrical potential, and those that are reduced will have a negative electrical potential. Aerobic microorganisms require positive *Eh* values and anaerobes require negative *Eh* values. Ranges of *Eh* required for growth of bacteria can be found in Table 2. The growth of microorganisms can also influence the *Eh* of a food product. Aerobes can reduce the oxidation-reduction potential of a food product and can reduce *Eh* through the following chemical reaction:

$$\tfrac{1}{2}O_2 + 2H^+ + 2e^- \rightarrow H_2O.$$

Foods from plant sources have *Eh* values around +400 mV. Cheeses will have a range of −20 to −200 mV (Banwart 1979). The *Eh* of meat will change during processing: Prerigor is +250 mV, postrigor +10 to −130 mV, and ground beef +300 mV (Jay et al. 2005).

Table 3 Water activity of various foods

Food	Water activity (a_w)
Fruits and vegetables	0.97–1.00
Meats, poultry, and fish	0.95–1.00
Bread	0.96
Cheese	0.68–1.00
Jams and jellies	0.75–0.94
Honey	0.54–0.75
Dried milk and crackers	0.10–0.20

Source: Banwart (1979)

Table 4 Minimum a_w for the survival of foodborne pathogens

Bacterial foodborne pathogen	Minimum a_w
Bacillus cereus	0.930
Campylobacter jejuni	0.987
Clostridium botulinum Types A and B	0.940
Clostridium perfringens	0.970
Escherichia coli	0.950
Listeria monocytogenes	0.920
Salmonella	0.940
Staphylococcus aureus	0.870[a] (0.830[b])

Source: ICMSF (1996)
[a]Toxin production
[b]Growth

Water activity (a_w) is a measure of the water available for microbial growth. It is defined as the equilibrium relative humidity of a food product; the scale for water activity is from 1.0 (pure water) to 0. Microbial growth is affected by a_w, since with a reduced water activity, less water is available in the cell for the microorganism to use for biochemical processes. The a_w of a food product depends upon the solutes present in a food system and their interaction with water. The most common solutes used for reducing water activity are sugar or salt, but other substances may also reduce water activity, such as polyols (e.g., xylitol). Therefore, the constituents of the food matrix play a large role on a_w. The water activity for various foods can be found in Table 3, while the minimum a_w for specific foodborne pathogens can be found in Table 4.

The minimum a_w for bacterial growth is generally 0.90–0.91, but *Staphylococcus aureus* can survive and grow in foods with an a_w as low as 0.83. Generally, yeasts must have an a_w above 0.87–0.94, but some osmotolerant species grow in an a_w as low as 0.60. Molds can grow in foods with the lowest water activity, 0.70–0.80; *Xeromyces* is one species of mold that can grow in an a_w as low as 0.60. Water activity can be used as an effective method for preventing the growth of microorganisms in foods.

Some foods also have inhibitory substances or naturally occurring antimicrobials present (Davidson et al. 2005). For example, egg albumin contains the enzyme lysozyme, which disrupts the cell walls of Gram-positive bacteria by hydrolyzing the glycosidic bonds of N-acetylmuramic acid and N-acetylglucosamine in the peptidoglycan layer, avidin, which ties up the vitamin biotin, conalbumin, which ties up iron, and protease inhibitors, which inhibit protein degradation. All of these compounds act in concert to inhibit microbial growth. Raw milk also contains lysozyme and the lactoperoxidase system, which requires the interaction of the enzyme lactoperoxidase, thiocyanate, and hydrogen peroxide to produce the antimicrobial hypothiocyanate. Lactoferrin is also present, which sequesters the iron necessary for microbial growth similar to conalbumin found in eggs.

Spices also contain antimicrobials, which are components of the essential oils and are often phenolic in nature. The most active spice-derived antimicrobial compounds are found in cinnamon (cinnamic aldehyde), cloves (eugenol), oregano (carvacrol), and thyme (thymol). Onions, garlic, and horseradish also contain many sulfur-containing antimicrobials. For example, garlic contains allicin (diallylthiosulfinate), which is generated from the interaction of alliin with allinase. Cranberries and other berries also contain many antimicrobial phenolic compounds, such as benzoic acid. Hops were first used in beer production for their antimicrobial properties from the compounds lupulone and humulone. Olives contain phenolic glycosides, and the antimicrobial chitosan can be extracted from shrimp, lobster, and mushrooms.

5 Interventions

When discussing interventions that may be applied for food systems, there are two major categories in which the systems fall: *inactivation* and *inhibition*. Inactivation is achieved when microbial death occurs, whereas inhibition occurs when microbial growth is slowed or stopped.

5.1 Inactivation

Heat is the most effective and one of the most commonly employed interventions to inactivate microorganisms in foods. Heat is used for commercial processes designed to preserve foods and used in the home generally to prepare foods. Commercial processes can be classified as those that produce commercially sterile products or those that pasteurize products. Commercially sterile products are either shelf-stable or can be stored at nonrefrigerated conditions without spoiling. Pasteurization creates products that generally need a second preservation condition (e.g., refrigeration) for extended shelf-life. Heat is also used in the food industry for other processes, such as blanching of vegetables or steam sanitization of meat carcasses, but these are much less defined processes. In the home, cooking with heat using stove tops, ovens, and microwave ovens also kills microorganisms. Just as with some commercial processes, the home methods of heating often are not well defined or described.

To determine the effectiveness of a heat preservation process, one must know the heat resistance of the target microorganism. An assessment of microbial resistance to heat is used to determine the amount of heat that must be applied for a given time to inactivate a specific microorganism. The decimal reduction time (*D-value*) is the time in minutes necessary to inactivate 90% of the microbial population at a constant temperature and is graphically represented by a survivor curve. These curves may be established by exposing a known population of microorganisms to a constant temperature, monitored by thermocouples in thermal death time tubes, cans, or capillary tubes, and by determining the number of survivors for various periods of time. A D-value is calculated from the survivor curve and is $DT = -1/\text{slope}$ of the survivor curve, which is the resistance to heat of a microorganism at a specific temperature. For instance, D_{250} is the time (min) necessary to reduce a population of a specific microorganism 1 $\log_{10}$ (90%) at 250°C. The D-values for a microorganism are a function of not only the temperature and time but also the heating and recovery conditions used for determining the survivors of a heat process.

The effect of temperature on survival can be shown by plotting the $\log_{10}$ of D-values versus temperature to create a thermal death time (TDT) curve. The TDT curve can then be used to calculate a *z-value*, which is the number of degrees (Fahrenheit or Celsius) to cause a 90% change in log D. The D-value allows one to express the heat resistance at only one temperature, while the TDT curve measures the resistance at various temperatures.

Table 5 Heat resistance of serovars of *Salmonella* in various foods

Serovar	Food	Temp (°C)	D-value (min)	Approx. z-value (°C)
Senftenberg	Beef bouillon	65.5	0.66	
	Pea soup	65.5	1.11	5.6
	Skim milk	65.5	1.11	
	Milk	65.6	34.00	4.4
	Milk chocolate	70.0	480.00	18.9
	Milk chocolate	71.0	276.00	
Typhimurium	Milk chocolate	70.0	1050.00	17.7
	Milk chocolate	71.0	396.00	
	TSB + 10% MS	55.2	4.70	4.5
	TSB + 30% MS	55.0	11.00	4.6
	TSB + 42% MS	55.1	18.30	
	Ground beef	57.0	2.13–2.67	
Eastbourne	Milk chocolate	71.0	270.00	
None specified	Ground beef	57.2	4.20	

Source: ICMSF (1996)

The *F-value* is the number of minutes at a given temperature needed to destroy a homogeneous population of microorganisms. The equation for the F-value is D-value × (log$_{10}$ initial population − log$_{10}$ final population). Since z-values as well as temperatures will vary, a reference F-value, termed F_0, is used to describe the sterilization value of a thermal process where the z-value is 18°F or 10°C and $T = 250°F$ or 121.1°C.

The heat resistance or D-values can be impacted by many inherent factors of a microorganism, by environmental factors before, during, and after heat exposure, by growth conditions (temperature and nutrients) and the phase of growth, and even by the pH and solutes or food components. The conditions for microbial growth after heating are also important. Microbial injury that results in the decreased ability of microorganisms to repair and grow on selective microbiological media may greatly alter the number of survivors. Therefore, the best recovery conditions must be utilized when estimating the survivor population. Spores are more heat-resistant than vegetative cells of the same organism. *Clostridium botulinum* spores have a historical $D_{250°F} = 0.204$ min, while *Staphylococcus aureus* vegetative cells have a $D_{140°F} = 7.8$ min. Fungi generally have a $D_{150°F}$ of 0.5–1.0 min. Examples of the heat resistance of vegetative cells of *Salmonella* serovars are shown in Table 5.

Some heat processes are applied to create *commercially sterile* products. The definition of a commercially sterile food is the absence of living microorganisms that could grow under nonrefrigerated conditions. Commercial sterilization produces food products that are free of foodborne pathogens and of spoilage microorganisms that could grow during storage. Foods that undergo commercial sterilization may be divided into three categories: low-acid, high-acid, and acidified foods. *Low-acid foods* are foods that have a pH > 4.6, while *high-acid* and acidified foods have a pH 4.6. Acidified foods will achieve a pH 4.6 after the addition of an acidulant (e.g., vinegar or acetic acid) to the product. The commercial sterilization of low-acid (pH > 4.6) foods in hermetically sealed containers is termed *canning,* and spores of the bacterium *Clostridium botulinum* are the target foodborne pathogen to be destroyed. Low-acid canned foods (LACF) must be processed to achieve a 12-decimal reduction of the most resistant *Clostridium botulinum* spores. This procedure is termed a *12D process,* which is achieved by heating products to 115–150°C (240–300°F) using steam under pressure in a pressure vessel or "retort." The minimum required processing time is 3 min at 121.1°C (250°F) to achieve a theoretical 10^{12} reduction in *C. botulinum* spores. This would translate to less than 1 surviving spore in 1,000,000,000,000 containers. A 12D process will allow for adequate safety of LACF but sometimes will not guarantee commercial sterility, as some spores from meso-

Table 6 D-values for inactivation of various foodborne pathogens in fluid milk

Microorganism	D-value (min)	Temperature (°C)
Mycobacterium tuberculosis	0.200–0.300	65.5
Brucella	0.100–0.200	65.6
Coxiella burnetti	0.500–0.600	65.5
Listeria monocytogenes	0.013–0.021	71.7
Mycobacterium avium ssp. *paratuberculosis*	0.800–1.170	65.0

Source: Klijn et al. (2001), Sung and Collins (1998), and ICMSF (1996)

philic spore-forming spoilage bacteria may be more heat-resistant and survive a minimum process. Processors therefore often use a scheduled process that goes above the minimum required time and temperature to kill spoilage bacteria and thus achieve commercial sterility. Heat processing of low-acid foods can also be done by sterilizing a food outside a container and filling a sterile container in a sterile filling machine; this process is termed *aseptic processing*.

High-acid foods can be naturally high in acidity or have been acidified to cause a drop in pH. These products can be made commercially sterile with lower temperatures, which can be achieved with an atmospheric heating process. At this time, there are no minimum time–temperature processing requirements for these products, but they are processed to inactivate the most resistant spoilage microorganism present. Acidified foods are naturally low-acid but have acidulants added in order to decrease the pH (e.g., pickled beets). For these products, the acidification process must be closely monitored to ensure the minimum pH is achieved to inhibit growth of *C. botulinum* spores, and like foods that are naturally high in acid, a thermal heat process is not specified.

There are some spore-forming bacteria that have been associated with the spoilage of thermally processed acid foods. *Clostridium pasteurianum* and *C. butyricum* can grow in foods with a pH as low as 4.2 and have been associated with the spoilage of canned tomato products. *Alicyclobacillus acidoterrestris* is a spore-forming bacterium that was first discovered in 1997 as a cause of spoilage in acidified foods. This organism can survive in pH environments <4.0 and as low as 3.0. The optimal temperature for this organism is 45–50°C, and it does not produce gas like flat-sour organisms. It is extremely heat-resistant, with a $D_{90°C}$ of 11–14 min in 16° Brix grape juice (Splittstoesser et al. 1994). *A. acidoterrestris* is also sorbate- and sulfite-resistant, making it a very difficult spoilage organism to inactivate. Heat-resistant fungi also cause economic losses to producers. Filamentous (vegetative) forms of these organisms are inactivated by heat, but they are capable of forming very heat-resistant ascospores that have a $D_{90°C}$ of several minutes. Some of the most common fungi observed in this type of spoilage are *Byssochlamys, Neosartorya, Talaromyces,* and *Eupenicillium*.

Pasteurization is another form of heat treatment that is designed to destroy vegetative cells of pathogens and most spoilage microorganisms but has little effect on spores. Therefore, pasteurized food products are typically combined with other preservation processes, such as refrigeration or reduced water activity, to inhibit the growth of spore formers. Milk is one of the food products most commonly associated with pasteurization. Some target D-values for the inactivation of historically important foodborne pathogens in milk are found in Table 6. In addition to fluid milk, eggs can also undergo a pasteurization process to inactivate foodborne pathogens. Some bacterial strains can be resistant to pasteurization.

Equivalent pasteurization processes for milk may be accomplished using three processes: low-temperature, long-time (LTLT); high-temperature, short-time (HTST); and ultrahigh temperature (UHT). LTLT is commonly done in batch processing in a vat where milk is heated to 62.7°C (145°F) for 30 min. HTST requires a processing temperature of 71.7°C (161°F) for 15 s, and UHT pasteurization occurs at 140–150°C (280–300°F) for 1–2 s. The U.S. Department of Agriculture (USDA) suggests a pasteurization process of 60°C (140°F) for 3.5 min for whole eggs and 61.1°C (142°F) for 3.5 min for egg yolk.

Table 7 Lethal doses of ionizing radiation for various organisms

Organism	Lethal dose (kGy)
Viruses	10–200
Vegetative bacteria	0.1–5
Bacterial spores	10–50
Insects	0.1–1
Humans	0.005–0.010

Source: Jay et al. (2005)

Microwave heating has been explored for use in commercial thermal processing situations. Microwaves are advantageous for heating foods because they cause the internal heating of foods. This occurs when the microwave field oscillates, causing a redirection of water molecules and friction that generate heat. However, heating is very inconsistent, with cold spots that are nearly impossible to model, which has greatly reduced the application of this technology by the food industry because it would be very hard to validate the inactivation of pathogenic microorganisms.

Irradiation is another process employed to inactivate foodborne pathogens and spoilage microorganisms. Two categories of irradiation are employed by the food industry, ultraviolet (UV) radiation (nonionizing) and ionizing radiation. Shortwave UV radiation that results in microbial inactivation falls between 240 and 280 nm, with 260 nm being the most effective microbiocidal wavelength. UV radiation causes damage to DNA by cross-linking thymine dimers, halting repair mechanisms. UV radiation is limited in its application because it cannot penetrate through materials and therefore is best suited for the inactivation of microorganisms on surfaces.

Ionizing radiation has wavelengths less than 200 nm, and the doses of ionizing radiation are a measurement of absorbed energy. Gamma, electron beam, and X-rays are the most common types of irradiation that could be utilized in commercial food processing. The term Gray (Gy) is the common term used for ionizing radiation doses. One Gy equals 1 J/kg of absorbed energy. For larger doses of radiation, the term kiloGray (kGy) is employed, where 1,000 Gy = 1 kGy. To put radiation exposure into context, a single chest X-ray will result in a dose of 0.001 Gy for the average adult. Lethal doses of irradiation for various organisms can be found in Table 7.

Gamma irradiation is generated by cobalt-60 (^{60}Co) or cesium-137 (^{137}Cs), with ^{60}Co being the primary source utilized in the food industry. Both of these sources are inexpensive byproducts of atomic fission, can be recharged, and produce no radioactive waste, with ^{60}Co and ^{137}Cs having half-lives of 5 and 30 years, respectively. Gamma irradiation is a batch process where a chamber is utilized to expose the food to the radioactive isotope. Chambers must be built to retain all gamma rays within and as a result may have 1.83-m- (6-ft) thick concrete walls. Generally, this process is conducted in an off-site location since it requires a very unique infrastructure. The dose of irradiation to the food product is determined by the speed of the conveyor the product is moving on. One major advantage to this process is that the food product is completely penetrated by gamma rays, and microorganisms will be inactivated that reside on the interior of food products. One property of gamma irradiation that can be viewed as a drawback is that the source is always emitting gamma rays and cannot be simply switched off.

Electron beam irradiation utilizes high-energy electrons that are produced by linear accelerators or "electron guns." Electron beam irradiation does not penetrate as deep as gamma radiation, but as the energy is increased, the penetration capacity is improved. Therefore, electrons at 10-MeV units are utilized to expose food products to the electron beam with a resulting 1–3 cm of penetration. This property makes electron beam irradiation best suited for thin layers of product that can be exposed relatively quickly due to a relatively high dose rate (10^3–10^6 Gy/s). Electron beams can also be powered on and off, unlike gamma radiation, which is more conducive to on-site or in-line food processing.

Processes for the inactivation of microorganisms using irradiation have been classified based upon the severity of the process (Jay et al. 2005). The term "radurization" has been used for irradiation processes that sanitize and/or disinfect foods. This process is designed to improve the shelf life by reducing the population of spoilage microorganisms present, or to kill insects or parasites in foods. Dose levels are 0.75–2.5 kGy. "Radicidation" is comparable to pasteurization, and inactivates nonspore-forming bacterial pathogens to undetectable levels. Applied doses for radiciation are 2.5–10 kGy. "Radappertization" is when a food product is processed with doses of 30–45 kGy to render it sterile. This process achieves a 12D reduction of *Clostridium botulinum* spores in a food and, coupled with packaging in hermetically sealed containers, produces a LACF.

The effect of irradiation on food quality has also been widely studied. Energy from irradiation is absorbed by foods and, as with any process that involves the movement of energy, including heat, changes in food components occur. The nutritional content can be altered due to irradiation; for instance, the B vitamin thiamine has been shown to be degraded by the irradiation of food products. Food quality attributes have also been affected by irradiation. Flavor and odor changes have been observed in irradiated products, but the degree of these effects is highly dependent upon the irradiation dose. Fats have been shown to undergo oxidation, which can be minimized with vacuum packaging. Proteins have also been shown to exhibit off-flavors due to functional changes of amino acids and the release of compounds such as ammonia. Texture changes in fruits have also been observed due to pectin breakdown, which causes increased softening of irradiated products.

Irradiation is considered a food additive rather than a food process by the U.S. government (Anonymous 2009), and irradiated foods must be labeled "treated with radiation" or "treated with irradiation." The U.S. Food and Drug Administration (FDA) and the USDA have approved many foods and packaging materials for treatment with irradiation. However, approval must be granted for each product prior to its sale in the United States. For example, the irradiation of iceberg lettuce and spinach was approved by the FDA in 2008 to reduce foodborne pathogens and extend shelf life with a maximum dose of 4.0 kGy.

Several "nonthermal" processes have also been studied and utilized by the food industry to inactivate microorganisms. *High-pressure processing (HPP)*, or pascalization, utilizes pressures in the range of 100–800 MPa (14,500–116,000 psi). When HPP is utilized to inactivate vegetative cells, variable effects have been observed in efficacy. In order to inactivate spores, heat (70–80°C) and pressure must be utilized together. HPP results in cell lysis due to heat plus the application of pressure and subsequent depressurization. This type of processing has been found most suitable for purees and liquids, as well as shellfish such as raw oysters. HPP is a batch process where foods are generally exposed following packaging in plastic films. Pulsed electric fields (PEF) use high-voltage pulses, 20–80 kV/cm, for durations up to 1 s but generally in microseconds. Foods commonly processed with PEF are liquids or purees.

5.2 Inhibition

Temperature depression is one of the most widely used processes for inhibiting microbial growth through the slowing or stopping of metabolic activities. Chilling foods by refrigeration to −1°C to 7°C results in the reducing or stopping the growth of mesophilic and thermophilic microorganisms. Additionally, psychrotrophic microorganisms will have an increased lag phase and generation time, and their growth will continue at a slow rate. Some foodborne pathogens are psychrotrophic and can grow slowly at refrigeration temperatures, for instance, *C. botulinum* Type E, *C. botulinum* nonproteolytic Type B, *Yersinia enterocolitica*, and *Listeria monocytogenes*. Freezing will generally halt all microbial growth, but some microorganisms have been reported to grow at temperatures below 0°C. Freezing of foods can be achieved rapidly or slowly, with the latter resulting in large ice crystals that can disrupt cell membranes, causing increased lethality. However, these crystals also disrupt cells in foods and can negatively impact the sensory properties. Gram-positive bacteria, with the exception of *Clostridium perfringens*, are generally more resistant to inactivation by freezing, whereas Gram-negative bacteria and protozoa are more sensitive (Jay et al. 2005). Freezing has no lethal impact on foodborne viruses, spores, or toxins. When thawing frozen foods, microbial growth can occur if the temperatures are conducive. Because frozen foods thaw from the outside in, thawing foods at room temperature could result in significant microbial growth on the surface of a product. It has also been shown that repeated freeze-thaw cycles will cause microbial inactivation by dehydration, oxidation, and ice nucleation. However, this process will also typically cause such negative impacts on food quality that it would not be employed as a means to inactivate pathogens and spoilage organisms.

The pH in a food system can be altered (generally reduced) to inhibit microbial growth. Generally, organic acids are used to reduce the pH. With the addition of organic acids, there will be a rise in protons (H^+) but also undissociated acids. The undissociated form of acids has the primary inhibitory effect on microorganisms and can more easily diffuse through the cell membrane to dissociate upon entry into the neutral cytoplasm. This will cause a drop in the internal pH of the microorganism. In order to survive, the microorganism will expend energy pumping the dissociated acid outside the cell membrane, where it will recombine. This futile process will cause the organism to expend a large amount of energy to maintain the internal pH and disrupt the proton motive force. Mineral acids, such as phosphoric acid, serve only as a source of hydrogen ions and therefore are not as effective as organic acids at similar pHs. In most cases they are added to beverages and acidified foods to impart an acidic flavor.

Modified atmosphere packaging (MAP) technology involves altering the atmosphere in an impermeable food package. In this process, an atmosphere is selected that slows microbial growth and thus extends shelf life. Impermeable plastic films are used as a packaging material to control the atmosphere of a food product. These films can be selected to fit the needs of the food product, with gas barrier properties that are appropriate for the desired package atmosphere. In MAP, the atmosphere is removed and replaced with another gas mixture. This type of process is also referred to as a vacuum/gas flush. In passive MAP, a selected gas is added to the package but no attempt is made to maintain the atmosphere over the storage life. The atmosphere may change due to respiration. Passive MAP is utilized primarily in fresh or minimally processed produce. Active MAP incorporates a defined gas mixture that will be continuously maintained by the addition of materials in the packaging or sachets that produce various gas mixtures. Examples of compounds used in packaging to help actively maintain the atmosphere are oxygen scavengers or sulfur dioxide–generating pads.

Modified atmosphere storage typically utilizes mixtures of carbon dioxide, oxygen, and nitrogen and inhibits aerobic microorganisms. High CO_2 levels (15–100%) will inhibit aerobic microbial growth. The oxygen content can be less than 1% and as high as 80%, but typically is in the range of 5–10%. Lowering the oxygen content will protect foods from oxidative damage, but some oxygen can be beneficial. For instance, meat requires oxygen to form oxymyoglobin to have a characteristic red color. A small amount of oxygen can also prevent the growth of strict anaerobes. Nitrogen is a filler gas and is found in 15–100% of the modified atmosphere. Modified atmosphere storage can be used in bakery products, produce, *ready-to-eat foods*, fish, meat, poultry, and fresh pasta and may increase the shelf life by 50–400%. Modified atmosphere storage does not prevent the growth of facultatively anaerobic bacteria or yeasts. Some foodborne pathogens that are able to grow in these environments include *Campylobacter*, *Clostridium*, *Salmonella*, and *Yersinia*.

Vacuum packaging is yet another type of modified atmosphere. As the name implies, in this process the air is removed and not replaced by any gases. This also requires the use of a barrier film to maintain the vacuum. Vacuum packaging will reduce the oxygen content, inhibiting fast-growing aerobic microorganisms, such as *Pseudomonas* and molds, which results in a shift of the spoilage microflora, favoring the ones that grow only under anaerobic conditions. For example, bacteria that cause spoilage of vacuum-packaged meats are species of *Lactobacillus* and *Brochothrix*. The major advantage of vacuum packaging is that while the aerobic microorganisms can cause spoilage in a few days of refrigerated storage, the shelf life of vacuum-packaged products is generally several weeks. *Sous-vide*, translated from the French as "under vacuum," is a preparation of foods in vacuum packaging followed by mild heat treatments and chilling. This process will not inactivate spore-forming bacteria, and *C. botulinum* growth and toxin production must be a concern for these products.

Dehydration, or reducing the a_w of foods, is another way of inhibiting microbial growth in foods. Dehydration is when the water content is reduced 75–92%. This will generally result in a very stable food product. However, the dehydration process, even using heat, may allow some pathogens to survive. For example, *Salmonella* has been found to survive in spray-dried foods, and *Cronobacter* (*Enterobacter*) *sakazakii* survives in powdered infant formula. Reducing water activity to below 0.85 can prevent the growth of all foodborne bacteria and significantly extend the shelf life of food products. Reducing the water activity can be achieved by adding solutes or humectants. Ingredients commonly used to decrease water activity include salt, sucrose, glycerol, fructose, galactose, hydrocollids, high fructose corn syrup, and polyols (xylitol, mannitol, sorbitol). Jams, jellies, fruit cake, candies, pet foods, some cheeses, and jerky are examples of foods that utilize reduced water activity to maintain stability. Microorganisms that might spoil reduced-a_w foods include osmotolerant yeasts and molds.

Antimicrobial food preservatives can be incorporated into foods to delay spoilage or inhibit the growth of foodborne pathogens (Davidson et al. 2005). As described above, weak organic acids can be used as food antimicrobials. Traditional regulatory-approved organic acid antimicrobials include acetic, benzoic, lactic, propionic, and sorbic acids or their salts. Acetic acid or acetates have a dissociation constant (pK_a) of 4.75 and are commonly used to inhibit bacteria. The dissociation constant is the pH where 50% of the acid is in a dissociated state, while the remaining 50% is undissociated. This is a very important parameter when applying organic acids to as antimicrobials because the undissociated form is primarily responsible for the observed antimicrobial properties. *Acetic acid* is the primary component of vinegar and imparts its typical odor and flavor. It is commonly used in high-acid foods due to the low pK_a. Sodium acetate is used in bakery products and sodium diacetate is used in meat products at 0.25%. *Benzoic acid* occurs naturally in berries and other fruits. This acid has a pK_a of 4.19. Benzoic acid is commonly used in acidic foods since its pK_a is relatively low. It is primarily used to inhibit yeast and mold growth. The yeast *Zygosaccharomyces bailii* is resistant to benzoic acid at concentrations up to 0.3% and some bacteria are not inhibited by benzoic acid at the concentration allowed in foods, 0.1%. Benzoic acid has been applied in beverages (0.03–0.05%); syrups (0.1%); olives, pickles, and relishes (0.1%); jams and jellies (0.1%); and fruit products (0.1%). *Lactic acid* and lactates have a pK_a of 3.79 and are produced by lactic acid bacteria during fermentations of foods. Lactic acid is effective at inhibiting bacterial growth in food products but is also used as a flavoring agent. Sodium lactate has been utilized at 1.8–4.0% in meat products to delay the growth of foodborne pathogens like *Listeria monocytogenes*. It has also been applied as a spray to meat carcasses for the inhibition of foodborne pathogens.

Propionic acid and propionates have a pK_a of 4.87 and are used in products to inhibit mold and bacterial growth at concentrations around 0.4%. Propionates are typically applied in baked goods. *Propionibacterium freudenreichii* ssp. *shermani*, which produces propionic acid, is also used in the fermentation of Swiss cheese to give it its characteristic flavor and the holes or "eyes" via CO_2 production. *Sorbic acid* and sorbates have a pK_a of 4.75 and are used to inhibit the growth of yeasts, molds, and bacteria. Sorbic acid was first isolated in 1959 by A. W. Hoffmann from rowanberry oil. Sorbic acid was studied as a food preservative as early as 1940 when C. M. Gooding used it in packaging to inhibit mold growth on the surface of foods. Sorbic acid and its salts may be applied to fruit, bakery, and dairy products and to beverages and fermented foods at concentrations up to 0.3%. Microorganisms that are resistant to sorbic acid include *Zygosaccharomyces bailii, Saccharomyces rouxii, Penicillium,* and lactic acid bacteria.

Sulfur dioxide (SO_2) and metabisulfite ($Na_2S_2O_5$) are effective against yeasts and molds and are commonly used in fruit products and wine. Sulfites are most effective in foods with low pH. However, high concentrations of sulfites are known to cause allergic reactions in some individuals, primarily those with asthma.

Parabens are alkyl esters of p-hydroxybenzoic acid with varying lengths of carbon chains (methyl 1C, propyl 3C, and heptyl 7C). Parabens are synthetically produced through the esterification of the carboxyl group of benzoic acid. This property also allows parabens to remain undissociated and be used in foods with a pH up to 8.0. Methyl-, propyl-, and heptyl parabens are permissible for use within the United States. Parabens inhibit the growth of yeasts, molds, and Gram-positive and -negative bacteria. They are commonly used in concentrations to 0.1% in bakery products, salad dressings, and soft drinks. However, parabens may have limited use because they can impart a metallic or bitter taste and are not readily soluble in foods.

Sodium nitrite is used for several purposes in cured meats, including for color formation, for flavor, as an antioxidant, and as an antimicrobial against the growth of *Clostridium botulinum*. The efficacy of nitrite against *C. botulinum* is dependent upon the pH and salt content. A drawback of using nitrites in foods is that they can react with secondary amines to form nitrosamines, which are known to be carcinogenic.

Nisin is a peptide produced by *Lactococcus lactis* ssp. *lactis* that was approved in 1988 in the United States for use in food products as an antimicrobial food preservative. Nisin is a bacteriocin that has activity against Gram-positive bacteria, including spore formers. Nisin interacts with the cell membrane of these organisms to cause pore formation to inhibit or inactivate microorganisms. Nisin is not normally effective against most Gram-negative bacteria; however, the addition of nisin with the chelator EDTA expands the inhibitory activity to this group. EDTA apparently destabilizes the outer membrane of Gram-negative bacteria, making them more susceptible to nisin. Nisin is generally recognized as safe (GRAS) and has been applied to pasteurized processed cheese and several other processed foods.

Natamycin (pimaricin) is an antibiotic-like compound from the organism *Streptomyces natalensis*. This polyene macrolide antibiotic is an antifungal agent, but it is not effective against bacteria. It was first approved in the United States in 1982 as an antifungal agent on the surface of cheeses. Natamycin at 200–300 mg/l solution inhibits mold growth.

Lysozyme is an enzyme that is present in avian eggs, tears, mammalian milk, insects, and fish, with dried egg white being the commercial source. As discussed previously, lysozyme acts on the peptidoglycan layer of Gram-positive bacteria, disrupting the bacterial cell wall and potentially causing cell lysis. As with nisin, lysozyme activity can be increased by adding a chelator, such as EDTA. Lysozyme is GRAS and was approved in the United States in 1998 for use in certain cheeses.

Lactoferrin is a glycoprotein found in milk that acts as a chelator to sequester iron. In this manner, microorganisms that require iron for growth are inhibited. Lactoferrin exists in milk primarily as a tetramer with calcium and has two iron-binding sites per molecule. For each Fe^{3+} bound by lactoferrin, one bicarbonate (HCO_3^-) is required. Naidu (2002) developed an antimicrobial system in which lactoferrin is immobilized to food-grade polysaccharides and is dissolved in a citrate/bicarbonate buffer with NaCl. The resulting product, activated lactoferrin, was reported to be effective at inhibiting the attachment of *E. coli* O157:H7 on meats. The lactoferrin product is approved by the Food Safety Inspection Services of the USDA to reduce bacterial contamination by the target microorganism on beef carcasses (Naidu 2001).

Novel antimicrobials will continue to be evaluated by those in research to help the food industry inhibit the growth of foodborne pathogens. Lauric arginate and bacteriophages are examples of antimicrobial inhibitors that have been more recently studied (Hudson et al. 2005; Luchansky et al. 2005). Lauric arginate is a derivative of lauric acid, L-arginine, and ethanol and has been certified GRAS and approval granted for use on ready-to-eat meats. Bacteriophages are viruses that infect bacteria, causing lysis. This interaction is very specific between the host (bacteria) and phage; therefore, a specific phage can be propagated to target the destruction of foodborne pathogens. These examples of antimicrobial agents, as well as many others, will continue to be researched so that the food industry can improve the safety of products offered to consumers.

Fermentation is yet another process that has been used for centuries to inhibit the growth of foodborne pathogens. The primary microorganisms involved in fermentation are lactic acid bacteria, such as *Lactococcus, Lactobacillus, Pediococcus,* and *Streptococcus,* as well as yeasts, such as *Saccharomyces cerevisiae.* Microorganisms can also be utilized in fermentations for flavor alteration, which is commonly used in cheese production. Some organisms that are utilized in this manner are *Brevibacterium linens, Propionibacterium freudenreichii* spp. *shermanii, Penicillium camemberti,* and *Penicillium roqueforti.* Chemical alterations occur in food products during a fermentation process. Homofermentative lactic acid bacteria will produce roughly 90% lactic acid during anaerobic metabolism. Heterofermentative lactic acid bacteria will produce lactic acid, ethanol, acetic acid, and carbon dioxide during metabolism. Yeasts, like *Saccharomyces cerevisiae,* also known as baker's yeast, have been utilized in baking and wine production due to their ability to produce ethanol and carbon dioxide.

Many types of food products undergo fermentation for preservation. Dairy products such as cheese, buttermilk, and yogurt are products of milk fermentation. These types of fermentations are carried out by lactic acid bacteria like *Lactococcus* and *Lactobacillus.* These products are preserved by lactic acid production during fermentation. Vegetable products like sauerkraut, pickles, and olives use *Leuconostoc* and *Lactobacillus* species during fermentation. These products are also preserved by lactic acid as well as microbial succession. Meats that undergo fermentation include summer sausage, salami, and pepperoni. *Pediococcus* is one microorganism that is commonly used for these fermentations; it preserves the product through lactic acid production.

Fermented meats are also preserved through the addition of nitrites and spices and the smoking process. Cereal products like bread, crackers, and pretzels can be fermented by *Saccharomyces.* These products are leavened through fermentation and the production of carbon dioxide. Alcoholic beverages like wine, beer, and distilled spirits also undergo a fermentation process. These alcoholic fermentations utilize organisms like *Saccharomyces* for flavor enhancement as well as ethanol production.

5.3 Interactions

Described within this chapter are many types of interventions that can be used to inhibit or inactivate foodborne pathogens and spoilage microorganisms. "Hurdle technology" utilizes multiple barriers or "hurdles" to preserve foods and is defined as the utilization of existing and/or novel preservation methods to establish a series of preservation factors (hurdles) that any microorganism should not be able to overcome. The *hurdle concept* was developed in 1978 by Lothar Leistner at the Federal Center for Meat Research in Kulmach, Germany, as a semisystematic approach to improve the safety of foods. This was first applied to shelf-stable foods and then extended to other food products. Hurdle technology is also known as combined methods, combined processes, combined preservation, combination techniques, barrier technology, and intervention technology. Targets for hurdle technology include pathogenic, spoilage, and fermentative microorganisms. Quality attributes impacted by the hurdle technology can improve sensory attributes as well as the nutritive value of foods in addition to food safety. Hurdles may be physical, physiochemical, or microbial in nature, and there are currently over 60 potential hurdles that may be used by the food industry (Leistner 2000).

6 Summary

Food preservation has long been used to safely maintain edible products for consumption. Preservation processes are designed to slow the growth of microorganisms or to inactivate them. The growth of microorganisms is affected by extrinsic (atmosphere, temperature, humidity) and intrinsic or food-related (pH, oxidation-reduction potential, water activity, natural antimicrobials, and barriers) factors. Microorganisms can be inactivated in foods using heat (commercial sterility, pasteurization), irradiation (ultraviolet, electron beam, gamma), or nonthermal processing. The inhibition of microorganisms may be accomplished using chilling, freezing, acidification, modified atmosphere packaging, dehydration, reduced water activity, or antimicrobial food preservatives. A combination of multiple methods, processes, and techniques, called the "hurdle" concept, can be used to make the food safer and to extend the shelf life of food products.

References

Anonymous. 2009. Irradiation in the production, processing and handling of food. Code of Federal Regulations 21CFR Part 179.

Banwart, G.J. 1979. *Basic food microbiology*. Westport: AVI.

Daniels, J.A., R. Krishnamurthi, and S.S.H. Rizvi. 1985. A review of effects of carbon-dioxide on microbial-growth and food quality. *Journal of Food Protection* 48: 532–537.

Davidson, P.M., J.N. Sofos, and A.L. Branen. 2005. *Antimicrobials in foods*, 3rd ed. Boca Raton: CRC Press.

Hudson, J.A., C. Billington, G. Carey-Smith, and G. Greening. 2005. Bacteriophages as biocontrol agents in food. *Journal of Food Protection* 68: 426–437.

ICMSF. 1996. *Microorganisms in foods*, Characteristics of microbial pathogens, vol. 5. London: Blackie Academic and Professional.

Jay, J.M., M.J. Loessner, and D.A. Golden. 2005. *Modern food microbiology*, 7th ed. New York: Springer.

Klijn, N., A.A.P.M. Herrewegh, and P. de Jong. 2001. Heat inactivation data for *Mycobacterium avium* subsp *paratuberculosis*: Implications for interpretation. *Journal of Applied Microbiology* 91: 697–704.

Leistner, L. 2000. Basic aspects of food preservation by hurdle technology. *International Journal of Food Microbiology* 55: 181–186.

Luchansky, J.B., J.E. Call, B. Hristova, L. Rumery, L. Yoder, and A. Oser. 2005. Viability of *Listeria monocytogenes* on commercially-prepared ham surface treated with acidic calcium sulfate and lauric arginate and stored at 4°C. *Meat Science* 71: 92–99.

Naidu, A. 2001. Immobilized lactoferrin antimicrobial agents and use. U.S. Patent 6,172,040.

Naidu, A. 2002. Activated lactoferrin – a new approach to meat safety. *Food Technology* 56: 40–45.

Setlow, P. 1995. Mechanisms for the prevention of damage to DNA in spores of *Bacillus* species. *Annual Review of Microbiology* 49: 29–54.

Song, Y., R. Yang, Z. Guo, M. Zhang, X. Wang, and F. Zhou. 2000. Distinctness of spore and vegetative cellular fatty acid profiles of some aerobic endospore-forming bacilli. *Journal of Microbiological Methods* 39: 225–241.

Splittstoesser, D., J. Churey, and C. Lee. 1994. Some factors affecting growth of aciduric sporeforming bacilli in fruit juices. *Journal of Food Protection* 57: 1080–1083.

Sung, N., and M. Collins. 1998. Thermal tolerance of *Mycobacterium paratuberculosis*. *Applied and Environmental Microbiology* 64: 999–1005.

Taniwaki, M.H., A.D. Hocking, J.I. Pitt, and G.H. Fleet. 2009. Growth and mycotoxin production by food spoilage fungi under high carbon dioxide and low oxygen atmospheres. *International Journal of Food Microbiology* 132: 100–108.

Food Regulation in the United States

Patricia Curtis

1 Introduction

The two major food regulatory agencies are the Food and Drug Administration (FDA), which is part of the Department of Health and Human Services, and the U.S. Department of Agriculture's (USDA) Food Safety and Inspection Service (FSIS). The FDA's food regulatory responsibilities include all foods except meat, poultry, and egg products. FSIS's regulatory responsibilities include meat, poultry, and egg products. There are other federal agencies such as the Environmental Protection Agency, the Occupational Safety and Health Administration, the Federal Trade Commission, and a few others that also periodically develop regulations associated with food, but food is not their primary focus. This chapter will review the different laws and responsibilities of the FDA and FSIS.

2 The Food and Drug Administration

Although the FDA is responsible for regulating drugs as well as foods, the present discussion will be limited primarily to their food regulatory responsibilities. Prior to the nineteenth century, states regulated foods that were locally produced and distributed. The FDA began as the Division of Chemistry and after 1901 became the Bureau of Chemistry. The 1906 Federal Food and Drug Act included regulatory authority in the agency's mission. In 1927, the Bureau of Chemistry name was changed to the Food, Drug, and Insecticide Administration. In 1930, the name was shortened to this present version, and the FDA remained a part of the Department of Agriculture until 1940 (FDA 2009a).

By the late nineteenth century, adulterated and misbranded foods were a growing concern. Deceptive practices included adding chalk to diluted milk and adding alum and clay to poor wheat flour. The Division of Chemistry began investigating the adulteration of agricultural commodities as early as 1867. Harvey Wiley, who was chief chemist at the Division of Chemistry, took an unprecedented action against adulterated foods. He created a volunteer group known as the "poison squad" who utilized able-bodied volunteers to consume varying amounts of questionable food additives to determine their impact on health. Over the years, the FDA has provided Americans with increasingly comprehensive, science-based protections that ensure the highest quality of products essential for health and survival. Today, these products represent almost 25% of all consumer spending and include

---

P. Curtis (✉)
Department of Poultry Science, Auburn University, Auburn, AL, USA
e-mail: pat_curtis@auburn.edu

> **Box 1** six amendments to the 1906 act
>
> There were only six amendments to the 1906 act before it was replaced by the FDCA in 1938 (Pina and Pines 2002):
>
> - The "Sherley Amendment" of 1912, which expanded the definition of misbranding for drugs to include "any statement, design, or device regarding the curative or therapeutic effect of such an article or (its components), which is false and fraudulent."
> - The "Gould Amendment" of 1913, which required the quantity of a packaged food's contents to be plainly and conspicuously marked on the outer package in terms of weight, measure, or numerical count.
> - The "Kenyon Amendment" of 1919, which overturned an adverse administrative decision by defining the word "package" as used in the act to include "wrapped meats enclosed in papers."
> - The first standard definition of a food product, in "An Act to Define Butter and Provide a Standard Therefore," introduced in 1923 by the Congress in response to requests from the dairy industry.
> - The "McNary–Mapes Amendment" of 1930, authorizing the Secretary of Agriculture to establish standards of quality, condition, and fill of container for each class of canned foods. Congress passed this amendment at the urging of the canned foods industry.
> - The "Shrimp Inspection Amendment" implemented by Congress in 1934, which authorized the Secretary of Agriculture to create a voluntary seafood inspection program that would allow a participant to market its shrimp as "USDA Approved."

80% of the national food supply and all human drugs, vaccines, medical devices, tissues for transplantation, equipment that emits radiation, cosmetics, and animal drugs and food (McDermott 2006). The FDA enforces the Food Drug and Cosmetic Act (FDCA) and several other public health laws. The predecessor of the FDCA was the 1906 Federal Food and Drug Act (FFDA). The 1938 FDCA and its many amendments are probably the single most far-reaching food regulation (Box 1).

In the 1920s and early 1930s, consumers were exposed to increasing numbers of processed food, drug, cosmetic, and medical products. But it was not until a drug-related tragedy occurred that a new food and drug law was passed. In the 1930s, a manufacturer prepared a liquid form of sulfanilamide using diethylene glycol. Though sulfanilamide itself is not harmful, diethylene glycol is toxic. As a result, 107 people, many of whom were children, died from the toxic ingredient in the product called Elixir Sulfanilamide. Before the elixir was marketed, the manufacturer tested the product for flavor, but not for safety. Diethylene glycol, which was used as a solvent to prepare the elixir, was not listed as an ingredient on the label. A recall was initiated based on a technical violation of the 1906 Act: "Elixir" could be used only to describe a solution that contained alcohol, and because Elixir Sulfanilamide did not contain alcohol, it was misbranded. The manufacturer's failure to test the drug for safety or to list a lethal ingredient on the label did not violate the 1906 Act (Curtis 2005; Meadows 2006; Pina and Pines 2002). The public became increasingly concerned about food safety, the inability of the USDA to adequately regulate foods and drugs under the 1906 law, and the horror of the sulfanilamide tragedy. These concerns motivated Congress to pass the FDCA (Box 2), which included the following new provisions (FDA 2009b):

- Providing that safe tolerances be set for unavoidable poisonous substances
- Expanding the definitions of adulteration and misbranding
- Authorizing standards of identity, quality, and fill-of-container for foods
- Authorizing factory inspections
- Adding the remedy of court injunctions to the previous penalties of seizures and prosecutions

Box 2 Food Drug and Cosmetic Act

To better understand the FDCA, you need to understand a little about the history of the act. The predecessor of the FDCA was the 1906 Federal Food and Drug Act (FFDA), which was the first comprehensive U.S. legislation relating to foods and drugs. The FFDA prohibited the interstate transport of unlawful food and drugs under penalty of seizure of the questionable products and/or prosecution of the responsible parties. The FFDA was designed as an "enforcement" statute intended to punish violations by criminal prosecution or by seizure of the offending goods. In concise and simple terms, the 1906 act commanded the federal government to search out adulterated and misbranded foods and drugs in interstate commerce. The government was not authorized to establish industry-wide standards or broad rules to protect the public health or to approve any products before they could be marketed (Pina and Pines 2002). The FDCA remains the basic law governing food and drug regulations. Since 1938, there have been numerous amendments to the FDCA. A list of significant amendments and other acts related to the FDCA can be found at http://www.fda.gov/opacom/laws/default.htm#amendments. The FDCA in detail can be found in Title 21, Chap. 9 of the *U.S. Code* (USC). The USC is the codification by subject matter of the general and permanent laws of the United States based on what is printed in the *Statutes at Large*, which is the official compilation of federal laws and is published annually. The USC has been published every 6 years since 1926. The USC is where one can find the details of the FDCA and its amendments. To conduct a search of the USC relative to some section of the FDCA, one should be aware that the section numbers for the USC and the FDCA do not match. A cross-reference of section numbers for the FDCA Act and the USC can be found at http://www.fda.gov/RegulatoryInformation/Legislation/FederalFoodDrugandCosmeticActFDCAct/ucm086299.htm. The new Federal Digital System makes searching much easier. To locate information about a law or regulation, simply type in the topic in the search box located at http://www.gpo.gov/fdsys/search/home.action. The search engine will provide a series of links related to the topic.

2.1 Food Code

The FDA publishes the Food Code, a model that assists food control jurisdictions at all levels of government by providing them with a scientifically sound technical and legal basis for regulating the retail and food service segment of the industry (restaurants and grocery stores and institutions such as nursing homes). Local, state, tribal, and federal regulators use the FDA *Food Code* as a model to develop or update their own food safety rules and to be consistent with national food regulatory policy. The Association of Food and Drug Officials reports that 49 of the 50 states and 3 of the 6 territories have adopted codes patterned after the *Food Code*. Those states and territories represent 96% of the U.S. population. The *Food Code* also serves as a reference for best practices for the retail and food service industries (restaurants, grocery stores, nursing homes) on how to prevent foodborne illnesses. Many foodservice operations apply *Food Code* provisions to their own food safety practices and policies. The most current information regarding adoption of the *Food Code* can be found online (FDA 2011a).

Between 1993 and 2001, the *Food Code* was published every 2 years. Publication intervals have now increased to every 4 years. The 2005 *Food Code* was the first full edition to be published since 2001. During the 4-year interval between complete revisions, a supplement is available that includes updates and modifications or clarifies certain provisions. The changes contained in the supplement reflect current science and emerging food safety issues. The supplement therefore offers the most up-to-date information regarding mitigation of risk factors that contribute to foodborne illnesses. The most up-to-date edition of the *Food Code* and/or supplement can be downloaded from the FDA website (FDA 2011b).

2.2 HACCP

The FDA has adopted a food safety program developed nearly 30 years ago for astronauts and is applying it to seafood and juice. The FDA intends to eventually use it for much of the U.S. food supply. The Hazard Analysis and Critical Control Points (HACCP) system has been endorsed by the National Academy of Sciences, the Codex Alimentarius Commission (an organization setting international food standards), and the National Advisory Committee on Microbiological Criteria for Foods (established in 1988 to provide impartial, scientific advice to federal food safety agencies). Many of the HACCP principles were already in place in the FDA-regulated low-acid canned food industry when the FDA established the HACCP for the seafood industry in a final rule on December 18, 1995, and for the juice industry in a final rule released on January 19, 2001. The HACCP involves seven principles (FDA 2001):

- *Analyze hazards.* Potential hazards associated with a food and measures to control those hazards are identified. The hazard could be biological, such as a microbe, chemical, such as a toxin, or physical, such as ground glass or metal fragments.
- *Identify critical control points.* These are points in a food's production – from its raw state through processing and shipping to consumption – at which potential hazards can be controlled or eliminated. Examples are cooking, cooling, packaging, and metal detection.
- *Establish preventive measures with critical limits for each control point.* For a cooked food, for example, this might include setting the minimum cooking temperature and time required to ensure the elimination of any microbial hazard.
- *Establish procedures to monitor the critical control points.* Such procedures might include determining how and by whom the cooking time and temperature should be monitored.
- *Establish corrective actions to be taken when monitoring shows that a critical limit has not been met.* An example is reprocessing or disposing of food if the minimum cooking temperature is not met.
- *Establish procedures to verify that the system is working properly.* An example is testing time-and-temperature recording devices to verify that a cooking unit is working properly.
- *Establish effective recordkeeping to document the HACCP system.* This would include records of hazards and their control methods, the monitoring of safety requirements, and action taken to correct potential problems.

Each of these principles must be backed by sound scientific knowledge, for example, published microbiological studies on time and temperature factors for controlling foodborne pathogens.

2.2.1 Fish Products

On December 18, 1995, the FDA published the final rule "Procedures for the Safe and Sanitary Processing and Importing of Fish and Fishery Products." This regulation required processors of fish and fishery products to develop and implement HACCP systems for their operations. The regulation became effective on December 18, 1997.

The FDA also published the Fish and Fishery Products Hazards and Controls Guide ("the Guide"), in September 1996, to assist processors in the development of their HACCP plans, to provide information on how to identify hazards that may be associated with fish and fishery products, and to formulate control strategies for those hazards. The FDA's HACCP system focuses on identifying and preventing hazards that could cause foodborne illnesses rather than relying on spot checks of manufacturing processes of finished seafood products to ensure safety.

The General Accounting Office (GAO) has released a report evaluating the FDA's seafood HACCP program. The GAO concludes that while the FDA has made progress in ensuring the safety of seafood through the HACCP, the program needs to be strengthened in order to reach its full objective. The FDA instituted a mid-course correction to its seafood HACCP program to focus on those products that presented the highest risk to consumers – those firms that needed to control for pathogens, those that needed to control for histamines (which can cause severe allergic reactions), and those firms that did not have HACCP plans. The FDA believed that seafood processed by these three categories of firms presented the highest risk to consumers. Therefore, the agency redoubled its efforts, resulting in *more frequent inspections* of noncompliant firms, *more extensive laboratory* testing for pathogens and histamines, and, ultimately, *enforcement* action where appropriate. In addition, the following actions were taken to strengthen the HACCP program for seafood:

1. *Improved guidance* and training to the industry and regulators on control of pathogens and histamine
2. Development of an *inspector certification program* that emphasizes knowledge of controls for pathogens and histamine
3. Development of *guidance for fishing vessel operators* to address proper handling of fish that can form histamine
4. Development of *guidance for aquaculture operators* to prevent pathogen contamination of aquaculture sites
5. Increased emphasis on *compliance by foreign processors* and increased surveillance of imports
6. Creation of a *National Seafood HACCP Inspection Database* that collects information on the details of preventive controls for safety applied by seafood processors

In the last report, issued in July of 2008 and covering the seventh and eighth years of the seafood HACCP program, it is clear that the industry performance has improved.

The FDA used the previous evaluations to make changes to the compliance programs and to focus inspection and training efforts on certain industry segments or specific food safety hazards. The report revealed that efforts have paid off, as evidenced by the steady improvements over the years. However, despite increased FDA efforts, certain industry segments continue to be slow to improve (FDA 2008).

2.2.2 Juice Products

A 1997 study by the FDA's Center for Food Safety and Applied Nutrition found that while the contamination of juice products most likely occurs during the growing and harvesting of the raw product, contamination may also occur at any point between the orchard and the table. Therefore, on January 19, 2001, the FDA published a final rule in the *Federal Register* that required processors of juice to develop and implement HACCP systems for their processing operations. The rule came after a rise in the number of foodborne illness outbreaks and consumer illnesses associated with juice products, including a 1996 *E. coli* O157:H7 outbreak associated with apple juice products and two citrus juice outbreaks attributed to *Salmonella* spp. The apple juice outbreak sickened 70 people in the western United States and Canada, including a child who died from hemolytic-uremic syndrome caused by the infection. The *Salmonella* Enteritidis outbreak in 2000 was caused by unpasteurized orange juice and resulted in 88 illnesses in six western states. The *Salmonella* Muenchen outbreak in 1999 was caused by unpasteurized orange juice and resulted in 423 illnesses in 20 states and three Canadian provinces, and one death (HHS 2001).

The juice HACCP regulation applies to juice products in both interstate and intrastate commerce. Juice processors are required to evaluate their manufacturing process to determine whether there are any microbiological, chemical, or physical hazards that could contaminate their products. If a potential hazard is identified, processors are required to implement control measures to prevent, reduce, or eliminate those hazards. Processors are also required to use processes that achieve a 5-log, or 100,000-fold, reduction in the number of the most resistant pathogen in their finished products, compared to numbers that may be present in untreated juice. Juice processors may use microbial reduction methods other than pasteurization, including approved alternative technologies, or a combination of techniques. Citrus processors may opt to apply the 5-log pathogen reduction on the surface of the fruit, in combination with microbial testing, to ensure that this process is effective. Processors making shelf-stable juices or concentrates that use a single thermal processing step are exempt from the microbial hazard requirements of the HACCP regulation. Retail establishments where packaged juice is made and only sold directly to consumers (such as juice bars) are not required to comply with this regulation (HHS 2001).

Establishments covered by the juice HACCP regulation are still subject to the current good manufacturing practices (CGMP) regulations. In fact, CGMPs are an essential foundation for a successful HACCP system. Importers of juice must either ensure that all juice entering into the United States has been processed in compliance with Part 120, or import such juice from a country that has an appropriate memorandum of understanding with the United States. In addition, importers must maintain records that document the performance and results of the steps taken to demonstrate compliance with the regulation.

2.2.3 Pasteurized Milk Ordinance

To assist states and municipalities in initiating and maintaining effective programs for the prevention of milkborne disease, the U.S. Public Health Service (USPHS) developed a model regulation in 1924 known as the *Standard Milk Ordinance* for voluntary adoption by state and local milk control agencies. To provide for the uniform interpretation of this ordinance, an accompanying code was published in 1927, which provided administrative and technical details to aid in compliance. This model milk regulation, now entitled the *Grade "A" Pasteurized Milk Ordinance* (Grade "A" PMO), incorporates the provisions governing the processing, packaging, and sale of Grade "A" milk and milk products, including buttermilk and buttermilk products, whey and whey products, and condensed and dry milk products. The USPHS alone did not produce the Grade "A" PMO. As with preceding editions, it was developed with the assistance of milk regulatory and rating agencies at every level of federal, state, and local government, including both health and agriculture departments; all segments of the dairy industry, including producers, milk plant operators, equipment manufacturers, and associations; many educational and research institutions; and with helpful comments from many individual sanitarians and others (Box 3).

The FDA provides oversight for the processing of raw milk into pasteurized milk, cottage cheese, yogurt, and sour cream under the National Conference on Interstate Milk Shipments Grade "A" milk program. Under this program, state personnel conduct inspections and assign ratings, and FDA regional milk specialists audit these ratings. The inspection program starts with the dairy farm and continues through the processing and packaging of products at milk plants. Products that pass inspection may be labeled "Grade A." This program includes pasteurized milk from cows, goats, sheep, and horses. Raw milk and raw milk cheeses cannot be labeled Grade A because they are not pasteurized and are not covered by the program (FDA 2004).

> **Box 3** Grade "A" Pasteurized Milk Ordinance
>
> The recommended Grade "A" PMO by the USPHS/FDA is the basic standard used in the voluntary Cooperative State–USPHS/FDA Program for the Certification of Interstate Milk Shippers (IMS). All 50 states, the District of Columbia, and U.S. Trust territories participate in the IMS program. The National Conference on Interstate Milk Shipments (NCIMS), in accordance with the Memorandum of Understanding with the FDA, has recommended changes and modifications to the Grade "A" PMO. These changes have been incorporated into the 2007 revision. The Grade "A" PMO is an ordinance to regulate the production, transportation, processing, handling, sampling, examination, labeling, and sale of Grade "A" milk and milk products; the inspection of dairy farms, milk plants, receiving stations, transfer stations, milk tank truck cleaning facilities, milk tank trucks and bulk milk hauler/samplers; the issuing and revocation of permits to milk producers, bulk milk hauler/samplers, milk tank trucks, milk transportation companies, milk plants, receiving stations, transfer stations, milk tank truck cleaning facilities, haulers, and distributors; and the fixing of penalties. The Grade "A" PMO has been widely adopted and used for many years and has been upheld by court actions. When it is adopted locally, its enforcement becomes a function of the local or state authorities (FDA 2007a).

It is a violation of federal law to sell raw milk packaged for consumer use across state lines. However, each state regulates the sale of raw milk within the state, and some states allow it to be sold. Therefore, in some states dairy operations may sell raw milk to local retail food stores, to consumers directly from the farm, at agricultural fairs, or at other community events depending on the state law. In states that prohibit intrastate sales of raw milk, some people have tried to circumvent the law by "cow sharing," or "cow leasing." They pay a fee to a farmer to lease or purchase part of a cow in exchange for raw milk, claiming that they are not actually buying the milk since they are part-owners of the cow. For more details on the hazards associated with raw milk, refer to Chap. 9.

3 Food Safety and Inspection Service

President Lincoln founded the U.S. Department of Agriculture in 1862. Following the U.S. Civil War, westward expansion and the development of refrigerated railroad cars spurred the growth of not only the livestock industry, but also meat packing and international trade. In response to the growing pressure from veterinarians, ranchers, and meat packers for a unified effort to eradicate livestock diseases in the United States, the USDA's Bureau of Animal Industry (BAI) was created in 1884; it was effectively the true forerunner of the FSIS. The BAI's function was to focus on preventing diseased animals from being used as food. The BAI gained further responsibility to enforce the newly approved meat inspection act to ensure that salted pork and bacon intended for export were safe. In 1891, the act was amended to cover the inspection and certification of all live cattle for export and live cattle that were to be slaughtered and their meat exported (FSIS 2006a). However, in 1905, the BAI faced its first challenge with the publication of Upton Sinclair's *The Jungle* (Sinclair 1906). This book exposed insanitary conditions in the meat-packing industry in Chicago, igniting public outrage, which eventually led to the establishment of continuous governmental inspection. President Theodore Roosevelt commissioned the Neill–Reynolds report, which confirmed many of Sinclair's horrid tales.

3.1 The Federal Meat Inspection Act

In response to both *The Jungle* and the Neill–Reynolds report, Congress passed the Federal Meat Inspection Act (FMIA) in June 1906. The FMIA required the mandatory inspection of livestock before slaughter as well as the mandatory postmortem inspection of every carcass and set explicit sanitary standards for slaughterhouses. Finally, the FMIA allowed the USDA to issue grants of inspection and monitor slaughter and processing operations, enabling the USDA to enforce food safety regulatory requirements.

Following World War II, the meat processing industry changed significantly. The rapid growth of the federal railroad system and the development of refrigerated trucks allowed packing houses to move away from expensive urban areas. Competition in the meatpacking business led to the building of sophisticated, mechanized plants in less expensive rural areas. During the 1950s and 1960s, inspection increasingly focused on wholesomeness and visible contamination. The prevalence of animal disease as a food safety problem was decreasing. However, there was an increase in the kinds of products, the complexity of operations, and the volume of processed products produced, which resulted in increased concerns about mislabeling and economic adulteration.

In 1958, in response to the public's concern about invisible hazards from chemicals added directly or indirectly to foods, the FDCA (Food, Drug and Cosmetic Act) was amended with the 1958 Food Additive Amendment to deal with the safety of ingredients when used in processed foods, including animal drug residues in meat and poultry products (FSIS 2007a).

3.2 Poultry Products Inspection Act

Prior to World War II, poultry was purchased from small local farms. As the consumer market developed and dressed poultry became more available, the need for regulation became more evident. The Poultry Product Inspection Act (PPIA) was passed by Congress in 1957 and established the mandatory inspection of poultry products. This act was modeled after the FMIA. The PPIA defined "poultry" as any live or slaughtered domesticated birds, such as chickens, turkeys, ducks, geese, or guineas. Game birds were not covered by the PPIA. The PPIA required a variety of types of mandatory inspection for poultry and poultry products in interstate and foreign commerce. This inspection include the inspection of (1) birds prior to slaughter (antemortem inspection), (2) each carcass after slaughter and before processing, (3) plant facilities to ensure sanitary conditions, (4) all slaughtering and processing operations, (5) imported poultry products at the point of entry, and (6) the verification of the truthfulness and accuracy of product labeling (NCR 1987).

3.3 Wholesome Meat Act of 1967 and the Wholesome Poultry Products Act of 1968

The Wholesome Meat Act of 1967 and the Wholesome Poultry Products Act of 1968 amended the FMIA and the PPIA in an attempt to ensure uniformity in the regulation of meat and poultry products. These acts provide the basis for the current meat and poultry inspection system. The attempt to ensure uniformity extended the federal standards addressed in these acts to intrastate operations and provided for the state–federal cooperative inspection programs, which required state programs to be "at least equal to" the federal system. Meat and poultry processing plants had the option of applying for state or federal inspection. However, operating under state inspection limited the distribution of products to only intrastate commerce (Box 4).

Food Regulation in the United States

> **Box 4** State Meat & Poultry Inspection PRograms
>
> Most of the nation's meat and poultry and all processed egg products are produced in approximately 6,200 plants that are under federal inspection. However, 27 states have also established meat and poultry inspection programs for products produced and sold in-state (FSIS 2011). State meat and poultry inspection (MPI) programs are an integral part of the nation's food safety system. About 2,100 meat and poultry establishments are inspected under state MPI programs. All of these establishments are small or very small. State MPI programs provide more personalized guidance to establishments in developing their food safety–oriented operations, and inspection requirements are waived for meat from custom slaughter facilities, poultry from small farms, and products processed in retail stores and restaurants. But these products must remain safe for human consumption. The FSIS provides up to 50% of the state's operating funds, as well as training and other assistance. Comprehensive reviews of state MPI programs are performed on a cycle of every 1–5 years, depending on review findings. Comprehensive reviews currently include the review of state performance plan and in-plant review directed to statewide inspection system assessment, which include (1) HACCP plan and implementation, (2) sanitation standard operating procedures and implementation, (3) findings of state *Salmonella* testing programs, (4) findings of *E. coli* in-plant testing program, (5) labeling, and (6) inspection procedures and the review of compliance programs, laboratories, resource management, budget and finance, and civil rights (FSIS 2008).

3.4 Meat and Poultry HACCP

A major impetus for risk-based inspection began in the mid-1980s through the 1990s, with studies conducted by the National Academy of Sciences of the then-General Accounting Office and FSIS to fundamentally change the meat and poultry inspection program (NRC 1987). The modernization of the inspection system focused on improving food safety through a reduction in foodborne illness caused by biological agents, mainly pathogenic bacteria such as *Salmonella, Campylobacter,* and *Listeria monocytogenes* (FSIS 2007b). In 1998, slaughter and processing establishments were required to adopt the HACCP system; to verify that the systems effectively reduce contamination with pathogenic bacteria, establishments have to meet the performance standards for *Salmonella* set up by FSIS. To verify that process control systems work as intended to prevent fecal contamination, slaughter establishments are required to test for generic *E. coli*. FSIS also required plants to adopt and follow written standard operating procedures for sanitation to reduce the likelihood that harmful bacteria would contaminate the finished product. This combination of HACCP-based process control, microbial testing, pathogen-reduction performance standards, and sanitation standard operating procedures would significantly reduce the contamination of meat and poultry with pathogenic bacteria and reduce the risk of foodborne illness.

The new approach to food safety requires all plants to develop, adopt, and implement a HACCP plan when hazards likely to occur are identified in their process. Plants are expected to reassess their HACCP plans annually or when any changes are made that impact the original hazard analysis. The purpose of the *Salmonella* performance standards is to provide incentives for producers of raw meat and poultry to reduce the prevalence of *Salmonella* on their products and to provide a

> **Box 5** Listeria Regulations for Ready to Eat Products
>
> The establishment is not required to comply with regulations if the RTE products produced in the establishment are not exposed postlethality to the environment. Examples of products that are **not** included are products that are fully cooked in a cook-in-bag that leave the official establishment in the intact cooking bag; products receiving a lethality treatment and hot-filled as long as the lethality temperature and sanitary handling are maintained during the period of time in which the product moves from the point of lethality to the point of packaging; or thermally processed, commercially sterile (canned) products (FSIS 2007c).

tangible measure to judge the effectiveness of HACCP plans. The FSIS uses the results of the *Salmonella* tests, record review, and direct observation as a part of the meat and poultry inspection and enforcement process.

3.5 *Listeria Regulations*

Testing of both raw and processed products is a major component of the HACCP. Under the Federal Meat Inspection Act and the Poultry Products Inspection Act, ready-to-eat (RTE) products are adulterated if they contain *L. monocytogenes* or if they come into direct contact with a surface that is contaminated with this pathogen. Based on inspection and the results of testing, the FSIS has successfully implemented major initiatives to control *Listeria*. The FSIS begin testing RTE products for *L. monocytogenes* in 1987 and published an interim final rule, "Control of *Listeria monocytogenes* in Ready-to-Eat Meat and Poultry Products," which became effective on October 6, 2003. Under this rule, establishments producing RTE products must address *L. monocytogenes* through a written program such as their HACCP systems, sanitation standard operating procedures, or other prerequisite programs. Establishments were also required to verify the effectiveness of these actions through testing and to share the results with the FSIS. Meanwhile, the FSIS continues to conduct its own random testing to verify each establishment's control program.

Establishments are required to choose one of three alternatives to control for *L. monocytogenes* (Box 5). The alternatives are as follows (FSIS 2003):

- *Alternative 1*. Employ both a postlethality treatment and a growth inhibitor for *L. monocytogenes* on RTE products. Establishments opting for this alternative are subject to FSIS verification activity that focuses on the effectiveness of postlethality treatments. Sanitation is important but is built into the degree of lethality necessary for safety.
- *Alternative 2*. Employ either a postlethality treatment or a growth inhibitor for the pathogen on RTE products. Establishments opting for this alternative are subject to more frequent FSIS verification activity than those in Alternative 1.
- *Alternative 3*. Employ sanitation measures only. Establishments opting for this alternative are targeted with the most frequent level of FSIS verification activity. Within this alternative, the FSIS places increased scrutiny on operations that produce hot dogs and deli meats, which were identified as high-risk products for *L. monocytogenes* by the FSIS and the FDA.

3.6 Egg Products Inspection Act

Congress passed the Egg Products Inspection Act (EPIA) in 1970, which provided for the mandatory continuous inspection of the processing of liquid, frozen, and dried egg products. For the next 25 years, the Poultry Division of the USDA's Agricultural Marketing Service inspected egg products to ensure they were wholesome, not adulterated, and properly labeled and packaged to protect the health and welfare of consumers. In 1995, the FSIS became responsible for the inspection of all egg products with the exception of those products exempted under the EPIA, which are used by food manufacturers, foodservice institutions, and retail markets. Officially inspected egg products bear the USDA inspection mark. The EPIA does not cover shell eggs, which are covered under the Agricultural Marketing Act of 1946.

The law requires that all egg products distributed for consumption be pasteurized. Certain commodities are not presently considered egg products and are exempt from this law. Freeze-dried products, egg substitutes, imitation eggs, and similar products that are exempted from continuous inspection under the EPIA (FSIS 2006b) are under the jurisdiction of the FDA. Inspected, pasteurized egg products are used to make these commodities, and companies may elect to repasteurize these products following formulation and before packaging (FSIS 2006b).

With the implementation of the Nutrition Labeling and Education Act in 1994, egg products sold at retail are also required to bear nutrition labeling. The "Nutrition Facts" panel describes the nutrient composition of products per serving and their contribution to the overall diet. In addition to nutrition information on consumer packages, all egg products must be labeled with (FSIS 2006b):

- The common or usual name and if the product is comprised of two or more ingredients, the ingredients listed in the order of descending proportions.
- The name and address of the packer or distributor.
- The date of pack, which may be shown as a lot number or production code number.
- The net contents.
- The official USDA inspection mark and establishment number.

The basic instructions that FSIS inspection program personnel have used to perform their inspection duties (related to EPIA) are in the *Egg Products Inspector's Handbook*, which was originally developed by the Agricultural Marketing Service. The FSIS has clarified many of these policies and has issued further instructions in FSIS notices and directives. The *Egg Products Inspector's Handbook* is not available in electronic form, and it does not follow the agency's format for issuing instructions to inspection program personnel. As a result, inspection program personnel may question the accuracy of the instructions and the policies contained in the manual. Because the FSIS recognizes the instructions contained in this document as policy, there is a need to update the information contained in the document. Therefore, the FSIS plans to issue FSIS notices and directives that convey, clarify, and update the information contained in the *Egg Products Inspector's Handbook*. The FSIS intends to expedite this process by issuing directives and notices.

3.7 Imported Products

Imported meat and poultry products must undergo inspection under systems that are equivalent to those of the USDA. The Foreign Programs Division of International Programs conducts system reviews to evaluate that the laws, policies, and operations of the inspection system in the foreign country are equivalent to those in the United States. Imported meat, poultry and egg products are not subject to the FDA's prior notice requirements unless they are being imported for use in animal feed. Products are required to be presented to the FSIS for port-of-entry reinspection. In addition, all meat brokers, poultry products brokers, official establishments, and carriers and importers of poultry or livestock carcasses are required to keep business records and make them available to FSIS employees upon request.

4 Where to Find Laws and Regulations

The *U.S. Code* is the codification by subject matter of the general and permanent laws of the United States based on what is printed in the *Statutes at Large*. It is divided by broad subjects into 50 titles and published by the Office of the Law Revision Counsel of the U.S. House of Representatives. Of the 50 titles, only 23 have been enacted into statutory law. The *General Index* contains an alphabetical listing of useful subject headings. Entries are also listed under agency names. The *U.S. Code* citation is given for each entry. The *Statutes at Large* contain all the public laws currently in effect and their corresponding citations. Statutes are cited in the *U.S. Code* as "Title U.S.C. Section (subsection)," for example, "21 U.S.C. Sec. 301 (*a*)." The *U.S. Code* supplement is issued during each of the years between printings of the *U.S. Code*. The *U.S. Code* does not include regulations used by executive branch, agencies, decisions on the federal courts, treaties, or laws enacted by state or local governments. The *U.S. Code* can be searched on the Internet, via the Government Printing Office website, which has a link to the "Federal Digital System" (http://www.gpo.gov/fdsys/search/home.action). A search can be done by subject or by citation (Box 6).

The *Code of Federal Regulations* (*CFR*) is the codification of the general and permanent rules published in the *Federal Register* by the executive departments and agencies of the federal government. It is divided into 50 titles that represent broad subject areas. Each title is divided into chapters, which usually bear the name of the issuing agency. Each chapter is further subdivided into parts that cover specific regulatory areas, and large parts may be subdivided into subparts. All parts are organized into sections, and most citations in the *CFR* are provided at the section level (Box 7). For example, current good manufacturing practices (CGMP) are described under 21 CFR Part 110. The laws related to food are contained in Titles 7 (Agriculture), 9 (Animal and Animal Products), and 21 (Food and Drug). Each volume of the *CFR* is updated once every calendar year and is issued on a quarterly basis. Titles 1–16 are updated as of January 1; Titles 17–27 are updated as of April 1; Titles 28–41 are updated as of July 1; and Titles 42–50 are updated as of October 1. The CFR volumes are added to the GPO access website concurrently with the release of paper editions.

Box 6 How to search for a government law and/or regulation

The Federal Digital System located on the GPO website is the easiest way to find the most current information about a topic. For example, searching "state meat inspection regulations" on the FDS site will produce links to information in the *Federal Register* as well as the *Code of Federal Regulations* (CFR). However, another way to locate the same information is via the FSIS website (http://www.fsis.usda.gov/regulations_&_policies/Federal_Meat_Inspection_Act/index.asp). In January 2008, the FSIS published the *FSIS Review of State Meat and Poultry Inspection Programs*, which summarizes the review results from the Food Safety and Inspection Service's comprehensive reviews of the 27 states that currently operate state meat and poultry inspection (MPI) programs. A copy of this report can be found on the FSIS website at http://www.fsis.usda.gov/PDF/Review_of_State_Programs.pdf. In July 2008, the FSIS published *"At Least Equal to" Guidelines for State Meat and Poultry Cooperative Inspection Programs*, which provides information to state cooperative inspection programs on the criteria that the FSIS uses to make its annual determination of whether state meat and poultry inspection (MPI) programs are "at least equal" to the federal inspection program. A copy of these guidelines can be found on the FSIS website at http://www.fsis.usda.gov/PDF/At_Least_Equal_to_Guidelines.pdf.

> **Box 7** Federal Register
>
> The Federal Register is an important document from a legal and historical perspective. Yet few people read it regularly due to its large volume and dry style of content. The Federal Register has been available online since 1994 and federal libraries within the United States receive copies of the text. Any agency proposing a rule in the Federal Register must provide contact information for people and organizations interested in making comments to the agencies, which in turn are required to give due diligence to the comments received when it publishes its final rule. As part of the Federal E-Government e-Rulemaking Initiative, the web site http://regulations.gov was established to enable easy public access to the Federal Register and the Federal Docket Management System (FDMS). Thus, the public can access the entire rulemaking dockets and submit online comments to those responsible for drafting the rulemakings.

5 Summary

The two major food regulatory agencies in the United States are the Department of Health and Human Services' Food and Drug Administration (FDA) and the U.S. Department of Agriculture's Food Safety and Inspection Service (FSIS). The responsibilities of the FDA include all foods except meat, poultry, and egg products. The FDA publishes the *Food Code*, a model that assists food control jurisdictions at all levels of government by providing them with a scientifically sound technical and legal basis for regulating the retail and foodservice segment of the industry. The responsibilities of the FSIS include meat, poultry, and egg products. A major impetus for risk-based inspection began in the mid-1980s, and in 1998 the FSIS established requirements for all meat and poultry plants to adopt the hazard analysis and critical control point (HACCP) system, which is a system of process controls to prevent food safety hazards. The FDA has adopted HACCP regulations for seafood and juices, and the intention is to use this system for much of the U.S. food supply.

References

Curtis, P.A. 2005. *Food laws and regulations*. Ames: Blackwell.
FDA. 2001. HACCP: A state of the art approach to food safety. FDA Backgrounder. http://www.cfsan.fda.gov/~lrd/bghaccp.html. Accessed 5 Dec 2010.
FDA. 2004. Got milk? Make sure it's pasteurized. FDA Consumer Magazine. September–October 2004. http://www.foodsafety.wisc.edu/assets/pdf_Files/milk_pasteurized.pdf. Accessed 5 Dec 2010.
FDA. 2007a. Grade A Pasteurized Milk Ordinance, 2007 Revision. http://www.michigan.gov/documents/mda/MDA_DP_07PMOFinal_251324_7.pdf. Accessed 5 Dec 2010.
FDA. 2008. FDA's evaluation of the seafood HACCP program for fiscal years 2004/2005. http://www.fda.gov/Food/FoodSafety/Product-SpecificInformation/Seafood/SeafoodHACCP/ucm111059.htm. Accessed 28 Feb 2011.
FDA. 2009a. FDA history website. http://www.fda.gov/oc/history/default.htm. Accessed 5 Dec 2010.
FDA. 2009b. History of the FDA, drugs and foods under the 1938 Act and its amendments. http://www.fda.gov/AboutFDA/WhatWeDo/History/Origin/ucm055118.htm. Accessed 28 Feb 2011.
FDA. 2011a. Real progress in Food Code adoptions. http://www.fda.gov/food/foodsafety/retailfoodprotection/federalstatecooperativeprograms/ucm108156.htm. Accessed 28 Feb 2011.
FDA. 2011b. Food Code. http://www.fda.gov/Food/FoodSafety/RetailFoodProtection/FoodCode/FoodCode2009/default.htm. Accessed 28 Feb 2011.
FSIS. 2003. FSIS strengthens regulations to reduce *Listeria monocytogenes* in ready-to-eat meat and poultry products. News release. June 4, 2003. http://www.fsis.usda.gov/oa/news/2003/lmfinal.htm. Accessed 5 Dec 2010.

FSIS. 2006a. About FSIS, celebrating 100 years of FMIA. http://www.fsis.usda.gov/About_FSIS/100_Years_FMIA/index.asp. Accessed 5 Dec 2010.

FSIS. 2006b. Fact sheets: Egg products preparation, egg products and food safety. http://www.fsis.usda.gov/Fact_Sheets/Egg_Products_and_Food_Safety/index.asp. Accessed 5 Dec 2010.

FSIS. 2007a. About FSIS, agency history. http://www.fsis.usda.gov/About_FSIS/Agency_History/index.asp. Accessed 5 Dec 2010.

FSIS. 2007b. The evolution of risk-based inspection. http://www.fsis.usda.gov/PDF/Evolution_of_RBI_022007.pdf. Accessed 5 Dec 2010.

FSIS. 2007c. Summary of *Listeria monocytogenes* compliance guideline for small and very small meat and poultry plants that produce ready-to-eat products. http://www.fsis.usda.gov/PDF/LM_Guidelines_for_SVSP_Ready_to_Eat_Products.pdf. Accessed 5 Dec 2010.

FSIS. 2008. State inspection programs. http://www.fsis.usda.gov/regulations_&_policies/state_inspection_programs/index.asp. Accessed 5 Dec 2010.

FSIS. 2011. List of states currently operating meat and poultry inspection programs. http://www.fsis.usda.gov/regulations_&_policies/Listing_of_Participating_States/index.asp. Accessed 28 Feb 2011.

HHS. 2001. FDA publishes final rule to increase juice safety of fruit and vegetable juices. HHS News. January 18.

McDermott, C. 2006. The food and drug administration celebrates 100 years of service to the nation. FDA News P06-02, January 4. http://www.fda.gov/bbs/topics/NEWS/2006/NEW01292.html. Accessed 5 Dec 2010.

Meadows, M. 2006. A century of ensuring safe foods and cosmetics. FDA Consumer Magazine. January–February 2006. http://www.fda.gov/AboutFDA/WhatWeDo/History/FOrgsHistory/CFSAN/ucm083863.htm. Assessed 28 Feb 2011.

NRC. 1987. *Poultry Inspection: The basis for a risk-assessment approach. Committee on public health risk assessment of poultry inspection programs*. Washington, DC: National Academy Press.

Pina, K.R., and W.L. Pines. 2002. *A practical guide to food and drug law and regulations*, 2nd ed. Washington, DC: Food and Drug Law Institute.

Sinclair, U. 1906. The Jungle. Prepared and Published by E-BooksDirectory.com.

Role of Different Regulatory Agencies in the United States

Craig Henry

1 Introduction

Ensuring food safety is the joint responsibility of the government (federal, state, and local); the food industry (production agriculture; food processors; food ingredient, equipment, and packaging manufacturers; food transporters; warehouse operators; food wholesalers; and food retailers and restaurants); and the public (food storage, handling, and preparation by the consumer). Authorities at the federal, state, and local levels of government have integral roles in regulating food and food processing facilities. The U.S. federal government has defined the following principles in support of the food safety system: (1) Only safe and wholesome foods may be marketed; (2) regulatory decision making in food safety is based on science; (3) the government has the responsibility to enforce regulations; (4) manufacturers, distributors, importers, and others are expected to comply and are liable if they do not; and (5) the regulatory process is transparent and accessible to the public (Anonymous 2000).

For the successful regulation of foods, three basic elements are required. First, a system of laws and regulations enforced through science-based standards must be in place so that foods and their packages can be judged safe. It is important that food be appropriately labeled so that consumers are not misled. Second, an organized infrastructure (industry and government) should be supported with sufficient funds to inspect the food before and after it enters commerce. Third, an effective communication program must be available to keep all stakeholders properly advised of current issues, such as regulatory status, new technologies, risk assessment, identification of foodborne illness outbreaks, and traceability of food products.

The importance of maintaining a strong food safety program cannot be overemphasized. Although many improvements have been made in the last century, the Centers for Disease Control and Prevention (CDC) estimates there are 47.8 million foodborne illnesses, 127,839 hospitalizations, and 3,037 deaths per year in the United States (Morris 2011). More specifically, they estimate that each year 31 major pathogens acquired in the United States cause 9.4 million episodes of foodborne illness, 55,961 hospitalizations, and 1,351 deaths (Scallan et al. 2011). Therefore, despite the checks and balances incorporated in our current food safety system, the public's confidence can be significantly affected by adverse events, such as the recall of potentially hazardous foods.

C. Henry (✉)
Grocery Manufacturers Association (GMA), Washington, DC 20001, USA
e-mail: cwhenry@deloitte.com

2 Managing a Global Food Supply Chain

It is paramount that federal, state, and local authorities work together because no single governing body has sufficient funding or manpower to execute the food safety roles assigned by the laws and regulations overseeing the food manufacturing facilities in the United States or in countries that export to the United States. At the international level, food safety professionals provide some of the science that assists U.S. regulators through international forums, such as the Codex Alimentarius Commission, the World Health Organization, the Food and Agriculture Organization of the United Nations, and the International Office for Epizootics. This proactive effort provides guidance to regulators around the world as they confront the huge challenge of minimizing risks and hazards to the global community.

Today, consumers can enjoy products from a global, highly complex food industry. Therefore, food manufacturers are totally dependent upon building and maintaining consumer confidence in the products they manufacture. Thus, the interactions among ingredient vendors, food contact packaging providers, re-packers, co-manufacturers, brokers, and other suppliers are critical to effective food management throughout the supply chain, or from "farm to fork."

It is the food manufacturer's responsibility to prevent contaminated or adulterated food from reaching consumers. Should adulterated products reach the market, the consumer will hold not only the retailer and the manufacturer responsible, but also the government entity that oversaw the product's manufacture, and in the case of imported products, the government of the exporting country. There is no better example of this than the multiple challenges presented to the U.S. and Chinese governments in 2007 and 2008 due to the intentional adulteration of wheat gluten and milk formula with melamine – an industrial compound used in the manufacturing of plastic that is not approved for use in foods. Melamine was used to illegally increase the nitrogen content and the subsequent market value of these two protein-based products. Melamine-adulterated pet foods killed numerous animals, and melamine-adulterated infant formula killed a number of young children and made thousands more ill. As a result, the federal government took action to prevent the entrance of such products into the United States (Anonymous 2008). By issuing an import alert, all milk products, milk-derived ingredients, and finished food products containing milk from China were detained at the ports of entry and were not released until test results demonstrated that they were not adulterated.

The melamine problem in China also reminded us that consumers react very quickly to the media coverage of tainted foods, especially foods intended for children. An example of this reaction was the abrupt decline in the consumption of milk from China after the article "At Least Nine Countries Ban Chinese Dairy Imports After Tainted Milk Kills Four Babies" was published by the Associated Press on Tuesday, September 23, 2008. Therefore, it is in the best interest of all stakeholders to implement programs that ensure the production of safe foods. This includes suppliers capable of providing safe ingredients, food contact packaging materials, and finished products that protect not only their brand but also, and more importantly, the consumer and the market.

Manufacturers must select business partners that are capable of supplying safe foods. The legal and financial implications for the supplier of food-related products are huge, and many companies have been put out of business after causing serious health consequences due to adulterated products. Too often a company loses sight of safety and quality in favor of "cheaper" ingredients and finished goods in hopes of increasing their profit margins. "Topps Meat Co. Shutters Business 6 Days After Second-Largest U.S. Meat Recall" was reported by Fox News and the Associated Press on Friday, Oct. 5, 2007. Topps' commercialized meat products were unfortunately adulterated with *E. coli* O157:H7, which resulted in 30 related cases of foodborne illness across seven states and over 21 million pounds of product being recalled.

Currently, there are five key federal regulatory agencies responsible for protecting the U.S. food supply and its consumers. They are the Department of Health and Human Services' (DHHS) Food and Drug Administration (FDA), the U.S. Department of Agriculture's (USDA) Food Safety and Inspection Service (FSIS) and Animal and Plant Health Inspection Service (APHIS), the Environmental Protection Agency (EPA), and the newest, the Department of Homeland Security (DHS), which was established as a result of the September 11, 2001, terrorist attacks on the United States.

3 Federal Agencies Regulating Food Safety and Key Statutes and Regulations

Before we can truly address the food regulatory process, it is important to understand the link between laws and regulations. The various regulatory branches of the U.S. government work with Congress to develop statutes (laws) and enact them through regulations that, when final, are codified in the Code of Federal Regulations (CFR) and are published in the *Federal Register*. Throughout this process, public comments are solicited. Typically, after a review of the comments received, the regulatory agency drafts a final rule, which is also published in the *Federal Register* and is eventually recorded or "codified" in the CFR.

Table 1 lists a number of key statutes and regulations that directly impact food manufacturing in the United States as well as the manufacturing of foreign products intended for export to the United States. For more details, refer to chapter "Food Regulation in the United States."

3.1 The Food and Drug Administration (FDA)

In 1906, Congress created the Bureau of Chemistry within the USDA. This was done to provide oversight for the newly enacted Food and Drugs Act. Since then, what is now known as the FDA lies with the DHHS and is completely separate from the USDA. Today, the FDA's inspectors and scientists ensure the safety of the vast majority of foods consumed in the United States. This agency is also responsible for appropriate product labeling that must be truthful and not misleading to the consumer. With the volume of imports expanding at an ever-increasing rate, the number of shipments of foreign-produced regulated products has grown from about 1.5 million in 1992 to over six million today. FDA inspectors must strategically allocate their resources through risk-based inspection to meet this challenge. Currently, the FDA is also responsible for the effectiveness of medical drugs and monitors the accuracy of labels and label-related advertising.

The Federal Food, Drug, and Cosmetic Act (FFDCA) provides authority for the FDA to regulate production and manufacturing of all foods (except for meat and poultry) in the United States (Anonymous 2004a). The FDA sets standards of the identity, labeling, quality, and fill of containers for food; approves food additives and colors before they can be used in food manufacturing or marketed as ingredients; oversees the components of packaging that come in contact with food; and has the authority to prohibit the interstate transport of adulterated food and false or misleading labeling of products.

3.1.1 Office of Food

Under the Obama Administration, FDA Commissioner Margaret Hamburg established the Office of Foods on August 18, 2009 (Anonymous 2011a). This may be considered the first real attempt to capture the key regulatory functions associated with the oversight of food under the FDA under

Table 1 Current laws governing the inspection and labeling of food products in the United States

Agency	Laws
Food and Drug Administration – Food Safety and Labeling	Compliance Policy Guides and Defect Action Levels
	Federal Food, Drug, and Cosmetic Act
	21 CFR[a] Parts 1–99 (Recalls, Approved Colors)
	21 CFR Parts 100–169 (Labeling Requirements, Standards of Identity, GMPs, HACCP)
	21 CFR Parts 170–199 (Food Additives, Additives Generally Recognized as Safe)
	Infant Formula Act
	Import Alerts
U.S. Department of Agriculture – Food Safety and Labeling	9 CFR Parts 1–199 (Animals and Animal Products, Slaughter Operations)
	9 CFR Parts 200–end, Further Processed Products, HACCP
	Agricultural Marketing Act of 1946, amended by the Farm Security and Rural Investment Act of 2002 and the Food Conservation and Energy Act of 2008
	7 CFR, Voluntary Grade Standards for food products Mandatory Country of Origin Labeling for beef, pork, lamb, chicken, goat meat, fish, shellfish, perishable agricultural commodities, macadamia nuts, peanuts, pecans, ginseng
	7 CFR, Animal and Plant Health Inspection Services
	Child Nutrition labeling
	7 CFR §800–899
	Federal Meat Inspection Act
	Egg Products Inspection Act
	Labeling Policy Book
	Poultry Products Inspection Act
Department of Homeland Security – Customs and Border Protection	Bioterrorism Act of 2002
	19 CFR Part 134 (Customs Duties), Country of Origin marking
	6 CFR, Domestic Security
	42 CFR, Public Health
Department of Transportation	Sanitary Food Transportation Act
	49 CFR, Transportation
	Shipping 46 CFR
Department of Environmental Protection (EPA)	Safe Water Drinking Act
	Protection of the Environment 40 CFR

[a]CFR = Code of Federal Regulations. Individual volumes of the CFR may be obtained from the Superintendent of Documents, U.S. Government Printing Office, Washington, DC 20402 (http://www.access.gpo.gov/nara/cfr/cfr-table-search.html#page1)

focused leadership. Commissioner Hamburg describes the organization under the Office of Foods as well as the purpose of the FDA Foods Program as follows:

- The FDA Foods Program includes three major operating units – the Center for Food Safety and Applied Nutrition (CFSAN), the Center for Veterinary Medicine (CVM), and the foods-related activities of the Office of Regulatory Affairs (ORA) – and draws on the resources and expertise of the FDA's National Center for Toxicological Research and key Office of Commissioner staff offices.
- The new Office of Foods is responsible, on behalf of the Commissioner, for providing all elements of the FDA's Foods Program leadership, guidance, and support to achieve the agency's public health goals. The office is also the focal point for planning implementation of the recommendations of the President's Food Safety Working Group and the new food safety authorities being considered by Congress.

The mission of the FDA Foods Program is to protect and promote the health of humans and animals by

- Ensuring the safety of foods for humans, including dietary supplements
- Ensuring the safety of animal feed and the safety and effectiveness of animal drugs
- Setting science-based standards for preventing foodborne illness and ensuring compliance with these standards
- Protecting the food and feed supply from intentional contamination
- Ensuring that food labels contain reliable information consumers can use to choose healthy diets

The FDA is charged with regulation of the manufacture and distribution of food additives and drugs for animals, including animals from which human food is derived (meat and milk). This is administered by the FDA's Center for Veterinary Medicine (CVM). The Infant Formula Act of 1980 gives the FDA clear oversight of the nutrient requirements, quality control, record keeping, reporting, and recalls of adulterated products from the marketplace. Infant formula is the only product subject to the mandatory recall authority of federal regulators. Under the authority of this act, the FDA has direct access to all manufacturing records and test results. In December 2010, Congress passed the new Food Safety Modernization Act (H.R. 2751) and President Obama signed it into law in January 2011 (Anonymous 2011b). This piece of legislation is considered by many as the most significant regulatory reform in over three decades. It now provides both the USDA and the FDA with mandatory recall authority for all other foods in the event a manufacturer refuses to execute a Class 1 recall.

The FDA executes a Class 1 recall whenever dangerous or defective products that predictably could cause serious health problems or death are found. Examples of products that could fall into this category are a food found to contain the toxin produced by *Clostridium botulinum* or foods with undeclared allergens (Anonymous 2010). The FSIS has a similar definition for meat, poultry, and egg products amenable to its jurisdiction (Anonymous 1998, 2006).

In the event that a food becomes adulterated and must be taken away before reaching consumers, the manufacturing company will conduct a voluntary recall to remove the product from the marketplace. The recall is executed in coordination with the FDA and/or USDA, depending upon which agency has inspectional jurisdiction on the compromised food. Most state agencies actually provide the vast amount of manpower to go into retail outlets and conduct evaluations to determine how quickly and effectively the recalled product has been removed from store shelves. This process is called a *recall effectiveness check*.

The Public Health Service Act allows the FDA to execute the Grade A Pasteurized Milk Ordinance via a Memorandum of Understanding in cooperation with the states through the National Conference on Interstate Milk Shipments. The focus of the ordinance is to prevent the interstate shipment of adulterated milk products, and it includes restaurants and retail market operations. The FDA also regulates fishery products, except farm-raised catfish, which is regulated by the FSIS as of 2008 under the FFDCA. Under 21CFR 123 Fish and Fishery Products (Seafood Hazard Analysis Critical Control Point), the FDA addresses thermally processed products like canned tuna or salmon, which are considered to be commercially sterile or "shelf-stable products." The FDA also manages the Fair Packaging and Labeling Act and the Low-Acid Canned Food program.

3.1.2 Center for Food Safety and Applied Nutrition (CFSAN)

The CFSAN is responsible for the foods under FDA jurisdiction, including dietary supplements, and all cosmetics. This is true whether products are domestically produced or imported. The CFSAN uses education, surveillance, and analysis and mandates through regulation of uniform food labels so that the U.S. consumer can avoid hazards such as allergens. It is estimated that $240 billion of food products enter U.S. commerce through 50,000 food manufacturers, processors, and warehouses annually. The United States also imports $15 billion of seafood, fresh produce, and other foods. In order for the CFSAN to do its job, the federal government must allocate resources such as their food scientists to develop rapid detection methods for microbial and viral food contaminants. This cannot be accomplished entirely by the CFSAN, and it therefore works closely with other public state and local regulators and private stakeholders. The rapid identification and control of outbreaks of food-borne diseases are vital to maintaining consumer confidence.

3.1.3 Center for Veterinary Medicine (CVM)

The CVM ensures that animal feed is safe and that animal drugs do not leave hazardous residues in human foods such as milk, meat, and eggs. The center approves safe and effective products for animals. The center also works to prevent the spread of bovine spongiform encephalopathy (BSE) also called mad cow disease, which can be spread through animal feed.

3.1.4 Office of Regulatory Affairs (ORA)

The ORA is responsible for inspection and enforcement for all of the food, drugs, and consumer packaged goods that are amenable to FDA inspection. The ORA comprises about 33% of the FDA's personnel. These inspectors are located in more than 160 sites throughout the United States. Most recently, the FDA Commissioner announced the opening of at least three FDA offices in China. The FDA conducts approximately 22,000 domestic and foreign inspections a year. In addition to their food inspection activities, FDA investigators also inspect and review clinical trials prior to the submission of a product for approval. Currently, the FDA has 13 ORA laboratories that analyze product samples to ensure compliance. This includes approximately 9.3 million import shipments as well as products that do not meet FDA standards, which are barred at the port of entry.

3.2 The Food Safety and Inspection Services of the U.S. Department of Agriculture

In 1884, the Bureau of Animal Industry was established to prevent diseased animals from being used as food. Ultimately, this bureau became the USDA FSIS. The name is somewhat misleading in that the FSIS is only responsible for regulating meat, poultry, and egg products, although as of 2008 it includes catfish under its jurisdiction as well.

The FSIS has the responsibility for ensuring that meat, poultry, and egg products for interstate commerce, or of foreign production, are safe, wholesome, and accurately labeled. It is also the only agency that requires a mark of inspection or an inspection legend. The FSIS ensures proper packaging and labeling of the products amenable to their inspection. It employs approximately 7,800 inspectors located in close to 6,200 federally inspected meat, poultry, and processed egg products plants. Inspectors must verify that the manufacturers comply with statutory requirements. The FSIS also inspects plants processing catfish and farms producing catfish. Similarly to the FDA, the FSIS verifies that foreign countries have equivalent inspection systems before products can be exported to the United States (Box 1).

Many consider FSIS inspection as extremely resource-intense because inspectors are present virtually all the time and challenge processes and documentation daily. The FSIS inspection system has undergone significant changes since 1996 with the enactment of the 1996 Pathogen

> **Box 1 Federal Meat, Poultry and Egg Inspection Acts**
>
> The Federal Meat Inspection Act (FMIA) requires mandatory inspection of livestock. The FMIA inspects livestock prior to slaughter and postmortem, and also defines the sanitation standards for these slaughter and processing facilities. The Poultry Products Inspection Act (PPIA) requires the continuous inspection of the processing of egg products in each facility that processes egg products for commerce. Interestingly enough, the FDA inspects shell egg facilities, while the FSIS inspects further processed egg plants. Today the biggest difference between the FSIS and the FDA as to how they inspect facilities is that the PPIA and the FMIA are the requirements for continuous FSIS government inspection of slaughtering and processing by the FSIS. Processed eggs and egg products are also regulated by the FSIS; however, they are not inspected as such. Further, processed facilities are also inspected at random throughout their production hours. The FDA inspects facilities more based upon risk, which results in an in-depth inspection as infrequently as once every 7 years.

Reduction: Hazard Analysis and Critical Control Point regulation otherwise known as the "Mega-Reg." The Mega-Reg mandates that the FSIS require all establishments to develop and implement sanitation standard operating procedures. Additionally, slaughterers are required to conduct generic *E. coli* testing to monitor "process controls." The FSIS also samples raw products for *Salmonella* at establishments and measures these results against a national baseline that is updated (but infrequently).

3.2.1 USDA Animal and Plant Health Inspection Service (APHIS)

Established in 1972, the APHIS is a relatively new agency, but much of the important work that falls under its mission today has been the responsibility of the USDA for more than 100 years. In fact, for most of the twentieth century, the early animal and plant health bureaus within the USDA operated independently of one another, but the creation of the APHIS consolidated these functions. The APHIS is divided into a collection of six operational programs units, three management support units, and two offices supporting federal government-wide initiatives. The APHIS is responsible for the prevention, detection, and response of animal and plant diseases. Working together, surveillance strategies are developed, coordinated, and integrated. To ensure rapid communications, the APHIS has the Emergency Operations Center, which works with states and tribal governments to enhance emergency preparedness, surveillance programs, and laboratory networks.

3.2.2 Office of Field Operations (OFO)

The OFO is directly involved with the regulatory management of the nationwide program of inspection and enforcement activities regarding meat, poultry, catfish, and egg products. The FSIS has 15 offices around the country that oversee the inspection staff assigned to the various establishments within each district. The FSIS also has regulatory field services laboratories, managed by the Office of Public Health Services (OPHS), that conduct scientific testing for the detection and prevention of foodborne outbreaks.

The Risk Assessment Division, which is also managed through the OPHS, develops and performs risk assessments of biological/chemical hazards in meat, poultry, and egg products to support policy development activities. These risk assessments are used to evaluate intervention strategies to reduce foodborne risks and to guide, support, and enhance the agency's overall decision-making process, risk management policies, outreach efforts, data-collection initiatives, and research priorities.

3.3 The Department of Homeland Security (DHS)

The DHS leads a "unified national effort to prevent and deter terrorist attacks in the United States." The DHS administers the Public Health Security and Bioterrorism Preparedness and Response Act of 2002 (the Bioterrorism Act, Public Law 107–188). The Bioterrorism Act was passed to enhance the security of the United States. Under the Bioterrorism Act, the FDA was required to develop sampling and testing programs for food to detect adulteration, especially intentional adulteration. These testing programs are also used not only for domestic inspections but also at ports of entry for foods entering the United States. Under 21 CFR Part 1, this law has four key parts: the registration of certain food facilities not including meat and poultry plants; prior notice of imported food shipments; maintenance and availability of establishment records; and administrative detention of food for human or animal consumption.

The Homeland Security Act of 2002 provides the primary authority for the overall homeland security mission. This act charged the DHS with the primary responsibility for developing a comprehensive national plan to secure the nation's Critical Infrastructure Key Resources (CIKR) and recommended "the measures necessary to protect the key resources and critical infrastructure of the United States." This comprehensive plan is the National Infrastructure Protection Plan (NIPP), published by the DHS in June 2006. The NIPP provides the unifying structure for integrating a wide range of efforts for the protection of the CIKR into a single national program.

Homeland Security Presidential Directive (HSPD) 7 established U.S. policy for enhancing CIKR protection by establishing a framework for security partners to identify, prioritize, and protect the nation's CIKR from terrorist attacks. The directive identified 17 CIKR sectors and designated a Federal Sector-Specific Agency (SSA) to lead CIKR protection efforts in each. The FDA and USDA are identified as joint SSAs for the Agriculture and Food CIKR. The agencies work with the DHS, the Federal Bureau of Investigation (FBI), state and local government representatives, including the National Association of State Departments of Agriculture and the Association of State and Territorial Health Officers, through a Government Coordinating Council, and interact closely with food industry representatives on the Food and Agriculture Sector Coordinating Council (FASCC) to develop programs directed toward protecting against a terrorist threat to ensure the safety of the nation's food supply.

These assessments support the requirements for a coordinated food and agriculture infrastructure protection program as stated in the NIPP, Sector-Specific Infrastructure Protection Plans (SSPs), and the Homeland Security Presidential Directive 9 (HSPD-9), entitled "Defense of U.S. Agriculture and Food."

The NIPP, Food and Agriculture SSPs, and HSPD-9 all call for federal, state, and industry partners to work together to protect the nation's infrastructure. Specifically, HSPD-9 establishes a national policy to defend the agriculture and food system against terrorist attacks, major disasters, and other emergencies. HSPD-9 directs the government to work with industry to identify and prioritize sector-critical infrastructure and key resources, establish protection requirements, develop awareness and early warning capabilities to recognize threats, mitigate vulnerabilities at critical production and processing nodes, enhance screening procedures for domestic and imported products, and enhance response and recovery procedures.

The Strategic Partnership Program Agroterrorism (SPPA) Initiative is a public–private cooperative effort established by the FBI, DHS, USDA, and FDA in partnership with state and industry volunteers. The intent of the initiative is to collect the necessary data to identify sector-specific vulnerabilities, develop mitigation strategies, identify research gaps and needs, and increase awareness and coordination between the food and agriculture government and industry partners. To accomplish this, the SPPA brings together federal, state, local, and industry partners to collaboratively conduct a series of assessments of food and agricultural industries.

SPPA assessments are conducted on a voluntary basis between one or more industry representatives for a particular product or commodity, their trade association(s), and federal and state government agricultural, public health, and law enforcement officials. Together, they conduct a vulnerability assessment of that industry's production process using the Criticality, Accessibility, Recuperability, Vulnerability, Effect, Recognizability plus Shock tool. As a result of each assessment, participants identify individual nodes or process points that are of highest concern, protective measures and mitigation steps that may reduce the vulnerability of these nodes, and research gaps/needs. Discussions of mitigation steps and good security practices are general in nature, focusing on physical security improvements for food processing facilities, biosecurity practices, and disease surveillance for livestock and plants. Participants also identified research gaps and needs during each assessment.

The FASCC also conducts tabletop exercises to demonstrate how government and industry can work together more effectively during a food or water contamination incident. The sector will continue to host tabletop exercises to include scenarios on the introduction of a foreign animal disease into the nation's agricultural systems. The exercises focus on response and recovery coordination among federal, state, tribal, local, and industry stakeholders.

3.3.1 Collaborative Efforts Among the DHS, FDA, and USDA

The Bio-Terrorism Act of 2002 provided the FDA with additional regulatory authority to require registration of all foreign and domestic food facilities that offer products for sale in the United States (except meat or poultry facilities, which are registered with the USDA). The FDA maintains records to trace products and ingredients, and administer the detention of potentially hazardous foods and prior notice of imported food shipments. To increase awareness of the potential for intentional adulteration of the food supply, both the FDA and the USDA have developed online Food Defense Awareness training material and guidance documents targeted to federal, state, and local regulators, local law enforcement, food program administrators, and the food industry.

The DHS has established the Office of Food Security and Emergency Preparedness, a dedicated office at the USDA, to better coordinate the infrastructure development and enhance the capacity to prevent, prepare for, and respond to an intentional attack on the U.S. food supply. The USDA, meanwhile, has the National Consumer Complaint Monitoring System, which is a surveillance and sentinel structure that monitors and tracks food-related consumer complaints around the clock and serves as a real-time, early warning system of a potential attack on the food supply. In addition, the DHS and the CDC monitor reported illnesses in the United States to enable identification of human illness originating from the food supply (Anonymous 2004b).

The USDA and DHS support the development of the National Animal Identification System (NAIS). The intent of the NAIS is to quickly identify the source of highly contagious animal diseases and their spread. This is critical whether the diseases are intentionally or accidentally introduced into the U.S. animal population.

3.4 *The Environmental Protection Agency (EPA)*

The EPA is responsible for all chemicals used to produce items such as cleansers, paints, plastics, fuels, industrial solvents, and additives. In 1976, the Toxic Substances Control Act (TSCA) was passed and authorized the EPA to regulate chemicals that pose unreasonable risks to human health or the environment. The TSCA requires the EPA to regulate those chemicals that are manufactured, imported, processed, and/or distributed in U.S. commerce. They are also responsible for how existing chemicals are used or disposed and for the assessment of new chemicals before they enter commerce. The TSCA does not include pesticides that are regulated by the EPA under the Federal Insecticide, Fungicide, and Rodenticide Act (FIFRA), nor does it address food, food additives, drugs, cosmetics, or devices that are amenable to regulation under the FFDCA. Under the Safe Drinking Water Act, the FDA works in concert with the EPA to establish regulations governing bottled water standards.

The mission of the EPA is to protect the public and the environment from risks posed by pesticides and other chemicals. The EPA also promotes safer means of pest management. No food or feed item may be marketed legally in the United States if it contains a food additive or drug residue not permitted by the FDA, or a pesticide residue without EPA tolerance or in excess of an established tolerance. The FDA, FSIS, APHIS, and EPA also use existing food safety and environmental laws to regulate plants, animals, and foods that arise from biotechnology. For instance, the APHIS has the role of protecting against plant and animal pests and diseases. In addition, the Food Quality Protection Act (FQPA) requires the EPA to regulate pesticides, as they may affect food safety. The FQPA resulted in an amendment to the FIFRA and the FFDCA, changing how the EPA regulates pesticides in the United States.

4 Challenges of Establishing a Single Food Agency

No architect of a food safety regulatory system starting from scratch would ever come up with this complex maze of oversight authorities that characterizes the current system. Today's food regulatory scheme is like a house built 100 years ago, which started small but over the years was remodeled here and had rooms added there as new needs arose. Past presidential administrations and congressional bodies have commissioned various studies to evaluate the idea of transforming this multi-jurisdictional hodgepodge into a single agency to oversee the safety of the nation's food supply. Yet there are different thoughts on whether or not a single food safety agency would be the most appropriate choice for the United States (Anonymous 2007; Pape et al. 2004). As noted previously, the new Office of Foods under the FDA may actually be a first step toward a single food agency. Funding remains a huge challenge for establishing any agency, so time will tell as to how much Congress allocates for the new Office of Foods within the FDA. Currently, approximately 80% of the U.S. food supply is overseen by the FDA and approximately 20% by the FSIS, with the engagement of at least three other key agencies depending upon the circumstances. There are a number of key laws or statutes in existence that have precluded the formation of a single food agency. With the formation of these laws, numerous congressional committees and subcommittees have been formed over the years to review and authorize funding and other resources to these various branches of the federal government. The willingness to relinquish such key authorities is not a common trait exhibited by House or Senate committee chairmen or members; thus, there is tremendous resistance to change. Approval for any recommendations for change within the agencies responsible for food safety ultimately lies with the Senate, the House of Representatives, and the White House. Such a political process is very time-consuming and fraught with disagreement and frustration for all involved. Therefore, it is not a simple task to create a new "single food agency." Perhaps consolidation would provide improvements if the entire food safety system were to be transformed by legislature and by providing additional funding and authority. But in our current circumstances, the simple combination of agencies will result in the reduction of resources for a system that can hardly afford it (Anonymous 2007). From the food security perspective, it has been suggested that a single food agency would not provide any benefit and moreover may compromise food security at the beginning of the consolidation process (Hammonds 2004).

That raises the important question of who is in charge. One might consider the president of the United States as having the ultimate responsibility for public health. In fact, everyone, including the consumer, plays an important role in properly managing food safety as food moves through the supply chain to the dinner table. Foodborne illnesses can occur wherever there is a weak link in the supply chain. Fortunately, there is considerable redundancy in the U.S. food safety system, which is considered a true strength by many since it does not provide an easy way to intentionally disrupt the flow of safe food across the system (Anonymous 2000).

5 The Regulatory Process

The American public relies primarily on federal and state laws and regulations to protect them from foodborne illnesses or other food hazards from imported or domestically manufactured products. Under the U.S. Constitution, specific aspects of this task are delegated to the executive, legislative, or judicial branches of government (Fig. 1). Through the legislative branch, food legislation or "bills" are introduced by the members of Congress into the Senate and the House of Representatives. These bills propose to create new or revise existing laws or "statutes" to authorize actions deemed necessary to protect public health. Once both houses of Congress have agreed to the legislative language and a bill is passed, it is ready to be signed into law by the president (or to be enacted by congressional override of a presidential veto). The regulatory agency(ies) designated to implement the law must then develop all required programs, including compliance regulations and enforcement criteria, through notice and comment rule making unless the law has self-enacting provisions. Congress is also responsible for funding the agency to enable enforcement of the prescribed programs.

Ideally, the executive branch, typically referred to as the "administration" or the "White House," works with Congress as these laws are developed. Should a dispute arise regarding regulations developed under a specific law, the judicial branch may be called upon to interpret the law and make

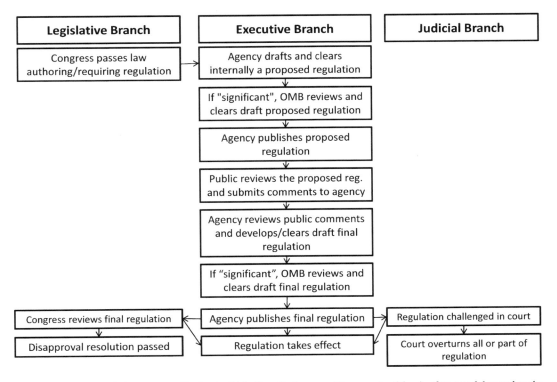

Fig. 1 The U.S. regulation process. Under the U.S. Constitution, specific aspects of food safety are delegated to the executive, legislative, or judicial branches of government

a fair and impartial judgment as to whether the proposed regulations accomplish the intent of the law. It is important to recognize the separation of powers among the three branches of government and that the process must employ science-based decision making and be openly transparent, such that public participation is solicited. This means anyone from the public can express his or her opinion regarding a particular law or regulation.

6 Example of the U.S. Regulatory Process: Making "Right Way Pizza"

An example of how the regulatory agencies work is provided in an example about the making of pizza. Pizza is a very popular food in the United States and around the world. However, making a pizza for sale in the United States as a finished, packaged product requires a full understanding of all of the steps in the manufacturing process, as well as the associated laws and regulations governing the process. This process will be used to illustrate how the U.S. regulatory process is involved in making "Right Way Pizza." Most manufacturers today typically do not rely on just a local market but are compelled to consider exporting their products to other markets to maximize production efficiencies and capture greater margins where demand is higher for safe, quality products (Fig. 2). Note that there are no less than four major U.S. federal agencies involved in regulating pizza production (Table 2).

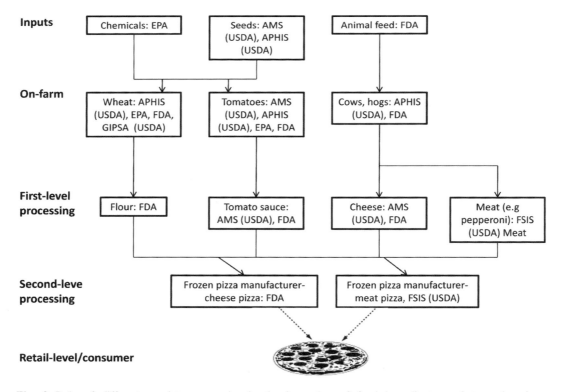

Fig. 2 Role of different regulatory agencies in the inspection of food ingredients used to make pizza. EPA = Environmental Protection Agency; FSIS = Food Safety and Inspection Services of the U.S. Department of Agriculture; FDA = Food and Drug Administration of the Department of Health and Human Services (Adapted from Anonymous 2004c)

Table 2 Summary of 21 CFR Part 110, current good manufacturing practice in manufacturing, packing, or holding human food[a]

Subpart	Section	Definitions	Definitions
A. General provisions	110.3		Acid foods/acidified foods
			Batter and blanching
			Critical control point
			Food and food-contact surfaces
			Microorganisms
			Pest and plant
			Quality control operation
			Rework
			Safe-moisture level
			Sanitize
			Water activity
	110.5	Current good manufacturing practice	Criteria for determining adulteration
			Food covered by specific GMPs is also covered by umbrella GMPs
	110.10	Personnel	Disease control
			Cleanliness
			Education and training
			Supervision of personnel with regards to these requirements
	110.19	Exclusions	Excluded operations (raw agricultural commodities)
			The FDA can issue special regulations to cover excluded operations
B. Buildings and facilities	110.20	Plant and grounds	Description of adequate maintenance of grounds
			Plant construction and design to facilitate sanitary operations and maintenance
	110.35	Sanitary operations	Requirements for
			Cleaning/sanitizing of physical facilities, utensils, and equipment
			Storage of cleaning and sanitizing substances
			Pest control
			Sanitation of food contact surfaces
			Storage and handling of cleaned portable equipment and utensils
	110.20	Sanitary facilities and controls	Requirements for
			Water supply
			Plumbing
			Sewage disposal
			Toilet facilities
			Hand-washing facilities
			Rubbish and offal disposal
C. Equipment	110.40	Equipment and utensils	Requirements for the design, construction, and maintenance of equipment and utensils
E. Production and process controls	110.80	Processes and controls	Delineates processes and controls for
			Raw materials and other ingredients
			Manufacturing operations
	110.93	Warehousing and distribution	Storage and transportation of food must protect against contamination and deterioration of the food and its container
Subpart G. Defect action levels	110.10		FDA has established maximum defect action levels (DALs) for some natural or unavoidable defects
			Compliance with DALs does not excuse violation of 402 (a)(4)
			Food containing defects above DALs may not be mixed with other foods

[a] *Source: Federal Register* 51, 1986. http://www.cfsan.fda.gov/~dms/gmp-1.html

Table 3 Production and regulatory considerations for "Right Way Pizza"

Production step	Governing agency	Compliance focus	Some potential concerns for Right Way Pizza managers
Inputs (ingredients)	EPA	Chemicals	Pesticides
	EPA, FDA/USDA	Water to the plant Water in the plant	High chemical, radiological, or bacterial levels
	USDA/Ag Marketing Service (AMS) USDA/APHIS	Seed	Unapproved genetically modified corn or wheat product
	HHS/FDA, CVM, DHS	Animal feed	Toxic chemicals like dioxin creating residues in meat
			Illegal drug residues
			Intentional contaminants like anthrax spores
On-farm	USDA/APHIS, EPA, HHS/FDA, USDA/GIPSA	Wheat	Unapproved genetically modified corn or wheat product
			Pesticide levels
	USDA/APHIS, EPA, HHS/FDA, USDA/GIPSA	Tomatoes	Foodborne pathogens, excessive rot or mold
			Pesticide levels
	USDA/APHIS, HHS/FDA, CVM	Cows and hogs	Toxic chemicals like dioxin creating residues in meat, animal disease (BSE)
			Illegal drug residues
		Milk	Bacterial toxins, drug residues
		Milk	Cross-contamination during shipping
First-level processing	USDA/FSIS, HHS/FDA	Facilities	cGMPs
	HHS/FDA	Flour	Foreign material, mycotoxins
			Unapproved genetically modified corn or wheat product
	AMS, APHIS, EPA, HHS/FDA	Tomato sauce	Foodborne pathogens arising from improper cooking/holding, excessive mold count
		Unapproved cleaning chemicals	Residues
	USDA/AMS, HHS/FDA	Cheese	Foodborne pathogens arising from improper milk pasteurization or improper curing
			Foreign material
			Allergen controls
	UDA/FSIS	Meat (pepperoni)	Foodborne pathogens arising from improper fermentation process, use of BSE risk material
			Foreign material
			Allergen controls
Second-level processing	HHS/FDA—Frozen cheese pizza, USDA/FSIS—Frozen meat and cheese pizza	Packaging	Approved materials
		Allergens	Proper approved labeling
		Cooking	Proper validated cooking instructions
		Storage	Appropriate temperatures, cross-contamination, off-odors
		Spices	Unapproved colors or dyes, mycotoxins, pathogenic bacteria, foreign material

In our example, we will assume the "Right Way Pizza" company has chosen to produce two types of packaged pizza products, a cheese-only pizza and a meat (over 2%) and cheese pizza. Based upon the U.S. regulations, this automatically means that both the FDA and the USDA will inspect the plant. In addition, the USDA will apply the "mark of inspection" to the meat and cheese pizza since it is amenable to its inspection process. This means that the procurement office at "Right Way Pizza" will need to be sure it secures only USDA-inspected beef products or imported beef products produced under an inspection process found to be equivalent by the FSIS.

7 Training

One area that is easily overlooked and is becoming more of a focus for regulators and inspectors is training. Training plays an important role in ensuring that safe and wholesome products are consistently produced for the marketplace. This is where the manufacturer must have "checks and balances" in place to monitor, verify, and validate the production process. Industry supports requirements for appropriate training for supervisors and workers to ensure that they have the necessary knowledge and expertise in food hygiene, food protection, employee health, and personal hygiene to produce safe food products. Such training must be accessible and delivered in a way that the workforce can easily understand. Food manufacturers must maintain a record of this training for each worker. Minimum training should include the following key areas:

- Proper food handling, storage, and distribution
- Maintenance and sanitation of facilities and equipment
- Environmental monitoring
- Personal hygiene
- Temperature control
- Food allergens management
- Pathogen control
- Good manufacturing practices applicable to certain agricultural operations (Table 3)

Training must be applied at all levels of the business, and top management must be visibly engaged and able to supply sufficient resources for proper training at all times.

8 Summary

The manufacturing of food involves a great deal of regulatory and management oversights to ensure products are safe for consumption. Although federal agencies have an important jurisdiction on the safety of foods, it should be noted that states also play an integral role in the inspection process. This is especially true during a recall, where state inspectors actually check all levels of distribution to ensure that adulterated product is removed from store shelves. Federal and state agents work closely with manufacturers and retailers to protect the consumer by rapidly removing potentially harmful products and properly disposing of them. It is this multifaceted regulatory structure that makes the U.S. food supply one of the safest in the world. Although mistakes can be made throughout the production and delivery of foods, these mistakes are minimized by effective checks and balances to ensure that consumers' confidence in our food supply remains high.

References

Anonymous 1998. Recall of meat and poultry products. U.S. Department of Agriculture, Food Safety and Inspection Service, Washington, DC. http://www.fsis.usda.gov/OPPDE/rdad/FSISDirectives/8080.1Rev5.pdf. Accessed 6 Feb 2011.

Anonymous. 2000. A description of the U.S. food safety system. http://www.fsis.usda.gov/oa/codex/system.htm. Accessed 6 Feb 2011.

Anonymous. 2004a. Federal food, drug, and cosmetic act. U.S. Food and Drug Administration. http://www.fda.gov/opacom/laws/fdcact/fdctoc.htm. Accessed 6 Feb 2011.

Anonymous. 2004b. Fact sheet: Strengthening the security of our nation's food supply. http://www.dhs.gov/xnews/releases/press_release_0453.shtm. Accessed 6 Feb 2011.

Anonymous. 2004c. Federal food safety and security system—fundamental restructuring is needed to address fragmentation and overlap (GAO-04-588 T). http://www.gao.gov/new.items/d04588t.pdf. Accessed 6 Feb 2011.

Anonymous. 2006. FSIS food recalls. http://www.fsis.usda.gov/factsheets/FSIS_Food_Recalls/index.asp. Accessed 6 Feb 2011.

Anonymous. 2007. Reforming the food safety system: What if consolidation isn't enough? *Harvard Law Review* 120(5): 1345–1366.

Anonymous. 2008. Melamine contamination in China (updated: October 18, 2008). http://www.fda.gov/ohrms/dockets/ac/08/briefing/2008-4386b3.pdf. Accessed 6 Feb 2011.

Anonymous. 2010. FDA 101: Product recalls. http://www.fda.gov/downloads/ForConsumers/ConsumerUpdates/UCM143332.pdf. Accessed 6 Feb 2011.

Anonymous. 2011a. FDA Office of food. http://www.fda.gov/AboutFDA/CentersOffices/OC/OfficeofFoods/ucm196720.htm.

Anonymous. 2011b. FDA FSMA. http://www.fda.gov/Food/FoodSafety/FSMA/default.htm.

Hammonds, T.M. 2004. It is time to designate a single food safety agency. *Food and Drug Law Journal* 59: 427–432.

Morris, J.G. Jr. 2011. How safe is our food? *Emerging Infectious Diseases* http://dx.doi.org/10.3201/eid1701101821.

Pape, S.M., P.D. Rubin, and H. Kim. 2004. Food security would be compromised by combining the Food and Drug Administration and the U.S. Department of Agriculture into a single food agency. *Food and Drug Law Journal* 59: 405–416.

Scallan, E., R.M. Hoekstra, F.J. Angulo, R.V. Tauxe, M.-A. Widdowson, S.L. Roy et al. 2011. Foodborne illness acquired in the United States – major pathogens. *Emerging Infectious Diseases* [Epub ahead of print].

Part IV
List of Other Food Safety Resources

Food Safety Resources

Omar A. Oyarzabal and Steffen Backert

There are several food safety resources available in the form of books, journals, trade magazines, and more recently though the Internet. This chapter briefly summarizes, in table format, most of the available resources that can provide complementary reading to the chapters described in this book. The summary starts with the books that deal mainly with food safety (Table 1), continues with other microbiology books with sections that deal with food safety (Table 2), then lists refereed journals related to food safety (Table 3), and finally provides some Internet resources related to food safety (Table 4).

The Internet has become an important repository of information. Many regulatory agencies now post their documents only on the Internet. However, the Internet can also be misleading by providing information that has been compiled and presented without scrutiny. Therefore, it is extremely important to review different resources when looking for information about a new topic on the Internet.

O.A. Oyarzabal (✉)
Department of Biological Sciences, Alabama State University, Montgomery, AL, USA
e-mail: oaoyarzabal@gmail.com

S. Backert
University College Dublin, School of Biomolecular and Biomedical Sciences,
Science Center West, Dublin, Ireland

Table 1 Books related to food safety organized by title, authors/editors, and publisher

Book title	Authors/editors	Publisher
Food safety	Jones, J. M.	American Association of Cereal Chemists, Eagan Press
Food safety and food quality (Issues in environmental science and technology)	Hester, R. E., and R. M. Harrison	Royal Society of Chemistry
Food safety: a guide to what you really need to know	Hemminger, J. M.	Iowa State University Press
Food safety: a practical and case study approach (Integrating safety and environmental knowledge into food studies towards European sustainable development)	McElhatton, A., and R. J. Marshall	Springer Science + Business Media
Food safety control in the poultry industry (Woodhead publishing in food science and technology)	Mead, G. C.	CRC Press
Food safety culture: creating a behavior-based food safety management system (Food microbiology and food safety)	Yiannas, F.	Springer Science + Business Media
Food safety for the twenty-first century: managing HACCP and food safety throughout the global supply chain	Wallace, C., W. Sperber, and S. E. Mortimore	Wiley-Blackwell
Food safety governance	Dreyer, M., and O. Renn	Springer Science + Business Media
Food safety handbook	Schmidt, R. H., and G. E. Rodrick	John Wiley and Sons
Food safety (Point/Counterpoint)	Sherrow, V.	Chelsea House Publications
Food safety: the implications of change from producerism to consumerism	Sheridan, J., M. O'Keeffe, and M. Rogers	Wiley-Blackwell
ISEKI 6: case studies in food safety and environmental health	Ho, P., V. Cortez, and M. Margarida	Springer Science + Business Media
Mass spectrometry in food safety	Zweigenbaum, J.	Springer Science + Business Media
Principles of food sanitation	Marriott, N., and R. Gravani	Springer Science + Business Media
Procedures to investigate foodborne illness	Todd, E., and J. Guzewich	Springer Science + Business Media
Safe or not safe	Pechan, P., O. Renn, A. Watt, and I. Pongratz	Springer Science + Business Media

Table 2 Microbiology books with sections that deal with food safety

Book title	Authors/editors	Publisher
Canned foods – principles of thermal process control, acidification and container closure evaluation	Weddig, L. M.	GMA Science and Education Foundation
Chemical food safety: a scientist's perspective	Riviere, J. E.	Wiley-Blackwell
Compendium of methods for the microbiological examination of foods	Vanderzant, C., and D. F. Splittstoesser	American Public Health Association
Compendium of the microbiological spoilage of foods and beverages	Sperber, W., and M. Doyle	Springer Science + Business Media
Control of foodborne microorganisms	Juneja, V. K., and J. N. Sofos	Marcel Dekker
Effective risk communication	Sellnow, T. L., R. R. Ulmer, M. W. Seeger, and R. Littlefield	Springer Science + Business Media
Food microbiology – the laboratory	Entis, P.	The Food Processors Institute
Food microbiology: fundamentals and frontiers	Doyle, M. P., L. R. Beuchat, and T. J. Montville	American Society for Microbiology Press
Foodborne parasites	Ortega, Y. R.	Springer Science + Business Media
Genomics of foodborne bacterial pathogens	Wiedmann, M., and W. Zhang	Springer Science + Business Media
HACCP – A systematic approach to food safety	Stevenson, K. E., and D. T. Bernard	The Food Processors Institute
HACCP and ISO 22000: application to foods of animal origin	Arvanitoyannis, I. S.	Wiley-Blackwell
Handbook of meat processing	Toldrá, F.	Wiley-Blackwell
Handbook of poultry science and technology (Vols. 1 and 2)	Guerrero-Legarreta, I.	Wiley-Blackwell
Handbook of seafood quality, safety and health applications	Alasalvar, C., K. Miyashita, F. Shahidi, and U. Wanasundara	Wiley-Blackwell
Handbook of meat, poultry and seafood quality	Nollet, L. M. L. et al.	Wiley-Blackwell
Listeria, listeriosis & food safety (Food science and technology)	Ryser, E. T., and E. H. Marth	CRC Press
Manual of clinical microbiology	Murray, P. R. et al.	American Society for Microbiology
Microbiological analysis of red meat, poultry and eggs	Mead, G. C.	CRC Press
Microbial hazard identification in fresh fruits and vegetables	James, J.	Wiley-Blackwell
Microbiologically safe foods	Heredia, N. L., I. V. Wesley, and J. S Garcia	Wiley-Blackwell
Microbial safety of fresh produce	Fan, X., B. A. Niemira, C. J. Doona, F. E. Feeherry, and R. B. Gravani	Wiley-Blackwell
Microorganisms in foods 7: microbiological testing of food safety management	International commission on microbiological specifications for foods	Springer Science + Business Media
Microorganisms in foods 8: use of data for assessing process control and product acceptance	International commission on microbiological specifications for foods	Springer Science + Business Media

(continued)

Table 2 (continued)

Book title	Authors/editors	Publisher
Modern food microbiology	Jay, J. M.	Kluwer Academic/Plenum Publishers
PCR methods	Maurer, J.	Springer Science + Business Media
Preharvest and postharvest food safety Contemporary issues and future directions	Beier, R. C., S. D. Pillai, T. D. Phillips, and R. L. Ziprin	Wiley-Blackwell
Principles of microbiological troubleshooting in the industrial food processing environment	Kornacki, J.	Springer Science + Business Media
Quantitative microbial risk assessment	Haas, C. N., J. B. Rose, and C. P. Gerba	John Wiley & Sons
Safety of meat and processed meat	Toldrá, F.	Springer Science + Business Media
The BRC global standard for food safety: a guide to a successful audit	Kill, R.	Wiley-Blackwell
The microbiology of meat and poultry	Davies, A., and R. Board	Blackie Academic & Professional
The microbiology of safe food	Forsythe, S. J.	Wiley-Blackwell
Viruses in foods	Goyal, S.	Springer Science + Business Media

Table 3 Refereed journals related to food safety

Journal	Publisher	URL
Applied and Environmental Microbiology	American Society for Microbiology	http://aem.asm.org/
European Food Research and Technology	Springer Science + Business Media	www.springer.com/food+science/journal/217
Food and Bioprocess Technology	Springer Science + Business Media	www.springer.com/food+science/journal/11947
Food and Environmental Virology	Springer Science + Business Media	www.springer.com/biomed/virology/journal/12560
Food Analytical Methods	Springer Science + Business Media	www.springer.com/food+science/journal/12161
Food Safety Trends	International Association for Food Protection	http://www.foodprotection.org/
International Journal of Food Microbiology	Elsevier	http://www.sciencedirect.com/science/journal/01681605
Journal of Applied Microbiology	Wiley-Blackwell, on behalf of the Society for Applied Microbiology	http://www.wiley.com/bw/journal.asp?ref=1364-5072
Journal of Microbiological Methods	Elsevier	http://www.sciencedirect.com/science/journal/01677012
Journal of Food Protection	International Association for Food Protection	http://www.foodprotection.org/
Journal of Food Safety	Wiley-Blackwell	http://www.blackwellpublishing.com/journal.asp?ref=0149-6085
Letters in Applied Microbiology	Wiley-Blackwell, on behalf of the Society for Applied Microbiology	http://www.wiley.com/bw/journal.asp?ref=1364-5072

Table 4 Internet resources related to food safety

Internet website	URL
Bad bugs	http://www.fda.gov/Food/FoodSafety/FoodborneIllness/FoodborneIllnessFoodbornePathogensNaturalToxins/BadBugBook/default.htm
Center for Food Safety and Applied Nutrition (CFSAN)	http://www.fda.gov/AboutFDA/CentersOffices/CFSAN/default.htm
Centers for Disease Control and Prevention	http://www.cdc.gov/
Division of Foodborne, Bacterial and Mycotic Diseases	http://www.cdc.gov/nczved/divisions/dfbmd/
Eating Outdoors, Handling Food Safely	http://www.fda.gov/Food/ResourcesForYou/Consumers/ucm109899.htm
Fight BAC!	http://www.fightbac.org
Food Irradiation – Key Topics (ERS)	http://www.ers.usda.gov/topics/view.asp?T=102818
Food Safety Actions Briefing Room (ERS)	http://www.ers.usda.gov/briefing/FoodSafety
Food Safety Research Information Office	http://www.nal.usda.gov/fsrio
FoodSafety.gov	http://www.foodsafety.gov
Foodborne illness	http://www.fda.gov/Food/FoodSafety/FoodborneIllness/default.htm
HACCP Information Page	http://www.fda.gov/Food/FoodSafety/HazardAnalysisCriticalControlPointsHACCP/default.htm
WHO Food Safety	http://www.who.int/foodsafety/en/

Glossary

Acetic acid (and acetates) An organic acid and corresponding salt commonly used as antimicrobial food preservatives. Acetic acid is the primary component of vinegar.
Actin A globular 42-kDa protein found in almost all eukaryotic cells. Monomeric actin (G-actin) can be converted into filamentous actin (F-actin) by actin-polymerization machineries. One of these polymerization machineries is Arp2/3 (actin-related proteins 2 and 3), which plays an important role in many cellular processes, such as signaling, motility, and muscle contraction, or infections.
Adhesin Cell-surface components of bacteria that facilitate bacterial adhesion or adherence to host target cells or to inanimate surfaces. Adhesins are classic host colonization factors of bacteria. For example, *Campylobacter* encode a variety of adhesins (including CadF, JlpA, and PEB1) as does *Listeria* (internalin A and B, called InlA and InlB).
A/E lesion Attachment and effacement lesions are characterized by effacement of microvilli and marked cytoskeletal changes beneath the attached bacteria. EPEC and some other pathogens cause characteristic A/E lesions on infected gut enterocytes.
Aflatoxin A group of carcinogenic, fungal toxins produced by *Aspergillus* spp., primarily *Aspergillus flavus* and *Aspergillus parasiticus*, especially in stored grains and nuts. After entering the human body, aflatoxins may be metabolized by the liver to produce harmful molecules.
Algae Algae represent a large and diverse group of simple, typically autotrophic organisms, ranging from unicellular to multicellular forms.
Allergens Compounds that can cause allergic reactions. They trigger the host immunity system to activate and fight against them.
Anaerobic The absence of oxygen. Feature of an organism that can grow under oxygen-free conditions.
Antemortem inspection The inspection of animals at the processing plant right before slaughtering.
Antibiotics Substances that kill or slow down the growth of bacteria and other microorganisms. Antibiotics have no known effects on viruses.
Antimicrobial food preservatives Chemical compounds that are synthetic or naturally occurring that are added to foods to inhibit the growth of or inactivate spoilage or pathogenic microorganisms.
Antimicrobial interventions Measures or treatments used to inhibit or destroy the growth of microorganisms.
Antioxidant A substance, such as vitamin E or beta-carotene, that is capable of counteracting certain damaging effects of oxidation in animal tissues.
Aseptic processing/packaging The processing and packaging of sterile (aseptic) products in a sterile container in a manner that maintains the product's integrity.
Asymptomatic A patient who is a carrier for a given disease or infection but shows no obvious symptoms.

ATP Adenosine-5'-triphosphate is a multifunctional nucleotide used in all living cells as a coenzyme. ATP plays a crucial role in intracellular energy transfer.

Attack rate The cumulative incidence of infection following exposure to a disease-causing agent.

Bacteriophage A widespread natural group of viruses that commonly infect bacteria.

Benzoic acid (and benzoates) An organic acid used as an antimicrobial food preservative. It occurs naturally in berries (e.g., cranberries) and some other fruits.

Berry A fleshy fruit produced from a single ovary that does not contain a stone or pit, but may contain seeds. The term commonly refers to any small edible fruit.

Botulism A rare and serious paralytic illness caused by the secreted toxin produced by *Clostridium botulinum*.

Bovine spongiform encephalopathy A fatal, spongy, chronic degeneration of the central nervous system in cattle that is caused by self-replicating infectious proteins called prions. Commonly referred to as "mad cow disease."

Buffering capacity The resistance of a given substance or food that changes the pH to a specific value.

CiaB The *Campylobacter* invasion antigen B (CiaB) is about 73 kDa in size and is encoded by the *ciaB* gene. CiaB is crucial for host cell entry of *C. jejuni* and shares similarity with type III secretion system (T3SS) proteins associated with cellular invasion in other bacterial pathogens.

Cladistics The methodology used to classify species of organisms into groups called clades, which consist of all the descendants of an ancestral organism and the ancestor itself. In biological systematics, a clade is a monophyletic group or a single "branch" on the "tree of life."

Class I recall The removal of a food product from the market because there is a reasonable probability that the use of or exposure to this product may cause serious adverse health consequences or death. Recalls may be conducted on a company's own initiative or by FDA request under its statutory authority.

Clustering methods A group of analytical techniques used to organize data into groups arranged by similarity or difference indices. Clustering is commonly used for the analysis of relationships among multiple bacterial strains. The patterns generated by clustering methods are usually arranged in hierarchical structures to generate dendograms.

Colony-forming units (CFU) A measure of viable microbial organisms in a given volume or amount of material. Unlike direct microscopic counts, where both dead and alive cells are counted, CFU determines the numbers of viable cells only. The results are commonly reported as CFU/ml for liquids or CFU/g for solids.

Colostrum Milk secreted immediately after a mammal has given birth. It contains high levels of nutrients and antibodies essential for the development of passive immunity.

Commercially sterile A food treated to inactivate all microorganisms that could grow under non-refrigerated conditions. The products are considered "shelf-stable" products.

Conjugation The transfer of genetic material between bacteria, such as circular DNA plasmids. This process is initiated by direct cell-to-cell contact and employs the so-called type IV secretion systems (T4SSs).

Creutzfeldt–Jakob disease A degenerative neurological disorder affecting humans and caused by prions.

Crohn's disease A serious chronic and progressive inflammation of the intestine. Also known as regional enteritis, the disease may affect any part of the gastrointestinal tract from mouth to anus, and is associated with a wide range of symptoms, such as abdominal pain, diarrhea, vomiting, and weight loss.

Cross-contamination The transfer of harmful microorganisms or substances through the improper handling of food.

Cyclosporiasis An infection caused by the species *Cyclospora cayetanensis*, a protozoan pathogen transmitted by feces or feces-contaminated products and water.

Cysts The larval form of tapeworms.

Cysticercus (plural: cysticerci) The larval form of any of the tapeworms from the genus *Taenia*.

Cytoskeleton The cellular "scaffolding" or "skeleton" localized within the cytoplasm of eukaryotic and prokaryotic cells; it is composed of multiple proteins.

Dendritic cells (DCs) Immune cells that form part of the mammalian immune system. The main function of DCs is to process antigen material and present it on the surface to other cells of the immune system. DCs also act as messengers between the innate and adaptive immunity in the host.

Dressed poultry A term used to describe a ready-to-cook, whole bird that has been slaughtered and whose feathers, viscera, head, and feet have been removed.

D-value The time in minutes necessary to inactivate 90% of a microbial population at a constant temperature and graphically represented in "survivor curves." This term is also called the "decimal reduction time."

E-cadherin Cadherins are a class of large transmembrane adhesive proteins that depend on calcium to function. E-cadherin plays important roles in cell adhesion, ensuring that neighboring cells within tissues are tightly connected with each other.

Electron beam irradiation High-energy electron-based technology used to inactivate microorganisms in thin films of food products.

Encephalitis A viral or bacterial infection that leads to inflammation of the brain.

Endemic An infection maintained in a given population without the need for external inputs.

Endocytosis It is the process by which host cells uptake molecules by engulfing them. This process is utilized by all cells to uptake large polar molecules that cannot passively go through cell membranes.

Enteric A general term referring to the intestine.

Enterohemorrhagic Eschericha coli (EHEC) Strains of *E. coli* that can cause hemorrhage in the intestines resulting in bloody diarrhea and colitis.

Enterotoxin A bacterially produced toxin specific to the mucous membrane of the intestine that can trigger vomiting and diarrhea associated with food poisoning.

Enzyme-linked immunosorbent assay (ELISA) Rapid antibody-based assays for the detection or quantification of a given pathogen, toxin, or substance. ELISAs involve an enzyme reaction to visualize the antibody–antigen reaction.

Epidemiology The branch of medicine studying the incidence and prevalence of diseases in large populations.

Etiology The cause of a given disease.

Eukaryote Eukaryotes are higher organisms that contain complex structures and specific cell organelles with membranes.

Facultative anaerobe A bacterium that produces ATP by aerobic respiration, if oxygen is present, but can also produce energy through fermentation. In contrast, obligate anaerobic bacteria do not survive in the presence of oxygen.

Federal Food, Drug, and Cosmetic Act A set of laws passed by Congress in 1938 giving the U.S. Food and Drug Administration the authority to oversee the safety of foods, drugs, and cosmetics.

Flagellum (plural: flagella) A tail-like projection protruding from the cell of certain prokaryotic and eukaryotic cells. The main function of flagella is cell motility.

Fomite Any inanimate object or substance capable of carrying infectious organisms and hence transferring them from an infected person to another.

Food atribution The identification of which foods are vehicles for specific cases of illnesses.

Foodborne illness A disease caused by an infectious or toxigenic agent where the agent enters the body through food.

Foodborne illness outbreak A situation where two or more cases of illness result from the consumption of contminated food.

Food Code A document that provides a scientifically sound technical and legal basis for regulating retail and foodservice operations. This document assists food control jurisdictions at all levels of government. It is published by the U.S. Food and Drug Administration.

FoodNet A collaborative project of the Centers for Disease Control and Prevention consisting of active surveillance for foodborne diseases and studies designed to understand the epidemiology of foodborne diseases in the United States.

Fruit The seed-bearing edible portion of a plant.

Fulminant hepatitis A massive hepatic necrosis within 8 weeks of onset. It is a very severe and life-threatening form of acute hepatitis.

F-value The exact time at a given temperature needed to destroy a homogeneous population of microorganisms.

Gamma irradiation Gamma rays generated by cobalt-60 or cesium-137 and used to inactivate microorganisms in foods.

Gastroenteritis A term describing the inflammation of the gastrointestinal tract (stomach and intestine) and resulting in acute diarrhea and vomiting.

Gene loss Gene elimination from the chromosome in any cell. Gene loss results in the disappearance of a chromosome or larger segments of DNA.

Genetic fingerprinting A group of methods based on DNA analysis used to characterize a group of strains at the subspecies level.

Good Agricultural Practices (GAP) The best-known methods of land use to achieve agronomic and environmental sustainability. These practices are summarized in various documents published by the government, producer organizations, retailers, and importers.

Growth curve The general growth pattern of microorganisms over time, including the classical lag, log, stationary, and death phases in the growth of bacterial cells.

GTPases A family of hydrolase enzymes that can bind and hydrolyze guanosine triphosphate (GTP). GTPases can regulate a wide variety of processes in the cell, including growth, differentiation, movement, lipid vesicle transport, and microbial infections.

Guillain–Barré syndrome (GBS) An acute inflammatory demyelinating polyneuropathy affecting the peripheral nervous system. The syndrome was named after the French physicians Guillain, Barré, and Strohl, who discovered this disease in 1916.

Halophiles Organisms that live in natural environments with very high concentrations of salt.

Hazards In the context of food, hazards are agents (physical, chemical, or microbial) or conditions of food that have the potential to cause adverse health impacts.

Hazard Analysis Critical Control Point (HACCP) A systematic preventive approach for ensuring food safety that addresses physical, chemical, and biological hazards. HACCP is used to identify potential food safety hazards so that actions are taken to reduce or eliminate these hazards.

Hemolytic anemia A condition in which there are not enough red blood cells in the blood of an individual due to the premature destruction of red blood cells.

Hemolytic-uremic syndrome (HUS) A disorder caused by toxic substances produced by a digestive tract infection that destroys red blood cells and results in kidney injury.

Heterofermentative A microorganism that produces several end products through fermentation.

High-acid foods Foods having a pH ≤ 4.6.

High-pressure processing The utilization of high pressures [100–800 MPa (14,500–116,000 psi)] to inactivate vegetative bacterial cells in foods for purposes of improving shelf life and/or food safety.

Homofermentative A microorganism that produces a single end product through fermentation.

Horizontal DNA transfer A process in which a bacterium incorporates genetic material from another bacterium. The mechanisms of horizontal DNA transfer include (i) uptake of DNA by genetic transformation, (ii) bacteriophage transduction, and (iii) conjugative DNA transfer mediated by so-called type IV secretion systems (T4SSs). Horizontal DNA transfer is a major driving force in the evolution of bacteria.

Human spongiform encephalopathy A fatal spongy degeneration of the human brain caused by self-replicating infectious proteins called prions.
Hurdle concept The utilization of multiple barriers to control microorganisms and improve the safety or shelf life of food products.
Ileocolitis An inflammation of the ileum (the lowest part of the small intestine) and the colon (the large intestine). It is the most common type of Crohn's disease, and symptoms include diarrhea and cramping or pain in the abdomen, especially after having meals.
Incubation period The time between the initial infection with a microorganism and the start of the symptoms.
Infectious dose The number of colony-forming units of a pathogenic bacterium required to cause an infection in a host.
Inoculum (plural = inocula) The intentional introduction of bacteria, viruses, or toxins to a food matrix in order to study some properties (survival, death, etc.) of the introduced agent.
Invasion The ability of a given pathogen to enter host cell tissues. Invasiveness encompasses mechanisms for colonization (adherence and initial multiplication), production of surface-exposed proteins (e.g., invasins), and the ability to circumvent host defense mechanisms.
Jaundice A yellowish pigmentation of the skin, the conjunctival membranes over the sclerae, and other mucous membranes caused by increased levels of bilirubin in the blood.
Koch's postulates Criteria designed to establish a causal relationship between a microorganism and a disease. These postulates were published by Koch in 1890, and include (1) the microorganism must be found in abundance in all organisms suffering from a given disease, but should not be found in healthy organisms; (2) the microorganism must be isolated from a diseased organism and grown in pure culture; (3) the cultured microorganism should cause disease when introduced into a healthy organism; and (4) the microorganism must be reisolated from the inoculated, diseased experimental host and identified as being identical to the original specific causative agent.
Lactic acid (and lactates) A three-carbon organic acid used as an antimicrobial food preservative. Lactic acid is naturally produced by *Lactococcus* spp. and other lactic acid bacteria in many food fermentations.
Lactoferrin A glycoprotein found in milk that acts as a chelator to sequester iron.
Lipopolysaccharide (LPS) A molecule consisting of a lipid and a polysaccharide joined by a covalent bond. LPS is found in the outer membrane of Gram-negative bacteria and acts as an endotoxin to elicit strong immune responses in infected persons.
Low-acid foods Foods that have a pH ≥ 4.6.
Lysozyme An enzyme that disrupts the peptidoglycan layer of the cell walls of Gram-positive bacteria and cause cell lysis. This enzyme is present in avian eggs, tears, mammalian milk, insects, and fish.
Mad cow disease See *bovine spongiform encephalopathy*.
Malaise A feeling of general discomfort or uneasiness, of being "out of sorts." It is often the first indication of an infection or other disease.
Mastitis The inflammation of the udder in cows caused by an infection, typically by *Staphylococcus aureus*.
Melamine Melamine is an organic compound based on a trimer cyanamide. Recent events involved the sale of melamine-tainted milk that resulted in the death of several children in China.
Meningitis The inflammation of the meninges, which are the membranes covering the brain and spinal cord.
Mesophile(s) (meso = middle; phile = love) Microorganisms with a minimum growth range of 5–15°C, optimum of 30–45°C, and maximum of 35–47°C. This group includes most human and animal pathogens and several spoilage microorganisms.
Methylmercury The methylated form of mercury that, after ingestion, can be rapidly accumulated in the body, where it can be toxic to cells.

Microaerobic Bacteria that require environments with much lower oxygen levels than the one preset in the atmosphere (~21%) to grow.

Microarray A chip-based technology that assays large amounts of DNA material using high-throughput screening methods. DNA microarrays allow the expression of many genes to be analyzed in a single experiment.

Microfold cells (M-cells) These are cells found in the follicle-associated epithelium of Peyer's patches. M-cells transport organisms and particles from the gut lumen to immune cells across the epithelial barrier and thus play important roles in stimulating mucosal immunity.

Microtubule One of the major components in the host cytoskeleton that serves as structural components within cells. Microtubules are involved in many cellular processes, including mitosis, cytokinesis, and vesicular transport.

Miller–Fisher syndrome A rare variant of Guillain–Barré syndrome that manifests as a descending paralysis, proceeding in the reverse order as compared to the more common form of Guillain–Barré syndrome.

Modified atmosphere packaging (MAP) The removal of the existing atmosphere of a food packaged in impermeable plastic film and/or the replacement with an alternate gas mixture. MAP was designed to slow microbial growth and thus extend the shelf life of food products.

Molecular biology A branch of biology that deals with the molecular basis of biological activities in diverse systems.

Monoclonal antibodies (mABs) Molecules that recognize a unique site on the surface of a given antigen and are highly specific, for example, for a given target pathogen. mABs are produced by the culturing of specific immortal hybridoma cell clones.

Morbidity The probability that a randomly selected individual in a population at some date and location would become seriously ill in some period of time.

Mortality The relative frequency of deaths in a specific population.

Most Probable Number (MPN) method A routine method used in laboratories for the quantification of bacterial cells in foods. A replicated dilution series of suspected sample extracts is produced and cultured in enrichment broth followed by plating on selective agar plates for enumeration of the colonies of the target pathogen.

Mycotoxin A nonenzymatic metabolite of fungi that is injurious to other organisms.

Mutation A change in the DNA sequence of a given gene. Any number of nucleotides can be mutated, from a single base to an entire piece of chromosome. Mutations in the DNA sequence can alter the amino acid sequence of proteins and therefore change the phenotype in the affected organism.

Natamycin A polyene macrolide antifungal agent produced by *Streptomyces natalensis* that is used to inhibit molds on certain foods.

Necrotizing enterocolitis A serious bacterial infection of the ileum or colon that can lead to death (necrosis) of intestinal tissue, holes in the intestine, and septicemia.

Nematodes Simple organisms that are parasites of insects, plants, or animals.

Nisin A peptide produced by certain strains of *Lactococcus lactis* ssp. *lactis* that acts as an inhibitor of Gram-positive bacteria, including spore formers. This product has been approved for use as an antimicrobial food preservative by many regulatory agencies.

Norovirus Single-stranded RNA viruses of the family *Caliciviridae* that contain the Norwalk virus.

Nucleic acid sequence-based amplification (NASBA) An alternative method to PCR that involves the amplification of RNA sequences in a DNA background to detect live microbe cells without the false-positive signals from dead cells.

Nucleotide hybridization probes Small nucleotide sequences (ca. 15–30 bases long) used to detect the presence of a microorganism in a sample by detecting the complementary sequences of segments of the microorganism's DNA or RNA.

Nuts A wide variety of products, including botanically defined nuts, seeds, legumes, and drupes.

Oocyst A form of the parasite represented by a thick-walled structure that is highly resistant to environmental stress and chemical sanitizers. Inside this structure, sporozoan zygotes develop and can transfer to a new host.

Operons A functional unit in a chromosome containing a cluster of genes that are under the control of a single regulatory signal or promoter. Operons are found primarily in prokaryotes. The genes are transcribed together into one mRNA strand and are either translated into proteins or undergo *trans*-splicing to create monocistronic mRNAs that are translated individually.

Organoleptic Relating to sensory evaluation, including appearance (e.g., color), odor, taste, and texture.

Osmophiles Microorganisms that are able to grow in environments with high sugar concentrations.

Oxidation-reduction potential (E_h) The ability of a substance to gain or lose electrons. E_h is commonly measured in millivolts (mV).

Parabens Alkyl esters of p-hydroxybenzoic acid with varying lengths of carbon chains. The methyl (1C), propyl (3C), and heptyl (7C) esters are approved for use as antimicrobial food preservatives in the United States.

Parasite An organism living in, with, or on another organism (called the host) that takes its nourishment from and usually injures the host. Parasites cannot live independently.

Pasteurization A form of heat treatment designed to destroy vegetative cells of pathogens and most spoilage microorganisms. Pasteurization has little effect on spores.

Pathogen Any microorganism capable of causing disease in a given host.

Pathogenicity island (PAI) A DNA segment acquired by a horizontal DNA transfer event. PAIs are about 10–200 kb in size and are present in the genome of pathogenic microbes. PAIs encode one or more genes that are associated with virulence, e.g., adhesins, invasins, toxins, or secretion systems. Two well-known PAIs are the *Salmonella* pathogenicity islands and the locus of enterocyte effacement pathogenicity island in EPEC.

Pathogen-associated molecular patterns (PAMPs) Molecules produced by certain groups of pathogenic microbes and that are recognized by the host innate immune system. Typical PAMPs are lipopolysaccharide (LPS) and flagellin. They are recognized by Toll-like receptors and other pattern recognition receptors present in animals and plants.

Perinatal septicemia An infection of the blood caused by bacteria or viruses that can be passed from the mother to the baby during pregnancy or birth.

Peyer's patches Organized aggregates of lymphoid tissue that are usually found in the lowest part of the ileum in humans.

Phase variation A form of gene regulation resulting in a heterogenic phenotype of a clonal bacterial population in which individual cells either express the phase-variable protein(s) or not, or express one of multiple antigenic forms of the protein. Phase variation has been identified in a wide range of surface structures in bacterial pathogens and is implicated as a major virulence strategy.

Phenetics A system to classify organisms based on the overall similarities. The traits usually studied are morphology or other observable traits, regardless of their evolutionary relationships.

Phosphatidylinositol 3-kinase (PI3-K) A family of enzymes capable of phosphorylating a hydoxyl group of the inositol ring of phosphatidylinositol. PI3-Ks are involved in many cellular functions, such as cell growth, proliferation, differentiation, motility, survival, intracellular trafficking, and pathogenesis.

Photosynthesis A metabolic process that converts carbon dioxide into organic compounds, especially sugars, by using the energy from sunlight. Photosynthetic organisms mainly include plants and some specialized bacteria.

Phylogenetics The study of the evolutionary relatedness among groups of organisms by analyzing genetic information and morphological data. Phylogenetics includes the reconstruction and analysis of evolutionary trees and networks based on inherited characteristics.

Phytochemical Any of the various bioactive chemical compounds found in plants.

Plasmid A distinct DNA molecule (circular or linear) that is separate from and can replicate independently of the chromosome. Plasmids can naturally occur in bacteria or sometimes in eukaryotes.

Polyclonal antibodies A combination of immunoglobulin molecules produced by different B cells against a specific antigen. Each immunoglobulin usually identifies one epitope.

Polymerase chain reaction (PCR) A technique used to amplify a specific segment of DNA or RNA and thus generate millions of copies of a particular sequence. This amplified segment can then be easily identified and further analyzed.

Postmortem inspection The inspection of food animals after they have been slaughtered.

Poultry This term refers to any domesticated bird species used for food purposes (e.g., meat or egg). The U.S. Department of Agriculture recognizes six kinds of poultry species: chicken, duck, goose, guinea, pigeon, and turkey.

Prerequisite programs The programs that provide a foundation for an effective HACCP system. Most prerequisite programs reduce the likelihood of certain hazards. According to the World Health Organization (WHO), prerequisite programs are the practices and conditions essential for food safety that are needed prior to and during the implementation of HACCP.

Prion An infectious protein form. Prions are believed to be the agents responsible for several degenerative diseases of the nervous system, such as Creutzfeldt–Jakob disease in humans and scrapie in sheep.

Prokaryote Microorganisms that lack a cell nucleus or any other membrane-surrounded organelles. Most prokaryotes are unicellular (single-celled) microorganisms.

Propionic acid (and propionates) A three-carbon organic acid used primarily as an antifungal agent in food products. It is produced naturally by *Propionibacterium freudenreichii* ssp. *shermani*, as a flavor component of Swiss cheese.

Protozoa Unicellular, eukaryote organisms that only divide within a host organism. Protozoa are grouped in the kingdom Protista, alone with some plant-like algae, fungus-like water molds, and slime molds.

Pruritus An itch or a sensation that makes a person want to scratch.

Psychrotrophic (psychrotroph; psychro = cold; trophic = food) Microorganisms that can multiply at low temperatures (4–7°C), although their optimum temperature is between 15°C and 20°C. Psychrotrophic microorganisms are important agents in the spoilage of foods.

Ready-to-eat (RTE) foods Foods that will not be cooked or reheated before they are eaten. These products include sandwiches, salads, cooked meats such as ham, cheese, cakes, and desserts.

Real-time PCR (Q-PCR, qPCR, or qrt-PCR) A variation of the traditional PCR technique in which there is a simultaneous amplification and quantification of the target DNA. This procedure is sometimes also called quantitative real-time PCR or kinetic PCR (K-PCR).

Reiter's reactive arthritis Also called Reiter's Syndrome, it is classified as an autoimmune condition that develops in response to an infection in a different part of the body. The disease exhibits symptoms similar to various other conditions collectively known as arthritis. There are indications that disease development involves an infection with *Campylobacter*.

Reverse transcription PCR (RT-PCR) A variant of the PCR method in which an RNA strand is first reverse-transcribed into its DNA complement (complementary DNA or cDNA) using an enzyme called reverse transcriptase (RT). The resulting cDNA is then amplified using traditional PCR methods.

RiboPrinter An automated ribotyping method based on PCR amplification and restriction profiling of the ribosomal operons in bacterial species.

Ribotyping A molecular technique used to type bacterial strains based upon the restriction profile of ribosomal operons. The method involves the restriction enzyme analysis of PCR amplified products containing the 16S and 23S rRNA genes.

Rickettsia A genus of Gram-negative, nonspore-forming, nonmotile bacteria. These bacteria are obligate intracellular parasites, and their survival depends on entry, growth, and replication within the cytoplasm of eukaryotic host target cells (typically in the endothelium).

Salmonellosis An infection in the small intestine caused by *Salmonella* bacteria. Salmonellosis is one of the most common sources of food poisoning.

Scrapie A transmissible disease found in sheep and goats. Scrapie belongs to a group of diseases termed "transmissible spongiform encephalopathies" (TSEs), which are believed to be caused by the occurrence of an abnormal form of the so-called "prion" protein.

Sepsis A severe illness in which the human bloodstream is overwhelmed by pathogenic microorganisms, such as bacteria. In acute sepsis, blood pressure drops, resulting in shock. Major organs and body systems, including the central nervous system, lungs, liver, and kidneys, stop working properly.

Septicemic infection The presence of pathogenic microorganisms in the bloodstream, leading to sepsis. It is also called septicemia or bacteremia.

Serotypes/serotyping A technique used to type some bacteria or viruses based on the reaction of antigenic components with specific antibodies. There are serotyping schemes that target the lipopolysaccharides of the cell wall or the flagella or even the capsule produced by some bacterial species. Serotyping depends on the testing of the microorganism with many different antibodies to determine which antigens are present.

Shigellosis Intestinal infection caused by *Shigella* species, also known as bacillary dysentery. These bacteria produce toxins and effector proteins that can attack the lining of the large intestine, thus causing swelling and ulcers on the intestinal wall. The associated symptoms can range from watery diarrhea to bloody diarrhea, fever, and abdominal pain.

Sodium-d-glucose cotransporter (SGLT) A family of glucose transporter found in the intestinal mucosa of the small intestine (SGLT-1) and the proximal tubule of the nephron (SGLT-1 and SGLT-2). These transporters contribute to renal glucose reabsorption. SGLT-1 is known to be targeted by EPEC virulence factors.

Sodium nitrite A compound used in cured meats to affect color formation and flavor. This compound acts as an antioxidant and as an antimicrobial against the growth of *Clostridium botulinum*.

Sorbic acid (also sorbates) A six-carbon unsaturated organic acid used as an antimicrobial food preservative against yeasts, molds, and bacteria. This compound may be found naturally in rowanberry oil.

Spoilage A condition resulting in changes in the odor, flavor, and appearance of a food due to microbial activity.

Spores, bacterial (or endospores) A dormant stage of a bacterium that is generally much more resistant to environmental stresses than vegetative cells. Spores are produced by some Gram-positive bacteria, including *Bacillus* and *Clostridium*. Spores also form part of the life cycles of many algae, fungi, and some protozoans.

Standard cultural methods Standard microbiological methods used to isolate a targeted bacterium from a food sample. This methodology is based on the growth of the target bacterium on nonselective or selective culture media.

Symbiosis The short- or long-term interaction between at least two different species for their mutual benefit. Symbiotic relationships have been described for both prokaryotic and eukaryotic organisms.

Systematics The discipline that studies biological diversity and its origins. It focuses on understanding evolutionary relationships among organisms, species, higher taxa, or other biological entities, and the evolution of the properties of taxa.

Taxon (plural = taxa) A taxonomic group or entity. It is defined by a set of characteristics common to each member of the same unit or group (e.g., genus, species).

Tephritid fly larvae One of the fly families known as "fruit flies." Tephritid fly larvae are also known as "sunflower maggot fly" and affect sunflowers and related plants by eating part of the flower receptacle while the larvae are growing. Human pathogenicity arises from their ability to host bacterial pathogens such as *Salmonellae*.

Thermoduric Bacteria that can survive exposure to high temperatures. In the dairy industry, the term is applied to those organisms that survive, but do not grow, at pasteurization temperature. They usually include species such as *Streptococcus, Bacillus, Micrococcus, Lactobacillus,* and occasionally some Gram-negative rods.

Thermophiles (thermo = hot; phile = love) Microorganisms that have a minimum growth temperature of 40–45°C, optimum of 55–75°C, and maximum of 60–90°C. The primary importance of these organisms is in the spoilage of foods; they are often spore formers.

T cell Cells that belong to a class of white blood cells known as lymphocytes (also called T lymphocytes) that play crucial roles in cell-mediated immunity.

Toll-like receptors (TLRs) A class of protein receptors that play key roles in the host innate immune system. TLRs are single, membrane-associated, noncatalytic factors that recognize structurally conserved molecules derived from microbes (so-called PAMPs, see above). Once a given microbe has passed physical barriers such as epithelial cells, they can be recognized by TLRs, which then activate signaling, leading to the activation of immune cell responses.

Toxins Poisonous substances produced during the metabolism and growth of certain microbes and some higher organisms. Toxins can be small biomolecules, short peptides, or large proteins that are capable of triggering disease upon contact with or absorption by body tissues.

Transcription factors Proteins that bind to specific DNA sequence motifs, thereby controlling the conversion of genetic information from DNA to mRNA (gene transcription). Important transcription factors discussed here include AP-1 (activator protein-1) and NF-κB (nuclear factor kappa B), which are major host components controlling pro-inflammatory responses.

Transmissible spongiform encephalopathy (TSE) Also known as prion disease, TSE is a progressive condition that affects the brain and nervous system of various animals and humans. The infectious agent is a transmittable prion.

Type III secretion system (T3SS) Pilus-like surface appendages protruding from the bacterial cell wall. T3SSs allow for the direct injection of bacterial proteins, called effector proteins, into the contacting host cell. These secretion systems are very important because pathogens with defective T3SSs are 1,000 to one million-fold attenuated in animal models of infection. T3SSs are evolutionarily related to the flagellar export system in bacteria.

Type IV secretion system (T4SS) Pilus-like surface appendages protruding from the bacterial cell wall. Like T3SSs, T4SSs allow for the direct injection of bacterial effector proteins into the contacting host cell, but can also transfer DNA such as plasmids to other bacteria. These secretion systems are very important because they can spread antibiotic-containing plasmids by conjugation, and pathogens with defective T4SSs are attenuated in animal models of infection.

Tyrosine kinases Enzymes that can transfer a phosphate group from ATP to a tyrosine residue in a given target protein. The phosphorylation of proteins by kinases can change the protein conformation and/or binding capabilities for certain molecules and thus is an important mechanism in signal transduction. Prime examples of tyrosine kinases in eukaryotes discussed here include the FAK, Src, Tec, and Abl families, as well as the hepatocyte growth factor receptor (c-Met). These kinases can be highjacked by certain pathogens during infection.

Value-added products Products that are subjected to a step in the production process that improves the commodity for the consumer and results in a higher net worth.

Vegetative cells The state of a bacterium that is actively growing or has the capability of growing.

Viral particle/virus Small infectious agents that can replicate only inside living cells in organisms. Viral particles are also called virions and consist of nucleic acids (DNA or RNA), proteins, and in some cases an envelope of lipids.

Water activity (a_w) This term refers to the vapor pressure of a liquid divided by that of pure water at the same temperature or the equilibrium relative humidity of a food product; it is a measure of the water available for microbial growth. The scale for water activity is from 1.0 (pure water) to 0. This term is a parameter for the amount of unbound water available in the close environment of a microbe.

Zoonosis (plural = zoonoses) An infectious disease in animals that can be transmitted to humans. The natural reservoir for the infectious agent is a given animal. As discussed in this book, there are a variety of zoonotic bacterial species with an impact on food safety. For example, *C. jejuni* is a commensal in birds and other animals, but when transferred to humans, it has an important impact on health.

Zoonotic species or agents Infectious agents (e.g., bacteria) that can be transmitted between animals and humans.

Z-value The number of degrees (Fahrenheit or Celsius) to cause a 90% change in log *D*. Z-values can be shown by plotting the $\log_{10}$ of D-values vs. temperature to create a thermal death time (TDT) curve.

Index

A

AFLP. *See* Amplified fragment length polymorphism
Ahmed, R., 112
Allergens, 160, 167
Amplified fragment length polymorphism
    (AFLP), 59
Andersson. Y., 112
Andrews, R., 112
Angulo, F.J., 130
Animal and Plant Health Inspection Service (APHIS),
    219, 223
Antibody-based assays
    biosensor, 51
    description, 49
    DNA-based assays, 50
    ELISAs, 49
    immunomagnetic beads, 49–50
    Listeria Rapid Test, 51
    mono and polyclonal, 49
    Pathatrix systems, 51
    traditional agglutination methods, 49
APHIS. *See* Animal and Plant Health Inspection
    Service
Aquaculture drugs, 160, 166–167
Aramini, J., 112

B

Bacillus cereus, 138
Backert, S., 13–26
Bacteria evolution, foodborne pathogens
    bacteriophages, 7
    binary fission, 6–7
    conjugation mechanism, 7–8
    Escherichia coli, 7
    gene loss, 7
    genetic competence, 7
    genetic transformation, 7
    mutations and horizontal DNA transfer, 7
    Mycobacterium tuberculosis, 7
    pathogenicity islands (PAIs), 8
    phase variation, 8
    pilus, 7

    S. flexneri, 7
    transduction, 7
Bacterial foodborne infections
    Campylobacter jejuni (*seeCampylobacter jejuni*)
    defined, pattern recognition receptors (PRRs), 13
    dendritic cells (DCs), 13–14
    E. coli (*seeEscherichia coli*)
    estimations, WHO, 14
    evolution, protein toxins, 14
    fewer infections, 14
    gastrointestinal (GI) tract, 13
    health and economic problem, 25
    international human microbiome project, 26
    invasive bacteria, 14
    Listeriosis, 24–25
    nonphagocytic host cells, bacterial invasion, 15
    proteomics techniques, 26
    research, genome sequencing and molecular
        infection, 26
    Salmonella spp., 15–18
    Shigella spp. (*seeShigella* spp.)
    signaling mechanisms, 26
    sophisticated immune system, 14
    successful infection, 14
    toll-like receptors (TLRs), 13
    "trigger" mechanism, 14
    T3SS, 14
    WHO, 25
Bagamboula, C., 118
BAI. *See* Bureau of Animal Industry
Banwart, G.J., 191, 192
Barendsz, A., 118
Barrett, L.J., 41
Barrett, T.J., 112
Bartleson, C.A., 76
Bartlett, C.L.R., 112
Baskaran, S.A., 33–43
Bean, N.H., 129, 130
Beczner, J., 118
Begg, N.T., 112
Beneficial bacteria
    Brevibacterium linens, 43
    dairy fermentations, *Leuconostoc*, 43

fermentation, LAB, 42
glucose metabolism, 42–43
Micrococcus, 43
souring milk, *Lactococcus lactis*, 43
Berries
bacterial contaminants, 120
cultivation, 116
food safety and quality hazards, 118
fungi, 120
high nutritional composition, 117
pesticides, 120–121
processing and storage methods, 121
Protozoa, 117, 118
US production, 117, 118
viruses, 119
Besser, R.E., 99
Bills, G., 118
Bishop, J., 118
Blau, D., 112
Bodager, D., 118
Böhm, H., 61
Botana, L.M., 163
Botulinum, 102
Bovine spongiform encephalopathy (BSE), 40–41
Brophy, I., 112
Brown, D.J., 112
Brustin, S., 112
Bryan, F.L., 129, 130
BSE. *See* Bovine spongiform encephalopathy
Bureau of Animal Industry (BAI), 209, 222

C

Calder, L., 118
Cameron, S., 112
Campbell, K., 112
Campylobacter jejuni
acute and long-term infection, 21
adhesins, 20
causes, 18
consequences, 18–19
contamination, 18
host cell invasion, 20–21
infections, 18, 20
intestinal biopsies, 20
mutation, 20
signal transduction and host cell invasion, 19, 20
symptoms, 20
Cantor, C.R., 60
Ceibin, B., 112
Center for Veterinary Medicine (CVM), 222
Chalmers, R., 118
Chamuris, G., 118
Chandran, A.U., 112
Chan, E., 112
Cheung, W., 118
Chou, J.H., 112
CJD. *See* Creutzfeldt–Jakob disease
Clark, C., 112

Clostridium botulinum, 102, 138–139
Collins, M., 195
Contamination
environmental/industrial, 96
foodborne illness, 75
food/water, incident, 225
green onions, 88
intentional, 221
pathogen, 99
product/ingredient, 97
routes and sources, 102–103
Conway, W., 102
Cook, N., 118
Crerar, S., 112
Creutzfeldt–Jakob disease (CJD), 140–141
Critzer, F.M., 189–202
Crozier-Dodson, B.A., 95
Curtis, P, 203–215
CVM. *See* Center for Veterinary Medicine

D

Dairy products
CJD, 140–141
Cronobacter sakazakii, 141–142
cryptosporidiosis, 141
dairy-related outbreaks, 128–132
foodborne outbreaks, 127
Johne's disease, 141
pasteurization, 132–133
pathogens, 138–140
Dairy-related outbreaks
consume raw milk, 127
FDA, 131–132
food safety, 131
percentage, 128, 130
product and cause, 128, 131
report, United States, 128–129
Dalton, C., 112
Danyluk, M., 109
Darwin, C., 5
Davidson, P.M, 189–202
Debevere, J., 118
Dehydration, 199
de Jong, P., 195
Delazari, I., 102
de Louvois, J., 112
Dendritic cells (DCs), 13–14
Densitometric curves, 65
Department of Homeland Security (DHS)
Bioterrorism Act, 224
collaborative efforts, 225
FASCC, 225
Homeland Security Act of 2002, 224
homeland security presidential directive (HSPD), 224
Strategic Partnership Program Agroterrorism (SPPA) Initiative, 224–225
DHS. *See* Department of Homeland Security

Dice index, 64
Diets
    production, 10
    safety, 10–11
DNA sequencing
    datas, 65–66
    epidemiology, foodborne
        developments, 61–62
        investigation, multilocus sequence, 61
        multilocus sequence typing, 62–63
        sequence data, 62
    segments, whole genome, 61
    technology, 51
Dolye, M.P., 99
Doyle, M.P., 104
Dubey, J., 118
Duckworth, G., 112
Dwyer, D.M., 112

E

Egg Products Inspection Act (EPIA), 213
EHEC. *See* Enterohemorrhagic *E. coli*
Ellis, A., 112
Emerging and reemerging infectious diseases
    by bacteria, 4
    factors, 3
    hemolytic-uremic syndrome, 4
    human, 4
    protozoa, 4
    transmission, 4
    WHO, 3
    zoonotic, 3–4
Enterohemorrhagic *E. coli* (EHEC), 22–24, 149–150
Enteropathogenic *E. coli* (EPEC), 22–24
Environmental Protection Agency
    (EPA), 84, 219, 225–226
Enzyme-linked immunosorbent assays (ELISAs), 49
EPA. *See* Environmental Protection Agency
EPEC. *See* Enteropathogenic *E. coli*
EPIA. *See* Egg Products Inspection Act
Epidemiological studies, foodborne pathogens
    amplification methods, PCR (*see* Polymerase chain reaction)
    analysis and interpretation
        band patterns, 64–65
        cladistic methods, 64
        cluster analysis, 66
        densitometric curves, 65
        DNA sequence data, 65–66
        epidemiology and typing method, 63
        fingerprinting techniques, 63–64
        mathematical algorithms, 63
        phenetic methods/numerical taxonomy, 64
        strains, 63
    definitions, 58
    DNA sequencing (*see* DNA sequencing)
    DNA techniques, 57
    fingerprinting, 57
    lipid profiles, 57
    molecular epidemiology (*see* Molecular epidemiology)
    molecular methods, 57
    PCR amplification and restriction (*see* Polymerase chain reaction)
    phenotypic tests, 57
    survival, 57
    typing methods, 58, 68–69
    whole genome restriction method
        advantage, PFGE, 61
        Kpn I, Sma I, and *Not I* enzymes, 61
        measurements, DNA molecules, 60
        PFGE, 60–61
        PulseNet, 61
Epidemiology
    hepatitis
        HAV transmission, 78
        nucleic acid sequencing, 79
    noroviruses
        enzyme immunoassay (EIA) kits, 77
        foodborne illness, 75
        pathogens indentified, 76
Erickson, M.C., 104
Escherichia coli
    attaching and effacing lesions (A/E lesions), 22–23
    diarrhea, 22
    EHEC and EPEC, 23
    enteric pathogens, 115
    F-actin, 24
    products consumption, 109, 121
    SGLT–1, 24
    Shiga toxins, 23
    strain K–12, 22
    Tir molecules, 23–24
    T3SS effectors triggering, 24
    types, 22

F

Fanning, S., 147–155
Farm-to-table continuum, 89
Farrar, J.A., 112
Farr, D., 118
FDA. *See* Food and Drug Administration
FDCA. *See* Food Drug and Cosmetic Act
Federal Meat Inspection Act (FMIA), 210, 212, 223
Feng, H., 102
Fermentation, 201
Fingerprinting
    methodology, 63
    profiles (FP), 65
    techniques, 63–64
Fiore, A., 118
Fish and seafood products
    allergens, 167
    aquaculture drugs, 166–167
    bacterial pathogens, 161
    chemical contaminants, 165–166

core programs, 159–160
description, 159
histamine and scombrotoxin, 164–165
natural toxins
Fish and seafood products (*continued*)
    ciguatoxin, 162–163
    description, 162
    gempylotoxin, 164
    planktonic toxins, 163
    tetrodotoxin, 164
parasites, 161–162
product and package
    battered/breaded, 169
    cooked/pasteurized, 168
    dried, cured and salted fish, 170
    raw/partially cooked finfish, 170–171
    ROP, 167–168
    shelf-stable low-acid/acidified canned seafood, 171
    smoked, 169
    stuffed seafood, 169
regulatory requirements, 159
vertebrates and invertebrates, 160
Fisher, I.S.T., 112
Flessa, S., 118
FMIA. *See* Federal Meat Inspection Act
Food and Drug Administration (FDA)
adulteration, 203
center for food safety and applied nutrition (CFSAN), 221–222
colony-forming units (CFU), 100
consumers, 204
CVM, 222
dairy products, 131–132
FDCA, 204
food code, 205–206
food safety hazards, 103
HACCP, 206–209
jalapeño peppers, 100–101
office of regulatory affairs (ORA), 222
provisions, 204–205
Foodborne bacteria
hazard impact, 39
lyse substrates, 39
oxygen requirements, 38–39
physiological tolerance levels, 39
temperature tolerance, 38
Foodborne disease
E. coli O157:H7, 99
risk minimization, 104
Foodborne illness outbreaks
gastroenteritis, 76
hepatitis A, 83
Foodborne pathogens, emerging and reemerging
"at-risk" populations, 10
bacteria evolution, 6–8
changes and expansion, diets
    food production, 10
    food safety, 10–11
description, 3
disease agents, evolution, and epidemiology

cholera outbreak, 5–6
Darwin theory, 5
Koch's postulates, 5
measures, 6
pasteurization method, 5
1670s, 5
emergence and spread, 3
food production and processing
    context, 9
    impact, population increases, 9
    Listeria monocytogenes, 9–10
    1950s and 1970s, 9
    traditional agricultural systems, 9
geographical distribution, human infect
    food commodities, 9
    genetic diversity, 8
    massive food production, 8
    Salmonella serotypes, 8
    viruses, 9
human pathogens, 4–5
infectious diseases (*see* Emerging and reemerging infectious diseases)
Foodborne viruses
bacteria-based approaches, 73
farm-to-table continuum, 89
hepatitis
    epidemiology, 78–79
    A and E viruses, 77–78
Mixed Bag, 79
noroviruses
    epidemiology, 75–77
    RT-PCR, 75
    scanning electron micrograph, 74
    spectrum, gastroenteritis symptoms, 74
prevention and control
    cleanup of infected materials, 84–88
    food-handling hygiene, 82–83
    hygienic food growing and harvesting, 83–84
    PEP, 81–82
    vaccines, 80–81
Food code, 82, 85
Food Drug and Cosmetic Act (FDCA), 204
Food microbiology and safety
bacterial cell wall and gram's reaction, 33
bacterial growth, 37
beneficial bacteria (*see* Beneficial bacteria)
characteristics, prokaryotes and eukaryotes, 33, 34
growth factors
    classification, 37
    compounds, 37–38
    determination, 37
    extrinsic foods, 38
    intrinsic foods, 37–38
    parameters and range, 38
host-pathogen interactions, 43
involvement, microorganisms
    foodborne molds, 41
    foodborne prions, 40–41
    gram-negative, 40
    gram-positive, 40

Index 257

parasites and protozoa, 41–42
living organism, 33
multiplicity, food organisms
 bacteria and associated foods, 35–36
 disease control and prevention, US, 36
 infection, 36
 intoxications, 36
 outbreaks, 34, 36
 perspective, 34
 toxicoinfections, 36–37
spoilage, 42
types, foodborne bacteria, 38–39
Food Regulation, United States
 FDA, 203–209
 FSIS, 203, 209–214
 laws and regulations, 214–215
Food risk analysis
 assessment modeling
  characterization, 177
  description, 175
  exposure, 176
  hazard characterization and identification, 176–177
  methods, 178–181
 characterization, 183, 185–186
 description, 175
 exposure assessment, 182–184
 hazard identification, 182–184
Food Safety Inspection Service (FSIS)
 APHIS, 223
 BAI, 209
 EPIA, 213
 Federal Meat Inspection Act (FMIA), 223
 FMIA, 210
 imported products, 213–214
 Listeria Regulations, 212
 meat and poultry HACCP, 211–212
 Office of Field Operations (OFO), 223–224
 PPIA, 210
Francis, G.A., 103
Fruits, 113–114
FSIS. *See* Food Safety Inspection Service
Fung, D.Y.C., 101
Fyfe, M., 112

G
Gal, N., 112
Gastroenteritis
 foodborne illnesses, 75
 public health measures, 76
 viral acute disease, 80
Genobile, D., 112
Gilbert, R.J., 112
Gill, O.N., 112
GMP. *See* Good manufacturing practices
Golden, D.A., 191
Good manufacturing practices (GMP), 159, 229, 231
Goulding, J.S., 129, 130
Gram-negative foodborne pathogens, 40

Gram-positive foodborne pathogens, 40
Grant, D., 118
Greening, G.E., 118
Green, M.S., 112
Gregory, J., 112
Greig, J.D., 76, 148
Griffin, P.M., 99, 129, 130
Gupta, A., 112
Guy, R., 112

H
HACCP. *See* Hazard analysis and critical control points
Hackney, C., 118
Hajmeer, M.N., 95, 101, 102
Hall, A.J., 73
Hammond, R., 118
Hänninen, M., 118
Harris, L.J., 112, 118
HAV. *See* Hepatitis A virus
Hazard analysis and critical control points (HACCP)
 FDA, 206
 fish products, 206–207
 juice products
  CGMP, 208
  FDA, 207
  juice processors, 208
 pathogen elimination, 103
 PMO, 208–209
 principles, 206
 and produce, 104
Hazard identification
 contamination, 182
 formulas, 178, 180
 incidence, 183–184
 number, 183, 184
 quantification, 176
Hedberg, C., 98
Henry, C., 217–231
Hepatitis A virus (HAV)
 effective and safe vaccine, 80
 enteric virus, 119
 foodborne transmission, 81
 genetic sequences, 88
 Picornaviridae family, 77
 quality hazard, 118
 single-antigen vaccine or IG 54, 82
Hepatitis E virus (HEV)
 foodborne transmission, 81
 Hepeviridae family, 77
Herrewegh, A.A.P.M., 195
HEV. *See* Hepatitis E virus
High-pressure processing (HPP), 197
Histamine and scombrotoxin, 164–165
Hjertqvist, M., 112
Howes, M., 112
HPP. *See* High-pressure processing
Hu, D.J., 73
Human pathogens
 chimpanzees, 4–5

definition, 5
hepatitis B viruses, 5
Hutchinson, N.D., 112
Hwang, P.H., 112

I
Identification methods, bacterial foodborne pathogens
  antibodies (*see* Antibody-based assays)
  culture-based methods
    advantage, 46
    chromogenic agar, 47
    commercial kits, 47
    description, 46
    most probable number (MPN) method, 47
    novel enumeration methods, 47
    selective and different plating medium, 48
    standard methods, *Listeria*, 46
    strains, 46
    sublethal cells, 47–48
  detection technologies, 46
  illnesses estimation, CDC, 45
  nucleic acid based method (*see* Nucleic acid based method)
  outbreaks and infectious dose, 45
  treatments, 53
  uneven distribution, 53
Inactivation
  acid foods, 195
  commercially sterile products, 194
  D-values, 195
  electron beam irradiation, 196
  heat, 193
  HPP, 197
  ionizing radiation, lethal doses, 196
  irradiation, 196
  pasteurization, 195
  Salmonella serovars, 194
  TDT, 193
Ingle, B., 112
Inhibition
  antimicrobial food preservatives, 199
  dehydration, 199
  fermentation, 201
  lactoferrin, 200
  lysozyme, 200
  MAP, 198
  natamycin, 200
  nisin, 200
  parabens, 200
  pH, 198
  propionic acid and propionates, 199–200
  sodium nitrite, 200
  sulfur dioxide and metabisulfite, 200
  temperature depression, 197–198
  vacuum packaging, 198–199
Interventions methods, bacterial pathogens in foods
  description, 189
  environmental factors, 190
  inactivation, 193–197
  inhibition, 197–201
  interactions, 201–202
  intrinsic factors
    oxidation-reduction potential ranges, 191
    pH ranges, microorganisms growth, 191, 192
    spices, 193
    water activity, 192
    a_w survival, 192
  microbial factors, 189–190
Irradiation
  effect, 197
  electron beam, 196
  gamma, 196
Isaacs, S., 112
Ivey, C.B., 129

J
Jamieson, F., 112
Jay, J.M., 191, 196
Johny, A.K., 33–43
Jolbaito, B., 112

K
Karch, H., 61
Kärenlampi, R., 118
Killalea, D., 112
Kirk, M.D., 112
Klijn, N., 195
Kniel, K., 118
Koch, R., 5
Kotloff, K., 41
Kruse, H., 98
Kurdziel, A., 118

L
Lactic acid bacteria (LAB)
  heterofermentative, 42–43
  homofermentative, 42
Lactoferrin, 200
Lalor, K., 112
Langton, S., 118
Lao, C., 130
Latham, S.M., 4
Ledet Muller, L., 112
Lem, M., 112
Lett, S.M., 99
Lightfoot, D., 112
Lim, Y.L., 112
Lindsay, D., 118
Lior, L.J., 112
Listeria monocytogenes
  cheeses, 137–138
  dairy farm environments, 137
  food contamination, 24
  host T cell responses, 25
  infection, 25
  internalization process, 25

Index 259

outbreaks, 24–25
RTE, 151
sepsis and *meningitis*, 25
survival, 25
zipper mechanism, 25
Lister, J., 5
Little, C.L., 112
Loessner, M.J., 191
Lum, K., 159–172
Luo, Y., 102
Lusk, D., 118
Lyi, H., 112
Lysozyme, 200
Lyytikäinen, O., 118

M
MacDonald, K.L., 130, 135
Mackenzie, E.F., 112
MacKinon, C.L., 130
Madden, J., 98
Malison, M.D., 112
Mandrell, R.R., 102
MAP technology. *See* Modified atmosphere packaging technology
Matas, A., 112
Maunula, L., 118
McEvoy, J.L., 102
McIntyre, L., 112
Mead, P.S., 112
Meat products
　foodborne disease and role, 147–148
　food chain, controls, 154
　hazards association
　　Campylobacter, 150
　　EHEC, 149–150
　　Listeria monocytogenes, 150–151
　　microbial toxins, 153
　　parasites and viruses, 151–153
　　Salmonella, 149
　production chain, 147, 148
Merom, D., 112
Michaels, B.S., 76
Microarray methods, 53
Microorganisms involved in food safety
　diarrhea, 41
　gram-negative, 40
　gram-positive, 40
　molds, 41
　parasites and protozoa, 41–42
　prions, 40–41
Middleton, D., 112
Miller, J.R., 41
Mitchell, E., 112
MLST. *See* Multilocus sequence typing
Modified atmosphere packaging (MAP) technology, 168, 198
Molecular epidemiology
　fingerprinting, 66–67
　reproducibility, 68

SID, 67
statistical methods, 67–68
strains relationship, 68
technique, 66
typability, 68
Morris, J.E., 112
Morris, J.G. Jr., 112
Morse, S.S., 4
Multilocus sequence typing (MLST), 62–63
Murray, C., 112

N
Naidu, A., 200
Nannapaneni, R., 45–53
Narciso, J., 109
Natamycin, 200
Nichols, G., 118
Nisin, 200
Norovirus, 98
Notermans, S., 118
Nucleic acid based method
　DNA sequencing technology, 51
　explanation, 51
　microarray methods, 53
　NASBA, 52–53
　PCR targeting, 51–52
　quantitative real-time PCR (qrt-PCR), 52
　reverse transcription PCR (RT-PCR), 52
Nucleic acid sequence-based amplification (NASBA) method, 52–53
Nuts, 114–116

O
O'Beirne, D., 103
O'Brian, S.J., 112
O'Brien, K., 112
O'Brien, S., 112
O'Grady, K.A., 112
Olsen, S.J., 130
Olsson, A., 112
O'Mahony, M., 112
Oscar, T.P., 175–186
Osterholm, M.T., 135
Oyarzabal, O.A., 3–11, 57–69, 235–239

P
Paccagenella, A., 112
Parabens, 200
Parasites and viruses
　cysticercosis and taeniosis, 152
　meatborne, 151
　norovirus, 152–153
　Trichinella, 152
Parish, M., 109
Pasteurization
　adoption, 133
　description, 132

milk ordinance, 132–133
purpose, 132
Pasteurized Milk Ordinance (PMO)
    FDA, 208
    raw milk, 208–209
    USPHS, 208
Pasteur, L., 5
Pathogens
    Bacillus cereus, 138
    bacterial
        centers for disease control and prevention (CDC), 98
        pathogenesis, 99
    Brucella, 138
    Campylobacter jejuni, 135–136
    Clostridium botulinum, 138–139
    contamination, 120
    enteric viruses, 121
    Escherichia coli, 136
    fruit-related outbreaks, 114
    Listeria monocytogenes, 136–138
    organic acids, 96
    produce contamination, 103
    public health concerns, 133
    Salmonella, 133–135
    Staphylococcus aureus, 139
    Yersinia enterocolitica, 139–140
Payne, D.C., 73
Payne, L., 112
PCR. *See* Polymerase chain reaction
Pearce, M., 112
Pettersson, H., 112
PFGE. *See* Pulsed-field gel electrophoresis
Pichette, A.S., 112
Pierson, M., 118
Plym Forshell, L., 112
PMO. *See* Pasteurized Milk Ordinance
Polymerase chain reaction (PCR)
    amplification and restriction
        AFLP, 59
        automated method, 60
        Campylobacter jejuni, 60
        C. coli, 60
        flaA and *flaB* genes, 60
        indentify DNA enzymes, 59
        Listeria and Streptococcus, 60
        protocols, 59
        RFLP, 59
        ribotyping, 59–60
        typing methods, 59
    description, 58
    DiversiLab Web Interface, 58
    REP-PCR, 58–59
Pönkä, A., 118
Potter, M., 98
Poultry Product Inspection Act (PPIA), 210
Powling, J., 112
PPIA. *See* Poultry Product Inspection Act
Pramukul, T., 112
Pritchard, K., 118
Produce safety
    disease prevention and control, 104
    foodborne illness
        bacterial pathogens, 98–102
        microbial contamination, 97
    measures, risk reduction
        health issues, 104
        WGA, 103
    pathogens, 95
    potential hazards
        chemical, 96
        physical, 96
    ready-to-eat (RTE) foods, 95
    routes and sources, 102–103
Pulsed-field gel electrophoresis (PFGE)
    BioNumerics, 67
    DNA restricted, *C. jejuni*, 61
    technique, 60–61
    whole genome, 60, 61

R
Ravel, A., 148
Ready-to-eat (RTE) foods, 137, 151, 162, 212
Reduced-oxygen packaging (ROP)
    cooked/pasteurized, 168
    raw/partially cooked finfish, 170–171
    raw shellfish, 170
    smoked, 169
Regulatory agencies, USA
    food safety, 217
    global food supply chain management
        EPA, 219
        Melamine, 218
    regulatory process, 227–228
    Right Way Pizza, 228–231
    single food agency, 226–227
    statutes and regulations
        DHS, 224–225
        EPA, 225–226
        FDA, 219–222
        food safety and inspection services, 222–224
        inspection and labeling, 219, 220
    training, 231
Reisfeld, A., 112
Resources, food safety
    books, 235–236
    internet, 235, 239
    journals, 235, 238
    microbiology books, 235, 237–238
Restriction fragment length polymorphism (RFLP), 59
Reverse transcription-polymerase chain reaction (RT-PCR)
    norovirus detection, 77
    sensitive molecular assay, 75
    viral genomic RNA, 79
RFLP. *See* Restriction fragment length polymorphism
Riemann, H.P., 102

Index 261

Risk assessment modeling
  description, 175
  exposure assessment, 176
  hazard characterization, 176–177
  hazard identification, 176
  methods
    discrete distribution, 178
    hazard and risk characterization, 178–180
    hazard identification and exposure, 178, 180
    logical function, 178
    "LOOKUP" function, 181
    "POWER" function, 180
    @risk functions, 181
    "ROUND" function, 180
    scenario analysis, 181
    transparency, 181
  risk characterization, 177
Risk characterization
  description, 177
  formula, 178, 180
  frequency distribution, 185, 186
  iterations, 185–186
  model, 178, 179
Roberts, D., 112
Rodgers, F., 112
Rodhouse, J.C., 112
Rohde, M., 13–26
Rooney, R., 118
ROP. *See* Reduced-oxygen packaging
Rossman, A., 118
Rowan, A., 118
Rowe. B., 112
RT-PCR. *See* Reverse transcription-polymerase chain reaction
Ryser, E.T., 127–142

S
Safety, products
  berries (*see* Berries)
  CDC foodborne outbreak online database (FOOD) analysis, 109, 110–111
  foodborne illness, 113
  fruits, 113–114
  nuts, 114–116
Salmonella spp.
  animal food, 16
  bacterial attachment, toxins and proteins, 18, 19
  categories, 149
  cheese, 135
  dairy products, 133–134
  disease agents, 110–111
  electron micrograph scanning, 16
  family and species, 15–16
  foodborne pathogens and zoonotic, 96
  genetic matches, 101
  HTST and LTLT pasteurization, 134
  infections, 17, 149
  infective dose, 101
  internalization, 114
  macropinocytosis, 18
  outbreaks, humans, 16
  pathogenicity islands, 17
  product consumption, 121
  role, 17–18
  SipA and SipC, 18
  sporadic food poisoning, 16
  T3SS–1 effector, 18
  typhoid and nontyphoid species, 17
Sanderson, P.J., 112
Scheil, W., 112
Schwartz, D.C., 60
Schwieger, M., 112
Scott, P., 118
Semmelweis, I, 5
SGLT–1. *See* Sodium-d-glucose cotransporter
Sharapov, U., 73
Shellfish, 82, 84, 220
Shigella spp.
  acute GI disorder, 21
  adherence factors, 21–22
  diagnosis, 21
  IcsA, 22
  IpaA-D binds, 22
  macrophages, 22
  pathogens and symptoms, 21
Shohat, T., 112
SID. *See* Simpson's index of diversity
Simmons, G., 118
Simpson's index of diversity (SID), 67–68
Sinclair, U., 9
Slater, P.E., 112
Slutsker, L., 130
Sneath and Sokal index, 64
Snow, J., 5
Sodium-d-glucose cotransporter (SGLT–1), 24
Stack, M., 118
Stafford, R., 112
Staphylococcus aureus, 76, 139
Stuart, J.M., 112
Sufi, F., 112
Sumner, S., 118
Sung, N., 195
Surnam, S., 112
Susman, M., 112

T
Tan, A., 112
Tauxe, R.V., 98, 112
Taylor, J.L., 112
Taylor, L.H., 4
Taylor, P., 118
TDT. *See* Thermal death time
Tegtmeyer, N., 13–26
Thermal death time (TDT) curve, 193
Thornley, C., 118

Threlfall, J., 112
Todd, E.C., 76
Tournas, V., 118
Toxin, 220
Tuttle, J., 112
Typing methods, 68–69

U

United States Public Health Service (USPHS), 208
U.S. Department of Agriculture-Food Safety and Inspection Service (USDA-FSIS), 100, 222
U.S. National Institute of Allergies and Infectious Diseases, 4
Uyttendaele, M., 118

V

Vacuum packaging, 198–199
Valcanis, M., 112
van Leeuwenhoek, A., 5
van Zandvoort-Roelofsen, J., 118
Venkitanarayanan, K., 33–43
von Bonsdorff, C.-H., 118
Vugia, D.J., 112

W

Wachsmuth, K., 98
Wall, P.G., 112
Wang, J., 118
Ward, K., 118
Ward, L.R., 112
Watchel, M.R., 102
Weber, J.T., 99
Wells, J.G., 99
Whitehand, L.C., 102
Whyte, P., 147–155
Wilkinson, N., 118
Wilson, D., 112
Wilson, M.M., 112
Wood, R.C., 135
Woolhouse, M.E.J., 4

Y

Yersinia enterocolitica, 76, 139–140

Z

Zhou, B., 102

Printed by Printforce, the Netherlands